Praise for *Dataproc Cookbook*

A fantastic resource for anyone getting started with Google Dataproc. The book walks you through the essentials and covers more advanced topics with clear, practical recipes. It's the kind of reference you'll keep coming back to if you work with Dataproc in any capacity.

—*Sushil Kumar, Principal Software Engineer, Twilio*

Dataproc Cookbook serves as a definitive and go-to guide, packed with practical, tested recipes that truly accelerate a user's ability to leverage Dataproc for high-performance big data pipelines on Google Cloud. The remarkable depth and breadth of the content are a testament to Narasimha and Anu's dedication and expertise.

—*Jerome Rajan, Data Architect, Google*

Dataproc Cookbook is packed with practical recipes that accelerate the journey of leveraging Spark and Hadoop on Dataproc. The authors' extensive experience as Google Cloud Data Engineer consultants shines through in this guide that bridges the gap between foundational theory and real-world applications. *Dataproc Cookbook* is your invitation to leverage their expertise and boost your Google Cloud proficiency.

—*Vitalii Fedorenko, Data Engineer Manager, Google*

This book makes working with Dataproc clear and manageable. The structure guides you step by step, and the examples are easy to follow, turning a complex topic into a smooth learning experience.

—*Daniel Villaseñor, Cloud Data Architect, Google*

Dataproc Cookbook is the missing ingredient for any team looking to rapidly translate Big Data theory into tangible results on Google Cloud. It cuts through the complexity, offering practical, battle-tested recipes that empower you to conquer common challenges and accelerate your Dataproc journey from day one.

—*Ryan McDowell, Data and Analytics Practice Lead,*
Google Cloud

Drawing on years of practical and migration experience, my colleagues Anu and Narasimha have authored this indispensable cookbook for anyone serious about mastering Google Cloud Dataproc. It brilliantly bridges the gap from foundational knowledge to advanced optimization and features, offering a wealth of best practices and insights into leveraging Dataproc's full potential. A truly well-done resource!

—*Blake DuBois, Sr. Staff Technical Solutions Consultant,*
Google Cloud

Dataproc Cookbook
Running Spark and Hadoop
Workloads in Google Cloud

Narasimha Sadineni and Anuyogam Venkataraman

O'REILLY®

Dataproc Cookbook

by Narsimha Sadineni and Anuyogam Venkataraman

Copyright © 2025 Narsimha Sadineni and Anuyogam Venkataraman. All rights reserved.

Printed in the United States of America.

Published by O'Reilly Media, Inc., 1005 Gravenstein Highway North, Sebastopol, CA 95472.

O'Reilly books may be purchased for educational, business, or sales promotional use. Online editions are also available for most titles (*http://oreilly.com*). For more information, contact our corporate/institutional sales department: 800-998-9938 or *corporate@oreilly.com*.

Acquisitions Editor: Aaron Black	**Indexer:** nSight, Inc.
Development Editor: Jill Leonard	**Cover Designer:** Susan Thompson
Production Editor: Christopher Faucher	**Cover Illustrator:** Karen Montgomery
Copyeditor: Shannon Turlington	**Interior Designer:** David Futato
Proofreader: Piper Content Partners	**Interior Illustrator:** Kate Dullea

June 2025:   First Edition

Revision History for the First Edition

2025-06-02:   First Release

See *http://oreilly.com/catalog/errata.csp?isbn=9781098157708* for release details.

978-1-098-15770-8

[LSI]

Table of Contents

Preface

Welcome! We're thrilled you're diving into the world of Google Cloud Dataproc. Why are we so excited? Because efficiently handling massive datasets is no longer just a baseline requirement—it's the core engine powering today's most significant innovations, from deep business analytics to the incredible breakthroughs happening in artificial intelligence. Even as AI captures headlines, the fundamental truth remains: the quality, structure, and accessibility of your data determine the success of any analytics, machine learning, or AI initiative. The cleaner and more readily available your data, the greater the insights and advantages you can unlock.

The evolution of distributed systems for data processing has progressed from the constraints of single VMs, through the power of specialized Massively Parallel Processing (MPP) systems, to the revolutionary breakthrough of Hadoop utilizing clusters of commodity hardware—a shift that fundamentally redefined the scale of data we could handle. Technologies like Apache Hadoop (MapReduce, HDFS, Hive) allowed us to tackle data problems at a scale previously unimaginable, and to do so within practical time frames. Spark, with its in-memory processing capabilities, pushed the boundaries even further, enabling large-scale data operations in mere seconds.

However, managing the underlying infrastructure for these powerful tools often presented significant hurdles—long hardware procurement cycles, heavy upfront investments, and complex maintenance. This is where the cloud, and specifically Google Cloud Platform (GCP), enters the picture, offering a paradigm shift. Imagine accessing cutting-edge hardware like the latest GPUs, scaling resources up or down in minutes instead of months, and adopting a flexible, pay-as-you-go cost model. This agility is revolutionary!

Google Cloud Dataproc sits right at the heart of this exciting intersection. It provides a managed service designed to let you run your familiar Hadoop and Spark workloads (and other tools like Flink and Presto) seamlessly on GCP's robust infrastructure. This means you can migrate existing applications with minimal-to-no code changes, shedding the burden of infrastructure management and focusing instead on extracting value from your data. Dataproc makes leveraging the power and flexibility of the cloud for big data workloads incredibly straightforward—and that's something to be genuinely excited about! Until now, practical, consolidated resources beyond official documentation have been scarce, and this book aims to be your definitive guide. Packed with practical, tested recipes, it's your go-to guide for exploring the real-world power of Dataproc. While Dataproc is our primary focus, the underlying Google Cloud fundamentals explored here—including resource organization, IAM, logging, monitoring, and security—provide valuable, transferable knowledge applicable across the GCP ecosystem. Let's dive into harnessing the capabilities of Google Cloud Dataproc for your data.

Who Should Read This Book

This is a handy cookbook on Dataproc that will help you accelerate your Hadoop migration and Dataproc learning journey and optimize your workloads. It is designed for data engineers, data scientists, cloud architects, and more:

Data engineers
> Professionals responsible for designing, building, and maintaining data processing pipelines using Dataproc. This book will help you learn about the various features, best practices, and optimization techniques for managing big data workflows.

Data scientists
> Researchers and analysts who work with large datasets and need to perform advanced analytics and machine learning tasks. This book will help you understand how to leverage Dataproc's capabilities to process and analyze data effectively.

Cloud architects
> Professionals responsible for designing and implementing data processing solutions on Google Cloud Platform. This book will help you understand how to integrate Dataproc with other services and architectures to create scalable and efficient data processing systems.

Data analysts
> Individuals who work with data to derive insights and make informed business decisions. This book will help you learn how to leverage Dataproc's capabilities to process and transform data for analysis and reporting.

Students and researchers
> People studying data engineering, data science, or related fields who want to gain a comprehensive understanding of data processing technologies and how to use Dataproc effectively.

IT managers and decision makers
> Executives and managers responsible for making decisions regarding data infrastructure and processing solutions. This book will help you understand the benefits, costs, and use cases of adopting Dataproc for your organization.

Why We Wrote This Book

Enterprises migrate their big data workloads to Google Cloud, often leveraging Dataproc as the crucial first step in this journey from on-premises environments. Through our direct experience helping large companies migrate and build big data solutions in Google Cloud Dataproc, we consistently recognized a distinct need beyond the existing resources. While foundational knowledge on Hadoop, Spark, and GCP is readily available, and official Dataproc documentation comprehensively details its features, we saw teams new to Dataproc on GCP frequently feel overwhelmed when trying to translate this information into practical solutions for common, immediate tasks. We often wished for a focused, hands-on guide ourselves during these projects, which directly inspired the "cookbook" approach of this book, focusing on working code recipes for the most frequent patterns. Our primary goal is to bridge this gap between theory and practice, enabling you to learn by doing, gain confidence through successful implementation, and ultimately accelerate your ability to effectively leverage Dataproc in your day-to-day work on Google Cloud.

Navigating This Book

This book is structured to guide you progressively from fundamental concepts to more advanced topics and real-world applications of Dataproc. Here's a brief overview of what each chapter covers:

Chapter 1, "Creating a Dataproc Cluster"
> Introduces the basics of creating Dataproc clusters on Compute Engine

Chapter 2, "Running Hive, Spark, and Sqoop Workloads"
> Walks you through submitting various types of jobs (Spark, MapReduce, etc.) to your cluster

Chapter 3, "Advanced Dataproc Cluster Configuration"
> Dives into advanced cluster configurations like autoscaling, custom machine types, and managing dependencies

Chapter 4, "Serverless Spark and Ephemeral Dataproc Clusters"
> Explores Dataproc Serverless for Spark, focusing on minimizing infrastructure management and optimizing costs

Chapter 5, "Dataproc on Google Kubernetes Engine"
   Details how to deploy and manage Dataproc workloads using Google Kubernetes Engine (GKE)

Chapter 6, "Dataproc Metastore"
   Covers options for managing metadata, including Hive Metastore and integration with services like Dataplex

Chapter 7, "Connecting from Dataproc to GCP Services"
   Provides practical examples of integrating Dataproc with key GCP services like BigQuery and Cloud Storage

Chapter 8, "Configuring Logging in Dataproc", and Chapter 9, "Setting Up Monitoring and Dashboards"
   Focus on the operational essentials of logging and monitoring your Dataproc clusters and jobs, skills applicable across GCP

Chapter 10, "Dataproc Security"
   Addresses critical security aspects, from managing secrets to network security using VPC Service Controls

Chapter 11, "Performance Tuning and Cost Optimization"
   Offers strategies for tuning Dataproc cluster and job configurations to maximize performance and cost-efficiency

Chapter 12, "Orchestrating Dataproc Workloads"
   Explains how to orchestrate Dataproc workflows using Cloud Composer

Chapter 13, "Using Spark Notebooks on Dataproc"
   Focuses on using notebooks (e.g., Jupyter, Vertex AI Workbench) with Dataproc for interactive analysis and AI/ML development

Chapter 14, "Migrating from On-Premises and Public Cloud Services to GCP"
   Discusses key considerations and strategies for migrating existing big data workloads to Dataproc from on-premises or other clouds

Conventions Used in This Book

The following typographical conventions are used in this book:

Italic
   Indicates new terms, URLs, email addresses, filenames, and file extensions.

`Constant width`
   Used for program listings, as well as within paragraphs to refer to program elements such as variable or function names, databases, data types, environment variables, statements, and keywords.

Constant width italic
> Shows text that should be replaced with user-supplied values or by values determined by context.

This element signifies a tip or suggestion.

This element signifies a general note.

This element indicates a warning or caution.

Using Code Examples

Supplemental material (code examples, exercises, etc.) is available for download at *https://github.com/anuyogamlab/Dataproc-Cookbook*.

If you have a technical question or a problem using the code examples, please send email to *support@oreilly.com*.

This book is here to help you get your job done. In general, if example code is offered with this book, you may use it in your programs and documentation. You do not need to contact us for permission unless you're reproducing a significant portion of the code. For example, writing a program that uses several chunks of code from this book does not require permission. Selling or distributing examples from O'Reilly books does require permission. Answering a question by citing this book and quoting example code does not require permission. Incorporating a significant amount of example code from this book into your product's documentation does require permission.

We appreciate, but generally do not require, attribution. An attribution usually includes the title, author, publisher, and ISBN. For example: "*Dataproc Cookbook* by Narsimha Sadineni and Anuyogam Venkataraman (O'Reilly). Copyright 2025 Narsimha Sadineni and Anuyogam Venkataraman, 978-1-098-15770-8."

If you feel your use of code examples falls outside fair use or the permission given above, feel free to contact us at *permissions@oreilly.com*.

O'Reilly Online Learning

O'REILLY® For more than 40 years, *O'Reilly Media* has provided technology and business training, knowledge, and insight to help companies succeed.

Our unique network of experts and innovators share their knowledge and expertise through books, articles, and our online learning platform. O'Reilly's online learning platform gives you on-demand access to live training courses, in-depth learning paths, interactive coding environments, and a vast collection of text and video from O'Reilly and 200+ other publishers. For more information, visit *https://oreilly.com*.

How to Contact Us

Please address comments and questions concerning this book to the publisher:

O'Reilly Media, Inc.
1005 Gravenstein Highway North
Sebastopol, CA 95472
800-889-8969 (in the United States or Canada)
707-827-7019 (international or local)
707-829-0104 (fax)
support@oreilly.com
https://oreilly.com/about/contact.html

We have a web page for this book, where we list errata, examples, and any additional information. You can access this page at *https://oreil.ly/dataproc-cookbook*.

For news and information about our books and courses, visit *https://oreilly.com*.

Find us on LinkedIn: *https://linkedin.com/company/oreilly-media*

Watch us on YouTube: *https://youtube.com/oreillymedia*

Acknowledgments

Writing this book has been a deeply rewarding journey, and we couldn't have done it without the support of many people.

Narasimha Sadineni

I extend my deepest gratitude to my family for their unwavering support throughout the writing of this book. To my wife, Manjeera Koduri, thank you for your incredible patience and sacrifice; you lovingly cared for our young daughters, Yeshika Sadineni and Jaanvika Sadineni, giving me the precious time needed to focus on this project. I am also profoundly thankful to my father, Rangarao Sadineni, and my brother, Sandeep Sadineni, whose emotional encouragement and handling of personal matters allowed me the necessary space to write. To my dear Yeshika and Jaanvika, thank you for your understanding during the many hours this book required—I cherish all our time together.

Anu Venkataraman

My heartfelt thanks to my husband, Vengatesh Parasuraman, and my daughter, Gravity, for their unconditional love and endless support. Vengatesh—my best friend and constant source of energy—stood by me through every high and low, and without his unconditional love and support, this book would not have been possible. I'm also deeply grateful to my mom, Rukmani Prabha, and dad, Venkataraman, for always believing in me and cheering me on with unwavering love.

We're grateful to our champions at O'Reilly. To Aaron Black, for believing in our project and getting our proposal approved, which was instrumental in this book coming into existence. To Jill Leonard, for over a year of invaluable guidance, critical early reviews, and keeping our spirits high. Her role as our first reader was invaluable, and her help was essential to this book's completion. Our sincere thanks to Kristen Brown, Chris Faucher, Shannon Turlington, Kate Dullea, Tom Dinse, and Kim Sandoval for their contributions across various phases, significantly refining and enhancing this book.

We're deeply grateful to our technical reviewers—Ryan McDowell, Vitalii Fedorenko, Blake DuBois, Shlok Karpathak, Sushil Kumar, Karthik Palaniappan, Jerome Rajan, Manuel Robles, Aravind Srinivasan, Daniel Villaseñor, and Shailendra Maktedar—for their expertise and thoughtful feedback, which greatly enriched every chapter and recipe in this book.

## CHAPTER 1

Creating a Dataproc Cluster

Dataproc is a paid Google Cloud service built on top of open source software Apache Hadoop, Apache Spark, and other big data technologies, such as Apache Kafka, JupyterHub, and Apache Solr. As a managed service, Dataproc abstracts creating, updating, managing, and deleting all the required cloud services and resources.

Dataproc offers three different environments for running it:

- Dataproc on Google Compute Engine (GCE)
- Dataproc on Google Kubernetes Engine (GKE)
- Dataproc Serverless

In this chapter, we focus on the first option: running Dataproc on GCE (see Figure 1-1).

Before you start using this product, you should understand the billing and charges. This service has two types of charges: a charge for the software and a charge for the underlying components (compute engine, disks, cloud storage, network, etc.). Dataproc's pay-as-you-go model allows you to pay for only the services you use. For more information on pricing, refer to the Dataproc pricing documentation (*https://cloud.google.com/dataproc/pricing*). Dataproc Serverless has a different pricing model that we will discuss in Chapter 4.

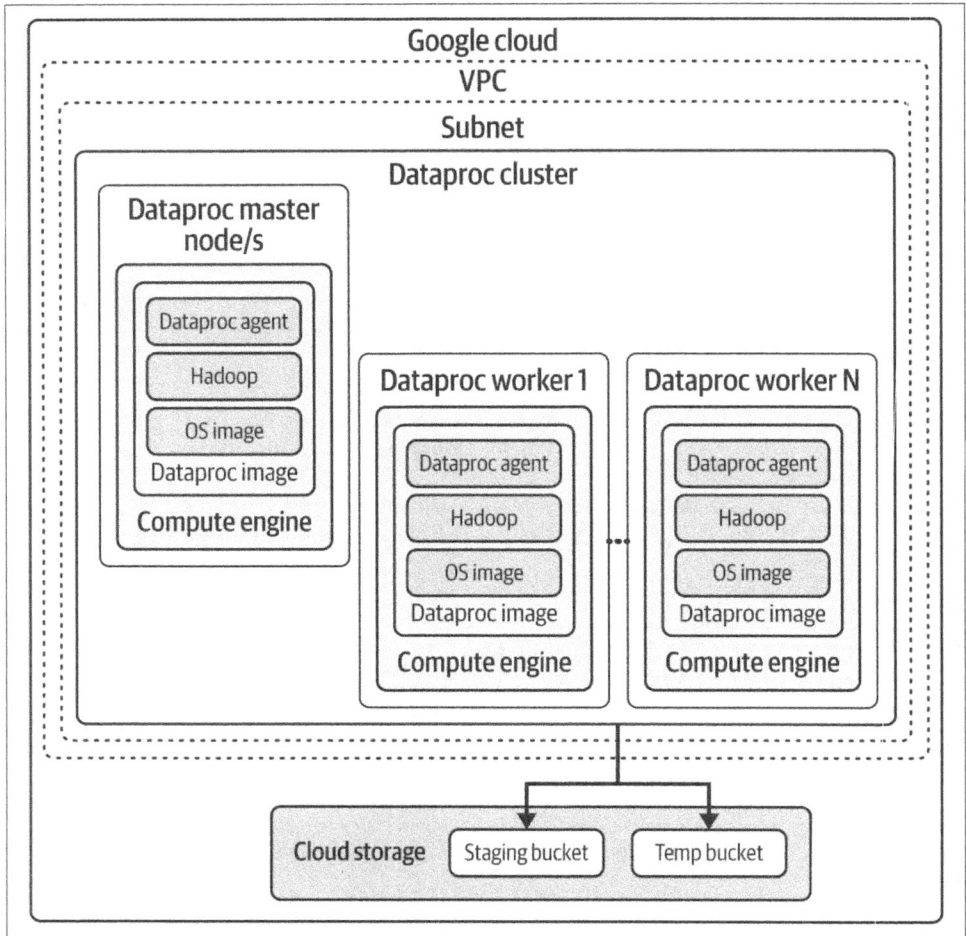

Figure 1-1. Dataproc on GCE high-level architecture diagram

The first step in the process of creating a Dataproc cluster is to secure a Google Cloud account. If you don't have one already, sign up for a new Google Cloud account at *cloud.google.com*.

> Google encourages new users to try out Google Cloud Platform (GCP) products by offering $300 in free credits across multiple services: Google's Free Tier products (*https://cloud.google.com/free*). Google can change or discontinue its Free Tier credits at any time. Refer to the Google Cloud documentation for the latest information.

Throughout your learning journey, the Dataproc product documentation (*https://cloud.google.com/dataproc/docs*) will help you gain knowledge along with this book. To keep updated about product releases, monitor the Dataproc release notes (*https://cloud.google.com/dataproc/docs/release-notes*). Google also offers a paid support model (*https://cloud.google.com/support*) to help you with any GCP-related issues. (If you are in an enterprise, your organization might already have purchased a support plan.) To participate in public discussions and ask questions, join the Google Group at *cloud-dataproc-discuss@googlegroups.com*.

This chapter provides a basic understanding of the prerequisites for creating a Dataproc cluster as well as the components that make up a cluster. We will discuss the various options for creating and customizing Dataproc clusters. Let's get started with the key components to install prior to working on Dataproc.

1.1 Installing Google Cloud CLI

Problem

You want to install Google Cloud (gcloud) CLI on your machine to interact with GCP services using the command line.

Solution

Download the gcloud CLI software from the Google Cloud SDK downloads repository (see the "Discussion" section that follows).

> Alternatively, you can use a browser-based cloud shell that comes with Cloud SDK, gcloud, Cloud Code, an online code editor, and other utilities preinstalled, fully authenticated, and up-to-date. To access the cloud shell, open *console.cloud.google.com?cloudshell=true* in your browser.

Discussion

Installation instructions for the gcloud CLI software differ based on your machine type (Mac, Windows, or Linux).

Mac users should follow these steps:

1. Open *cloud.google.com/sdk/docs/install#mac* in your browser.
2. Check that the supported version of Python is installed on your machine. At the time of writing, the minimum required Python version is Python 3. Run the commands `python3 -V` or `python -V` to determine which version of Python you have.

3. If Python is missing, download Python and install it (*https://www.python.org/downloads/macos*).

4. Determine your Mac architecture by running the command `uname -m`.

5. Download the package that matches your Mac's machine architecture (x86_64, ARM64, or x86) from the link in step 2 of *gCloud CLI installation instructions* (*https://oreil.ly/qIkn1*).

6. The downloaded file is in compressed (*tar.gz*) format. Extract the compressed archive file using the following command:

   ```
   tar -xvzf downloaded-file-name
   ```

7. Install the CLI with the command `./google-cloud-sdk/install.sh`.

8. After installation is successful, run the command `gcloud init` to authenticate with Google Cloud. You will be asked to configure the default user, project, and region/zone to use.

Windows users should follow these steps:

1. Download the installer (*https://oreil.ly/0Khh7*).

2. Run the installer by double-clicking it, and follow the prompts.

Linux users should follow these steps:

1. Determine your Linux machine architecture (32-bit or 64-bit) by using the command `getconf LONG_BIT`.

2. Check the Python version using the commands `python3 -V` or `python -V`. The minimum required version is Python 3.

3. If you are using a 64-bit machine, Python is bundled with the installer, and there is no need to install it manually. Otherwise, download (*https://python.org/downloads*) and install it.

4. Download the CLI package that matches your Linux machine architecture (32- or 64-bit) from the link in step 2 of *gCloud CLI installation instructions* (*https://oreil.ly/3jPrp*).

5. The downloaded file is in compressed (*tar.gz*) format. Extract the compressed archive file using the following command:

   ```
   tar -xvzf downloaded-file-name
   ```

6. Install the CLI with the command `./google-cloud-sdk/install.sh`.

7. Restart the cloud shell/terminal.

8. After installation is successful, run the command `gcloud init` to authenticate with Google Cloud. You will be asked to configure the default user, project, and region/zone to use.

Now you have successfully downloaded, installed, and set up authentication for gcloud CLI and are ready to run the commands on your machine. To verify the installation, issue the command `gcloud` at the command line. Successful execution will display the help screen.

1.2 Granting Identity and Access Management Privileges to a User

Problem

You want to use gcloud commands to grant the necessary identity and access management (IAM) permissions to a user or service account for creating a Dataproc cluster.

Solution

Run the following commands in your terminal, replacing the parameters enclosed in <> with your actual values.

Use this command to grant an IAM role for a specific user account:

```
gcloud projects add-iam-policy-binding <PROJECT_ID> \
  --member="<user:EMAIL_ADDRESS>" \
  --role=roles/dataproc.editor
```

Use this command to grant permission for a service account:

```
gcloud iam service-accounts add-iam-policy-binding \
  <compute_engine_default_account> \
  --member="<user:EMAIL_ADDRESS>" \
  --role=roles/iam.serviceAccountActor
```

Discussion

A *user* in this context refers to a human individual who authenticates with their Google account to manage and interact with GCP services. A *service account* is a special type of Google account designed for nonhuman entities, such as applications running on GCE or other GCP services.

IAM is the service that controls who can access which resource in Google Cloud. IAM privileges are required for users or service accounts that create Dataproc clusters. An IAM *role* is a collection of permissions defining the actions that a principal (user, service account, or group) is allowed to perform on Google Cloud resources.

Accessing IAM also requires an IAM role (viewer, editor, or owner). If you own the project, you will get the owner role for accessing IAM, or check with your project or platform admin to get the required access.

Resources in Google Cloud are organized in hierarchical format with Organization as the parent, followed by Folders, Projects, Services, and Resources (Google Cloud Storage bucket, Compute Engine, Dataproc cluster, BigQuery table, etc.), as shown in Figure 1-2.

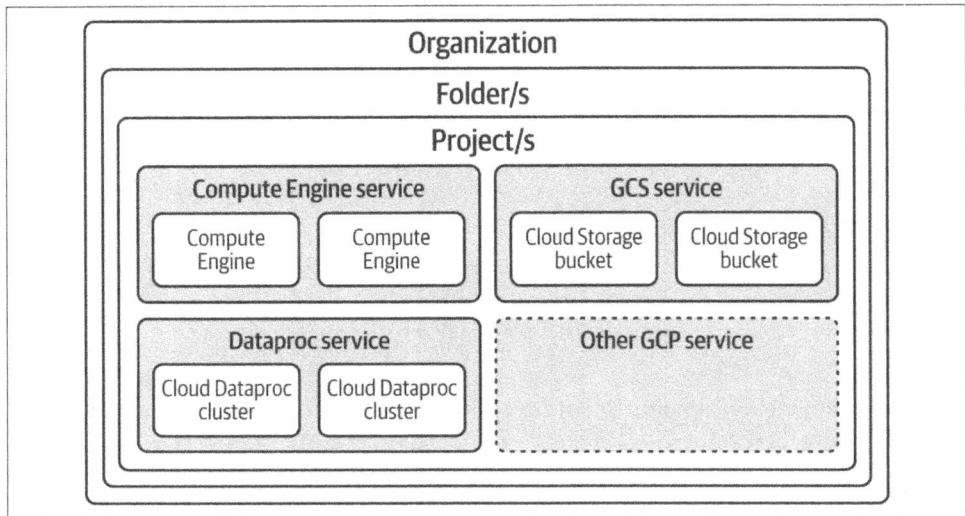

Figure 1-2. GCP resources hierarchy for IAM policy inheritance

IAM policies created at the parent level (within a hierarchy) are inherited by child components. For instance, the editor role assigned at the project level will be inherited by all services and resources created within that project. Similarly, an editor role assigned at the Dataproc service level will be inherited by all clusters within that Dataproc service.

For most of the resources (Folders, Projects, and Services like Dataproc or GCS), IAM offers basic roles that can be applied across all components of that service, which are listed in Table 1-1.

Table 1-1. Basic IAM roles for Google Cloud services

IAM role	Description
Viewer	Read-only access
Editor	All viewer permissions plus access to create, modify, and delete resources
Owner	All editor permissions plus additional high-level administrative permissions, such as managing IAM permissions, setting up billing accounts, and deleting projects

The gcloud command for creating editor access at the project level is as follows:

```
export PROJECT_ID=<PROJECT_ID>
export SERVICE_ACCOUNT_EMAIL=dataprociamtest0827
gcloud projects add-iam-policy-binding ${PROJECT_ID} \
  --member="user:${SERVICE_ACCOUNT_EMAIL}" \
  --role=roles/editor
```

Granting editor access at the project level gives editor access to all services in the project. To limit the user to having access only to Dataproc services, Dataproc has predefined roles, which are listed in Table 1-2.

Table 1-2. Predefined IAM roles for Dataproc services

IAM role	Description
Dataproc administrator	Grants full control over Dataproc resources
Dataproc editor	Grants permission to create and manage clusters and view the underlying resources
Dataproc viewer	Read-only access to Dataproc resources
Dataproc worker	Assigned to Compute Engine machines for performing cluster tasks

To assign the Dataproc editor role to a user, run the following gcloud command:

```
export PROJECT_ID=<PROJECT_ID>
export SERVICE_ACCOUNT_EMAIL=dataprociamtest0827
gcloud projects add-iam-policy-binding ${PROJECT_ID} \
  --member="user:${SERVICE_ACCOUNT_EMAIL}" \
  --role=roles/dataproc.editor
```

Each IAM role is a collection of permissions. Granting the Dataproc editor role to users will give them permission to create clusters as well as additional privileges to manage and delete clusters. For fine-grained permissions specific to cluster creation, you can create a custom role.

Here is the gcloud command to create a custom role:

```
export PROJECT_ID=<PROJECT_ID>
gcloud iam roles create custom.dataprocEditor \
  --project=${PROJECT_ID} \
  --title="Custom Dataproc Editor" \
  --description="Custom role for creating and managing Dataproc clusters" \
  --permissions=dataproc.clusters.create
```

Let's assign this custom role to a user:

```
export PROJECT_ID=<PROJECT_ID>
export SERVICE_ACCOUNT_EMAIL=<SERVICE_ACCOUNT_EMAIL>
gcloud projects add-iam-policy-binding ${PROJECT_ID} \
  --member=${SERVICE_ACCOUNT_EMAIL} \
  --role=roles/custom.dataprocEditor
```

Dataproc internally uses two types of service accounts:

- Dataproc VM service account
- Dataproc Service Agent service account

The Dataproc VM service account is used to create underlying resources like Compute Engine instances and to perform dataplane operations, such as reading and writing data to Google Cloud Storage (GCS). Dataproc uses the Compute Engine default service account as the Dataproc VM service account, but this can be customized using the `--service-account` option.

Users creating clusters require access to the Dataproc VM service account as a service account user:

```
export SERVICE_ACCOUNT_EMAIL=<SERVICE_ACCOUNT_EMAIL>

gcloud iam service-accounts add-iam-policy-binding \
  <COMPUTE_ENGINE_DEFAULT_SERVICE_ACCOUNT>  \
  --member="serviceAccount:${SERVICE_ACCOUNT_EMAIL}" \
  --role=roles/iam.serviceAccountActor
```

> The project's Compute Engine default service account can be listed using the gcloud command:
>
> ```
> gcloud iam service-accounts list \
>   --filter="displayName:Compute Engine default \
>   service account" \
>   --project=<PROJECT_ID_HERE>
> ```
>
> Dataproc automatically creates a service account known as the Dataproc Service Agent. This account is responsible for performing control plane operations on your behalf, such as creating, updating, and deleting Dataproc clusters. This automatically created service account cannot be replaced with a custom service account.

1.3 Configuring a Network and Firewall Rules

Problem

You want to create a new virtual private cloud (VPC) network for hosting virtual machines (VMs), and to attach firewall rules for allowing communication between the machines.

Solution

Create a VPC network:

```
gcloud compute networks create <NETWORK_NAME> \
  --subnet-mode auto \
  --description "VPC network hosting dataproc resources"
```

Attach a firewall rule to the network:

```
gcloud compute firewall-rules create <FIREWALL_NAME> --network dataproctest \
  --allow [PROTOCOL[:PORT]] --source-ranges <IP_RANGE>
```

Discussion

Compute Engines that are part of a Dataproc cluster must reside within a VPC network to communicate with one another and with external resources when necessary. The Dataproc service mandates that all nodes of the cluster be able to communicate with one another using the ICMP, TCP (all ports), and UDP (all ports) protocols.

The default project network typically has subnets created in the range of 10.128.0.0/9. It also includes the "default-allow-internal" firewall rule, permitting communication within this subnet range. Table 1-3 shows an example of this rule. If you are creating a custom network, ensure that you establish a rule aligned with Dataproc's requirements to enable internal communication.

Table 1-3. Firewall rule requirements for a Dataproc cluster network

Direction	Priority	Source range	Protocols:ports
ingress	65534	For default network: 10.128.0.0/9 For a custom VPC or subnet, use the custom subnet range.	tcp:0-65535, udp:0-65535, icmp

To create a VPC network in auto mode, run the following command:

```
gcloud compute networks create dataproc-vpc \
  --subnet-mode auto \
  --description "VPC network hosting dataproc resources"
```

Configuring the subnet mode as `auto` creates multiple subnets (one for each GCP region). This VPC creates subnets in all available regions, allowing you to create a Dataproc cluster in any region.

To prevent the automatic generation of too many (40+) subnets in auto mode, you can just create subnets in the required regions. This is a two-step process: first create a VPC and then add a subnet to it. Creating a VPC with custom subnet mode results in an empty VPC without any subnets:

```
gcloud compute networks create dataproc-vpc \
  --subnet-mode custom \
  --description "VPC network hosting dataproc resources"
```

The following code creates a subnet in the us-east1 region with a range of 10.120.0.0/20:

```
gcloud compute networks subnets create dataproc-vpc-us-east1-subnet \
  --network=dataproc-vpc \
  --region=us-east1 \
  --range=10.120.0.0/20
```

Understanding Subnet Ranges for Dataproc Clusters

A subnet range defines the pool of IP addresses available within a network. When planning for a certain number of nodes in your network, you need to choose a subnet range large enough to accommodate them, including the network and broadcast addresses.

To accommodate around 256 usable IP addresses, you need to consider that a subnet also requires a network address and a broadcast address. A subnet with $2^8 = 256$ total addresses would have 254 usable addresses (256 − 2). Therefore, a /24 subnet mask is suitable. An example range would be 10.120.1.0/24. This subnet has 256 total addresses (from 10.120.1.0 to 10.120.1.255), with 10.120.1.0 being the network address and 10.120.1.255 being the broadcast address, leaving 254 usable IP addresses for your nodes.

To accommodate around 512 usable IP addresses, you need a subnet with $2^9 = 512$ total addresses, providing 510 usable addresses (512 − 2). A /23 subnet mask will provide this capacity. An example range would be 10.120.2.0/23. This subnet has 512 total addresses (from 10.120.2.0 to 10.120.3.255), with 10.120.2.0 as the network address and 10.120.3.255 as the broadcast address, leaving 510 usable IP addresses.

To choose a suitable subnet range for the maximum number of hosts on a Dataproc cluster, you will need to consider the expected number of hosts and allow room for growth.

Resources in a VPC are not reachable until you create a firewall rule to allow communication. Attach the firewall rule matching the Dataproc service requirements. To create a firewall rule for a custom subnet with an IP range 10.120.0.0/20, use this command (output shown in Figure 1-3):

```
gcloud compute firewall-rules create dataproc-allow-tcp-udp-icmp-all-ports \
    --network dataproc-vpc \
    --allow tcp:0-65535,udp:0-65535,icmp \
    --source-ranges "10.120.0.0/20"
```

Figure 1-3. Successful firewall rule creation output

See Also

Refer to the Google Cloud public documentation (*https://oreil.ly/G9xfW*) to learn more about VPC network and firewall rules.

1.4 Creating a Dataproc Cluster from a Web UI

Problem

You want to create a Dataproc cluster using a web UI.

Solution

Google Cloud console is a web-based UI for accessing Google Cloud services that offers the option of creating and managing Dataproc clusters.

Discussion

To create a Dataproc cluster using a web UI, first log in to Google Cloud console (*https://console.cloud.google.com*), as shown in Figure 1-4.

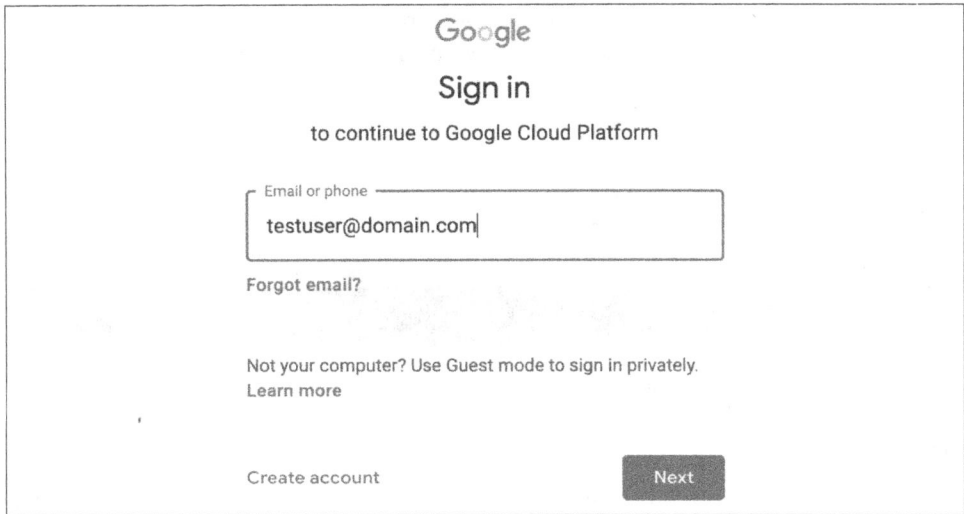

Figure 1-4. Sign in to Google Cloud console

Once you are logged in to Google Cloud console, you will see the dashboard or home page, as shown in Figure 1-5.

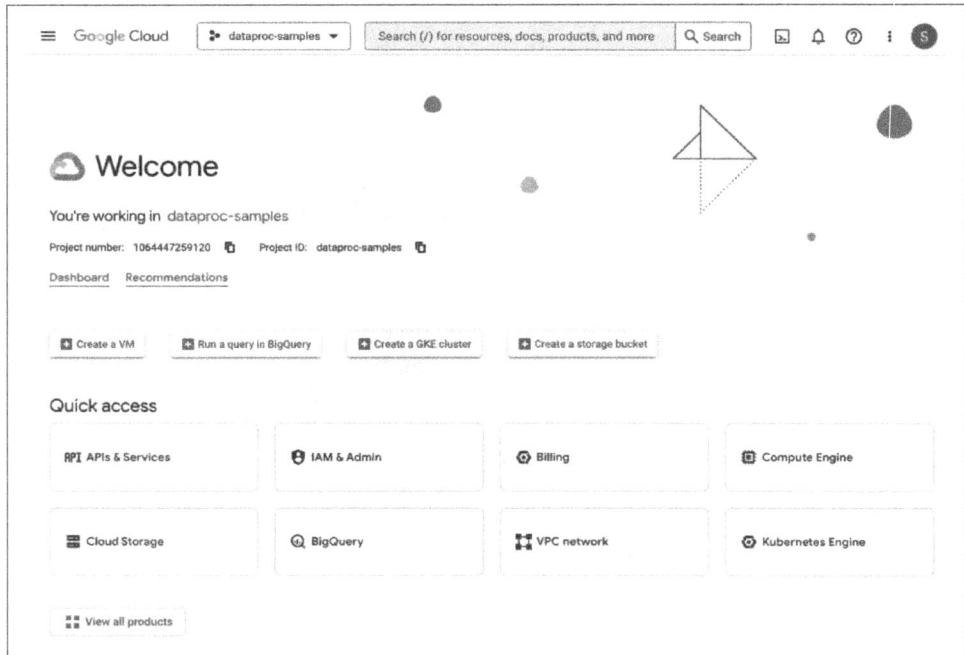

Figure 1-5. Google Cloud console home page

In the search bar, enter the keyword "dataproc" and select the Dataproc service, as shown in Figure 1-6.

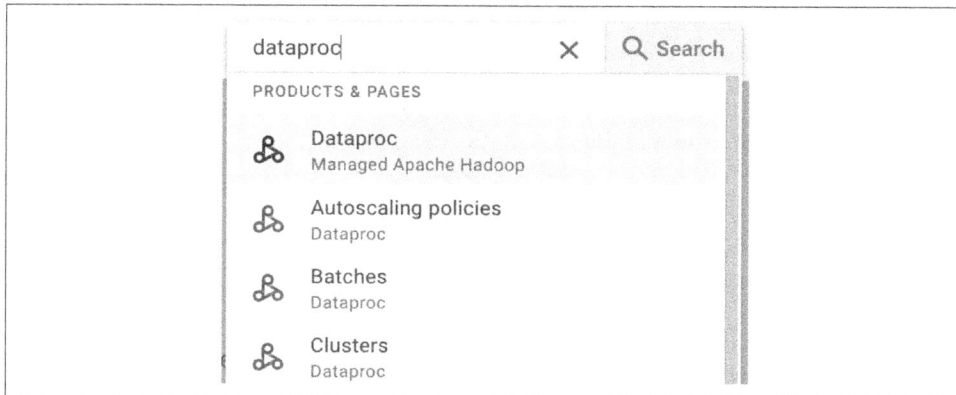

Figure 1-6. Searching for the Dataproc service in Google Cloud console

If the Dataproc API is not enabled in the project, you might be prompted to enable the service. The Dataproc service is not enabled by default. Click the Enable button to enable the service, as shown in Figure 1-7.

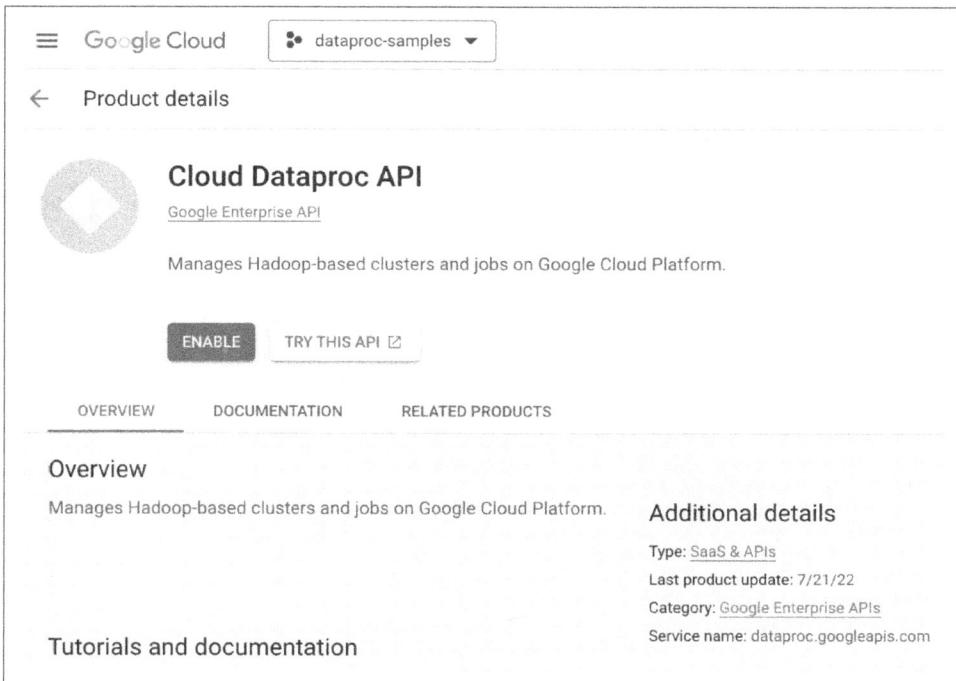

Figure 1-7. Enabling the Dataproc API service

> If the billing account is not linked to the project, you might be asked to link an existing account or to create a new billing account.

Selecting Dataproc from the search takes you to the Dataproc service home page. Click Create Cluster, shown in Figure 1-8, to create a new cluster.

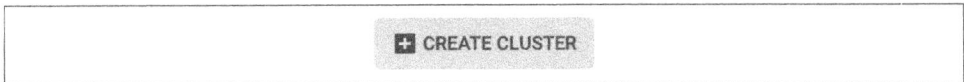

Figure 1-8. Create Cluster button for creating a Dataproc cluster

Dataproc clusters can be created on Compute Engine or GKE. Here, we will select the "Cluster on Compute Engine" option, as shown in Figure 1-9.

Figure 1-9. Console UI showing the different options for creating a Dataproc cluster

To create a basic cluster with all default values, just enter the cluster name and select the region, as shown in Figure 1-10. Then, click the Create button.

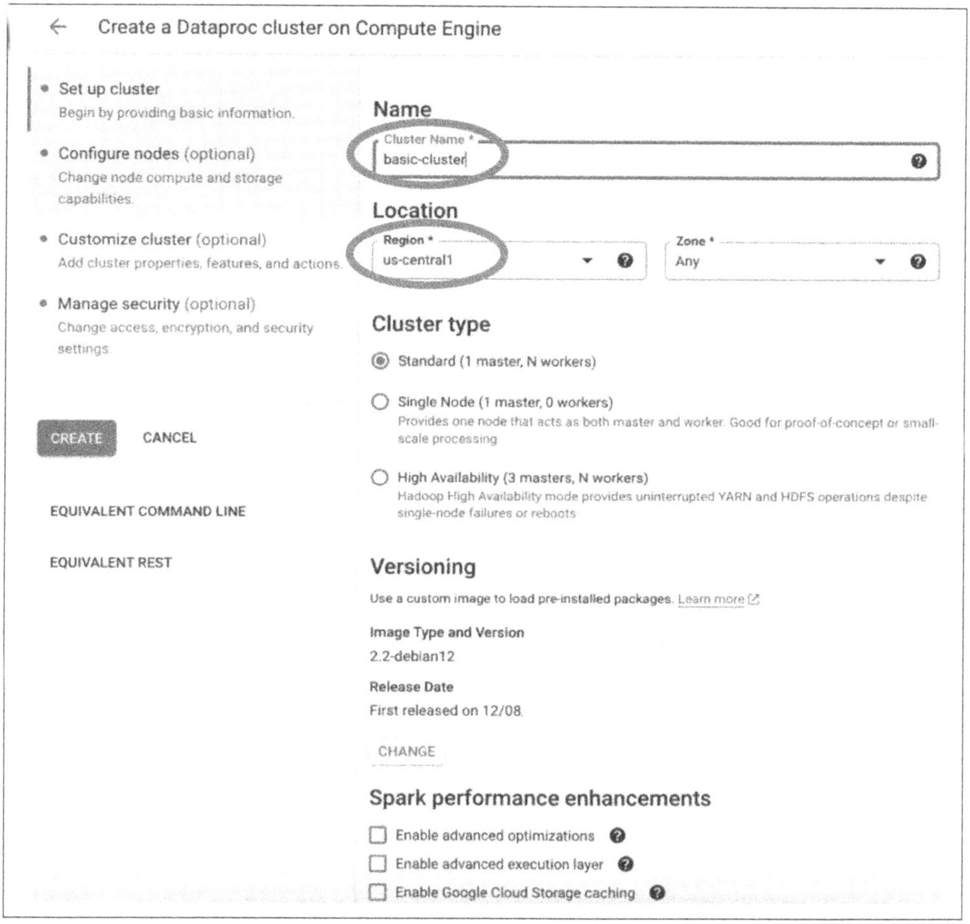

Figure 1-10. Entering the cluster name and region

Creating a cluster takes up to 90 seconds. Once the cluster is successfully created, you will see it in the Dataproc service home page, as shown in Figure 1-11. Click the linked cluster name to see the details of the cluster.

	Name ↑	Status	Region	Zone	Total worker nodes	Flexible VMs?	Scheduled deletion
☐	basic-cluster	✔ Running	us-central1	us-central1-c	2	No	Off

Figure 1-11. Dataproc service page listing available clusters

1.5 Creating a Dataproc Cluster Using Gcloud

Problem

Manually creating multiple clusters from the web UI is time-consuming. How can you accelerate development and testing from a local machine by creating clusters from the command line?

Solution

Install a gcloud CLI on your machine and run a `gcloud dataproc clusters create` command. This is the command to create a cluster with a basic configuration:

```
gcloud dataproc clusters create basic-cluster --region us-central1
```

This is the command to create a cluster with a custom configuration (machine types, disk, network, etc.):

```
gcloud dataproc clusters create basic-cluster \
  --region us-central1 \
  --zone "" \
  --image-version 2.0-debian10 \
  --master-machine-type n1-standard-4 \
  --worker-machine-type n1-standard-8 \
  --master-boot-disk-type pd-ssd \
  --master-boot-disk-size 100 \
  --worker-boot-disk-type pd-ssd \
  --worker-boot-disk-size 200 \
  --num-worker-local-ssds 2 \
  --network default \
  --enable-component-gateway
```

Discussion

The minimum required arguments to create a Dataproc cluster are the cluster name and the region. The following command creates a Dataproc cluster with the name `basic_cluster` in the region `us-central1`:

```
gcloud dataproc clusters create basic-cluster --region us-central1
```

The cluster created from the command will use the defaults listed in Table 1-4.

Table 1-4. Default values when creating a cluster with a name and region

Property	Default value
Number of master nodes	One
Number of primary workers	Two
Machine type	Chooses a machine type based on the Dataproc internal configuration

Property	Default value
Network	Uses the network with the name "default" that is available in the project when no network is specified
Zone	Intelligently picks a zone within the region specified
Dataproc version	Defaults to the latest version available
Disk type	Standard persistent disk
Disk size	1,000 GB
Component gateway	Disabled

> The default values could be changed by Google over time. For the latest values, refer to the documentation (*https://oreil.ly/oaECw*).

Additional customizations, such as machine types, secondary workers, disk types and sizes for primary and secondary workers, high availability, and component gateways, can also be configured using gcloud. Let's look at the command to customize a few more components in the cluster:

```
gcloud dataproc clusters create basic-cluster \
    --region us-central1 \
    --zone "" \
    --image-version 2.0-debian10 \
    --master-machine-type n1-standard-4 \
    --worker-machine-type n1-standard-8 \
    --master-boot-disk-type pd-ssd \
    --master-boot-disk-size 100 \
    --worker-boot-disk-type pd-ssd \
    --worker-boot-disk-size 200 \
    --num-worker-local-ssds 2 \
    --network default \
    --enable-component-gateway
```

From this command, the following components were customized:

region
> The region is where your cluster will be created.

zone
> Google Cloud has multiple zones in each region. For example, the us-central1 region has zones us-central1-a, us-central1-b, and us-central1-c. If you do not specify a zone when creating a Dataproc cluster, Dataproc will choose a zone for you in the specified region.

image-version

A combination of the operating system and Hadoop technology stack. Image version 2.0-debian10 comes with the Debian 10 operating system and Apache Hadoop 3.x and Apache Spark 3.1. It is recommended to explicitly specify the image version when creating Dataproc clusters to ensure consistency in the cluster configuration. Refer to the Dataproc cluster image version lists (*https://oreil.ly/xbwlD*) for available and supported Dataproc versions.

master-machine-type

Compute Engine type for installing master node services.

master-boot-disk-type

Disk type to be attached to the master node. Accepted values are pd-ssd, pd-standard, and pd-balanced.

master-boot-disk-size

Size of the master boot disk. By default, the values given are assumed to be in gigabytes. A value of 100 refers to a 100 GB disk to be attached to the master node.

worker-boot-disk-type

Disk type to be attached to the worker node. Accepted values are pd-ssd, pd-standard, and pd-balanced.

> A Dataproc cluster consists of VMs provided by Google Cloud's Compute Engine service. Choose a VM machine type that suits your data-processing needs. GCP offers a variety of machine types:
>
> - General purpose (N1, N2, N2D, T2D, T2A, etc.)
> - Cost optimized (E2)
> - Memory optimized (M1)
> - CPU optimized (C2, C2D, C3)
> - Custom machine types that can be created with custom memory and CPU configurations
>
> For data pipelines, general-purpose machine types N2D and N2 are popular choices. High-performance compute workloads use C3 machine types or GPUs.

worker-boot-disk-size

Size of the worker boot disk. By default, the values given are assumed to be in gigabytes. Value 100 refers to a 100 GB disk to be attached to the master node.

Dataproc clusters require storage attached to compute nodes for storing persistent or temporary data. GCP offers the following storage options:

- PD (persistent disk—either standard or balanced)
- PD SSD (persistent disk solid-state drive)
- Local SSD

Local SSDs are a recommended choice for Dataproc worker nodes as they offer greater performance than PD standard and are less expensive than PD SSD.

num-worker-local-ssds

Local SSDs are the recommended storage for worker nodes. They offer higher performance than standard disks with a good price-to-performance ratio.

num-worker

Number of primary worker nodes. You can also add secondary worker nodes that do compute only with no storage (HDFS).

Primary workers are the only worker machine types that have a DataNode component for storing Hadoop Distributed File System (HDFS) data. Choose the primary workers based on the amount of HDFS storage needed. Not all of your data goes to HDFS. In later chapters, we will cover what gets stored in HDFS versus the local filesystem versus GCS.

network

Virtual network that pools cloud resources together. When you create a project, it comes with a default network.

enable-component-gateway

Creates access to web endpoints for services like Google Cloud Resource Manager, NameNode web UI, and Spark history server.

We will learn more about these customizations in later chapters.

To view the list of clusters available in a region and project, use this command:

```
gcloud dataproc clusters list --project {project_name_here} \
  --region {region_name_here}
```

To delete a cluster, use this command:

```
gcloud dataproc clusters delete basic_cluster --region=us-central1
```

1.6 Creating a Dataproc Cluster Using API Endpoints

Problem

You want to create a cluster using the REST API to make the process of cluster creation platform independent.

Solution

The following `curl` command creates a new Google Cloud Dataproc cluster by making a POST request to the Dataproc API endpoint. It authenticates using your current gcloud credentials and sends the cluster configuration defined in the *request.json* file:

```
curl -X POST \
  -H "Authorization: Bearer $(gcloud auth print-access-token)" \
  -H "Content-Type: application/json; charset=utf-8" \
  -d @request.json \
  "https://dataproc.googleapis.com/v1/projects/[project_name]/regions/
<region-name>/clusters"
```

Discussion

Create a JSON request file, *request.json*, with the required configuration:

```
{
  "projectId": "dataproctest",
  "clusterName": "dataproc-test-cluster",
  "config": {
    "gceClusterConfig": {
      "networkUri": "default",
      "zoneUri": "us-central1-c"
    },
    "masterConfig": {
      "numInstances": 1,
      "machineTypeUri": "n2-standard-4",
      "diskConfig": {
        "bootDiskType": "pd-standard",
        "bootDiskSizeGb": 500,
        "numLocalSsds": 0
      }
    },
    "softwareConfig": {
      "imageVersion": "2.1-debian11",
    },
    "workerConfig": {
      "numInstances": 2,
      "machineTypeUri": "n2-standard-4",
      "diskConfig": {
        "bootDiskType": "pd-standard",
        "bootDiskSizeGb": 500,
```

```
            "numLocalSsds": 2,
            "localSsdInterface": "SCSI"
        }
      }
    },
    "labels": {
      "billing_account": "test-account"
    }
}
```

Making a REST API call to Google Cloud services requires the user to provide authorization tokens. Authenticate from the command line prior to executing the curl command.

To authenticate with a personal account, run the command:

```
gcloud auth login
```

To authenticate as a service account using a credential JSON file, run the command:

```
gcloud auth activate-service-account --key-file=<credential-json-file-location>
```

Execute the following curl command by replacing *project_name* and *region_name* (e.g., us-central1):

```
curl -X POST \
  -H "Authorization: Bearer $(gcloud auth print-access-token)" \
  -H "Content-Type: application/json; charset=utf-8" \
  -d @request.json \
  "https://dataproc.googleapis.com/v1/projects/{project_name}/regions/
{region-name}/clusters"
```

Successful execution of the curl command will give the following output:

```
{
"name": "projects/{project-name}/regions/{region-name}/operations/b5706e31.....",
  "metadata": {
    "@type": "type.googleapis.com/google.cloud.dataproc.v1.ClusterOperation
Metadata",
    "clusterName": "cluster-name",
    "clusterUuid": "5fe882b2-...",
    "status": {
      "state": "PENDING",
      "innerState": "PENDING",
      "stateStartTime": "2019-11-21T00:37:56.220Z"
    },
    "operationType": "CREATE",
    "description": "Create cluster with 2 workers",
    "warnings": [
      "For PD-Standard without local SSDs, we strongly recommend provisioning
1TB...""
    ]
  }
}
```

1.7 Creating a Dataproc Cluster Using Terraform

Problem

You want to automate the provisioning and managing of clusters with an infrastructure-as-code (IaC) framework.

Solution

Terraform is an IaC tool that enables users to create and maintain cloud infrastructure using declarative configuration language.

Discussion

Terraform is a widely used IaC tool for creating, maintaining, and managing cloud platform resources. This tool supports multiple cloud vendors, including Amazon Web Services (AWS), GCP, Microsoft Azure, and so on.

Install Terraform following the instructions in the public documentation (*https://oreil.ly/ITWBt*). Terraform code execution involves the following:

init
: Initializes a state of resources

plan
: Runs a preview and lets you know all changes to be applied on top of the current state of resources

apply
: Applies the changes on resources

destroy
: Deletes all the resources

The following is sample Terraform code to create a basic Dataproc cluster with no customizations:

```
provider "google" {
 credentials = file("service-account-credentials-file.json")
 project     = "project-id"
 region      = "us-central1"
}


resource "google_dataproc_cluster" "clusterCreationResource" {
 provider = google
 name     = "basic-cluster"
 region   = "us-central1"
```

```
  cluster_config {
    gce_cluster_config {
    network = google_compute_network.dataproc_network.name
  }

    master_config {
      num_instances     = 1
      machine_type      = "n1-standard-4"
    }
    worker_config {
      num_instances     = 2
      machine_type      = "n1-standard-8"
    }
    endpoint_config {
      enable_http_port_access = "true"
    }
 }
}
resource "google_compute_network" "dataproc_network" {
  name                    = "basic-cluster-network"
  auto_create_subnetworks = true
}
resource "google_compute_firewall" "firewall_rules" {
  name    = "basic-cluster-firewall-rules"
  network = google_compute_network.dataproc_network.name
  // Allow ping
  allow {
    protocol = "icmp"
  }
  //Allow all TCP ports
  allow {
    protocol = "tcp"
    ports    = ["1-65535"]
  }
  //Allow all UDP ports
  allow {
    protocol = "udp"
    ports    = ["1-65535"]
  }
  source_ranges = ["0.0.0.0/0"]
}
```

This Terraform configuration defines the necessary GCP infrastructure for a basic Dataproc cluster. The provider "google" block configures the connection to GCP, using a service account file for authentication and setting the project ID and default region. The configuration then defines three resources: a google_com pute_network named "basic-cluster-network" with auto-created subnetworks; associated google_compute_firewall rules allowing all ICMP, TCP, and UDP traffic from any source; and finally the google_dataproc_cluster itself named "basic-cluster". This cluster is placed in the specified region and network, configured with one n1-standard-4 master node, two n1-standard-8 worker nodes, and enabled HTTP port access for web UIs.

Save the Terraform sample code in the *main.tf* file. Then, navigate to the folder that has your *main.tf* file and run a command to initialize Terraform:

```
terraform init
```

Run a plan command:

```
terraform plan
```

Run `terraform apply` to review the execution plan and apply the changes needed to create the cluster:

```
terraform apply
```

To destroy the cluster, run a `destroy` command:

```
terraform destroy
```

> Terraform maintains the state of all resources it created in a file named *terraform.tfstate*. When running multiple times, Terraform compares the configuration (*.tf* files), the recorded state (state file), and the actual target environment state and applies updates only where needed. The `destroy` option will delete all the resources it created and maintained in the state.

1.8 Creating a Cluster Using Python

Problem

You want to automate creating a cluster using the Python programming language.

Solution

Google Cloud Dataproc offers Python client libraries for interacting with Dataproc services. Here is the code for creating a Dataproc cluster:

```python
from google.cloud import dataproc_v1

def create_dataproc_cluster(project_id, region, cluster_name):
    """Creates a Dataproc cluster."""
    dataproc_cluster_client = dataproc_v1.ClusterControllerClient()

    # Create the cluster config.
    cluster = {
        "project_id": project_id,
        "cluster_name": cluster_name,
        "config": {
```

```
            "master_config": {
                "num_instances": 1,
                "machine_type_uri": "n1-standard-2",
            },
            "worker_config": {
                "num_instances": 2,
                "machine_type_uri": "n1-standard-2",
            },
        },
    }


    operation = dataproc_cluster_client.create_cluster(
        project_id=project_id,
        region="us-central1",
        cluster=cluster
    )


    #result = operation.result()

    print(f"Created Dataproc cluster: {cluster.cluster_name}")


if __name__ == "__main__":
    project_id = "PROJECT-ID"
    region = "REGION"
    cluster_name = "CLUSTER-NAME"

    create_dataproc_cluster(project_id, region, cluster_name)
```

Discussion

Running a Python-based SDK requires installing the Python package google-cloud-dataproc. To install google-cloud-dataproc using pip, execute the following command:

```
pip install google-cloud-dataproc
```

Let's understand how the Python code creates a Dataproc cluster step by step.

The code first imports the dataproc_v1 module from the google.cloud package. This module provides the Python client library for the Google Cloud Dataproc API:

```
from google.cloud import dataproc_v1
```

The create_dataproc_cluster() function takes the three arguments project ID, region, and cluster name:

```
def create_dataproc_cluster(project_id, region, cluster_name)
```

The function first creates a client object for the `ClusterControllerClient` class. This class provides methods for creating, managing, and monitoring Dataproc clusters:

```
dataproc_cluster_client = dataproc_v1.ClusterControllerClient()
```

The function then creates a cluster configuration object. The cluster configuration object specifies the configuration of the cluster, such as the number of master and worker nodes, the machine types for the nodes, and the software that should be installed on the nodes:

```
# Create the cluster config.
    cluster = {
        "project_id": project_id,
        "cluster_name": cluster_name,
        "config": {
            "master_config": {
                "num_instances": 1,
                "machine_type_uri": "n1-standard-2"
            },
            "worker_config": {
                "num_instances": 2,
                "machine_type_uri": "n1-standard-2"
            },
        },
    }
```

The function then calls the `create_cluster()` method on the client object. The `create_cluster()` method creates a new Dataproc cluster and returns an `operation` object. The `operation` object can be used to track the progress of the cluster creation:

```
operation = dataproc_cluster_client.create_cluster(
        project_id=project_id,
        region="us-central1",
        cluster=cluster
    )
```

The function finally prints a message that the cluster has been created:

```
print(f"Created Dataproc cluster: {cluster.cluster_name}")
```

The `if __name__ == "__main__":` statement at the end of the code defines the main entry point for the program. When the program is run, this statement will be executed first. The statement then assigns the values to the variables `project_id`, `region`, and `cluster_name`. The `create_dataproc_cluster()` function is then called with these values:

```
if __name__ == "__main__":
    project_id = "project-id"
    region = "us-central1"
    cluster_name = "basic-cluster"

    create_dataproc_cluster(project_id, region, cluster_name)
```

To run the code, you can save it as a Python file and then run it from the command line. For example, if you save the code as *create_dataproc_cluster.py*, you can run it by typing the following command into the command line:

```
python create_dataproc_cluster.py
```

1.9 Duplicating a Dataproc Cluster

Problem

Users have reported an issue in the production environment, but you do not have direct access to it. To diagnose and verify the problem, you need to create an exact replica of the production cluster. This will allow you to reproduce the issue in a controlled environment and investigate further without affecting the live system.

Solution

Export the existing cluster configuration to a file:

```
gcloud dataproc clusters export <source-cluster-name> \
    --destination prod-cluster-config.yaml
```

Create a new cluster using the YAML configuration file:

```
gcloud dataproc clusters import <target-cluster-name> \
    --source prod-cluster-config.yaml \
    --region=region
```

Discussion

When working with existing clusters, you may want to view cluster details, such as worker details, labels, custom configurations, and component gateway URLs. The gcloud command for viewing the existing cluster configuration is as follows:

```
gcloud dataproc clusters describe <cluster-name-here> --region <region>
```

Creating a new cluster with the same configuration as the existing cluster is a two-step process. First, we have to export the existing cluster configuration to a file. Dataproc offers a gcloud command option to export the configuration in YAML file format. At the time of this writing, this option of configuration export is only supported from gcloud and can't be done from the web UI.

Run a command to export the configuration:

```
gcloud dataproc clusters export prod-cluster\
    --destination prod-cluster-config.yaml
```

Upon successful execution of the command, the cluster configuration will be stored in a file named *prod-cluster-config.yaml*. The cluster name and region are not included in the export because the name must be unique. When creating a new cluster using this configuration, the cluster name and region must be provided.

Run a command to create a new cluster using the configuration present in the YAML file:

```
gcloud dataproc clusters import prod-cluster-duplicate \
  --source prod-cluster-config.yaml \
  --region=region
```

## CHAPTER 2

Running Hive, Spark, and Sqoop Workloads

In the world of big data processing and analysis, Google Cloud's Dataproc simplifies managing and executing large-scale data workloads. In this chapter, we will cover the essential steps for running various big data jobs on your Dataproc cluster. A *job* in this context represents a specific task or workload to be executed on the Dataproc cluster. This can be a Hive query for structured data processing, a Spark application for distributed computation, or a Sqoop data transfer for moving data between databases and Hadoop.

To effectively follow along with this chapter, you will need the following prerequisites:

Dataproc API
Ensure that the Dataproc API is enabled for your project. This API is essential for interacting with your cluster.

Existing Dataproc cluster
You will need a Dataproc cluster that has already been created and is running on GCP. If you haven't set one up yet, Chapter 1 provides guidance on cluster creation.

We will explore the different methods you can use to submit these jobs to your Dataproc cluster. This includes using the Dataproc console UI as well as the gcloud CLI tool. Throughout the chapter, we'll provide practical examples to illustrate these concepts.

Let's get started!

2.1 Adding Required Privileges for Jobs

Problem

You need to grant users the necessary permissions to submit jobs to your Dataproc cluster.

Solution

Use Google Cloud's IAM to assign appropriate roles to users. At the service level, predefined roles include the privilege to submit jobs to the Dataproc cluster. The predefined Dataproc roles are as follows:

roles/dataproc.worker

> The most restrictive option, this role specifically grants the ability to submit jobs to the Dataproc cluster. Use the following gcloud command to give the dataproc.worker role to a user:

```
gcloud projects add-iam-policy-binding <PROJECT_ID> \
  --member="user:EMAIL_ADDRESS>" \
  --role=roles/dataproc.worker
```

roles/dataproc.editor

> This role includes permission to submit jobs along with other actions like creating or deleting clusters. To grant the dataproc.editor role to a user, use the following command:

```
gcloud projects add-iam-policy-binding <PROJECT_ID> \
  --member="<user:EMAIL_ADDRESS>" \
  --role=roles/dataproc.editor
```

roles/dataproc.admin

> This role offers admin access and gives full control over the Dataproc resources within the project. To grant the dataproc.admin role to a user, use the following command:

```
gcloud projects add-iam-policy-binding <PROJECT_ID> \
  --member="<user:EMAIL_ADDRESS>" \
  --role=roles/dataproc.admin
```

Discussion

You can grant IAM roles (*https://cloud.google.com/iam/docs*) at different levels for the user to submit jobs to a Dataproc cluster, as listed in Table 2-1.

Table 2-1. IAM roles at different levels

Approach	Description	Example roles and permissions
Project-level roles	Broad permissions across an entire project	roles/editor (project editor)
Service-level roles	Grants permission to the Dataproc service	roles/dataproc.worker roles/dataproc.editorroles/dataproc.admin
	Grants permission to GCS	roles/storage.objectAdminroles/storage.admin roles/storage.objectViewer
Custom roles	Fine-grained control	dataproc.jobs.create dataproc.jobs.get

Project-level roles

Project-level IAM roles in Google Cloud grant permissions across an entire project. If a user's task is simply to submit and manage jobs on the Dataproc cluster, granting the broad project-level role is not necessary as it violates the principle of least privilege in security.

Service-level roles

Service-level (or cluster-level) roles provide permissions focused on individual services or resources within the project. When submitting jobs to the Dataproc cluster, a combination of permissions at both the Dataproc service level and the GCS level is typically required. This is because jobs running on Dataproc typically:

- Read input data from GCS
- Write logs and output data to GCS

If your Dataproc jobs need to interact with other Google Cloud services (beyond Dataproc and GCS), additional service-specific permissions will be required. For our discussion here, we will include Dataproc-level and GCS-level permissions:

Dataproc-level permissions
> `roles/dataproc.worker` is specifically designed to provide the minimum permissions needed to submit jobs on a specific cluster. Consider `roles/data proc.editor` or `roles/dataproc.admin` if the user needs additional Dataproc management capabilities. For automated Dataproc job submission and related tasks, create a dedicated service account and assign the Dataproc worker role.

GCS permissions

    If your job has to interact with a storage bucket, add additional roles such as storage object viewer or storage object creator to allow read/write access to the relevant storage buckets.

Custom roles

When the predefined service-level IAM roles are too broad, you can create custom roles to allow more granular access for users or service accounts. For a user who only needs to submit and retrieve information about Dataproc jobs, the `data proc.jobs.create` and `dataproc.jobs.get` permissions are appropriate.

The `gcloud iam roles create` command is a valid template for creating custom roles:

```
gcloud iam roles create <dataproc_custom_role_name_here> \
  --project={project_name_here} \
  --title={dataproc_job_title_here} \
  --description="Allows submission and retrieval of Dataproc jobs" \
  --permissions="dataproc.jobs.create,dataproc.jobs.get" --stage="GA"
```

To assign a custom role that you created, first navigate to the IAM page on the Google Cloud console and locate "Grant Access," as shown in Figure 2-1.

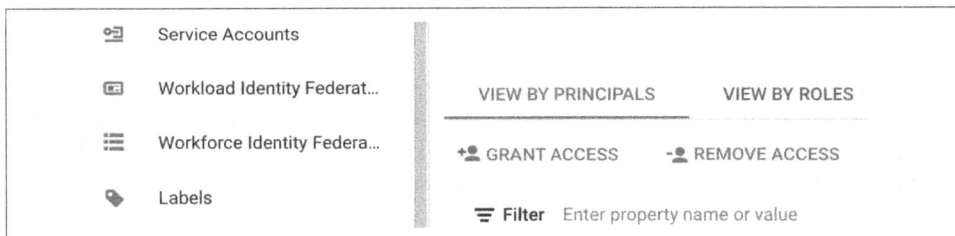

Figure 2-1. Google Cloud console IAM page

In the "Add principals" section, enter the user or service account email address, as shown in Figure 2-2.

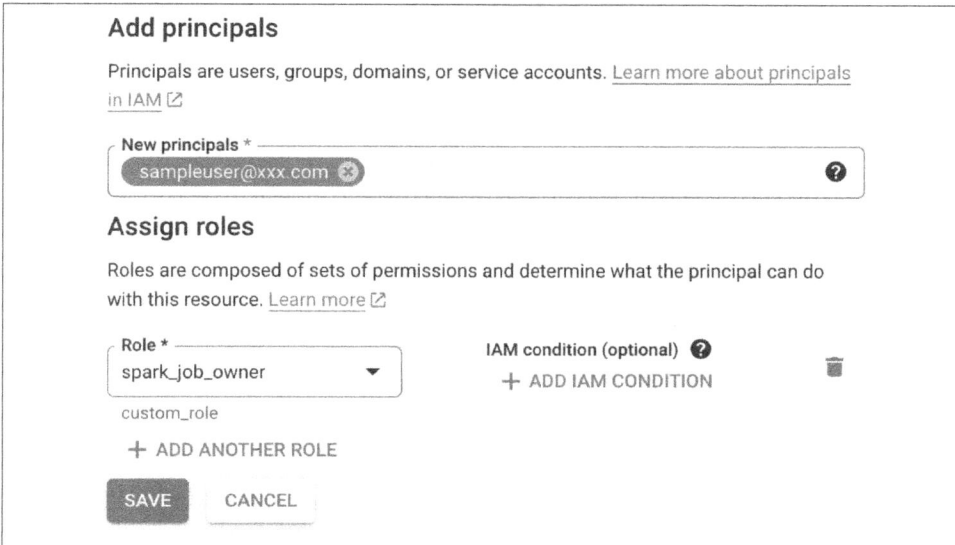

Figure 2-2. Adding the user or service account email address in the "Add principals" section

Select the custom role that you previously created and assign it to the user, as shown in Figure 2-3. In Figure 2-3, we picked the "spark_job_owner" custom role and assigned the user "sampleuser@xxx.com."

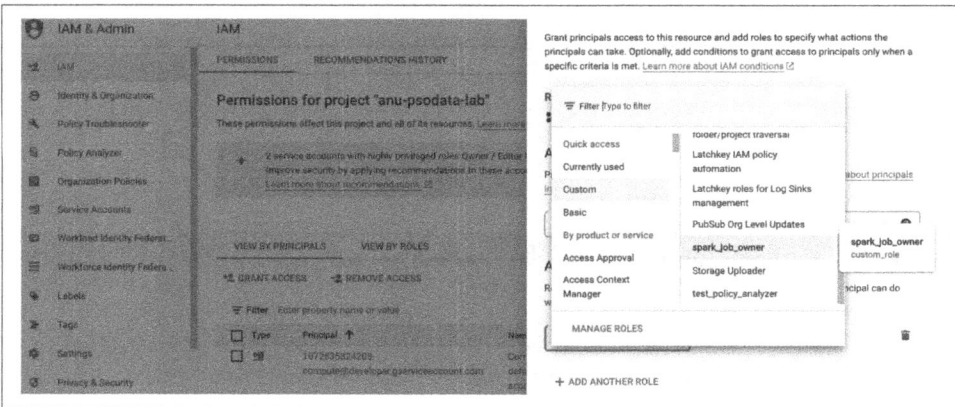

Figure 2-3. Assigning a custom role to the user

Staging and temporary buckets

When you run a Dataproc job, it interacts with GCS. Submitting and running jobs in the Dataproc cluster involves staging job dependencies (JAR files) and writing job logs. Dataproc automatically creates a staging bucket for job dependencies and a temporary bucket for the Spark and MapReduce job logs. The staging bucket will have the prefix "dataproc-staging-" and a temporary bucket will have the prefix "dataproc-temp-" by default in GCS. You can also specify a staging and temporary bucket when creating the Dataproc cluster.

A job's actual output data will go to the specified location (GCS, BigQuery, etc.). Users submitting jobs need to have read/write permissions for all source and target locations.

The temporary bucket has a 90-day time to live (TTL). The staging bucket doesn't have a TTL; it persists even after the cluster gets deleted.

Does every cluster have default staging and temp buckets? Clusters in the same region and using default staging and temp buckets will share the same buckets.

Users or service accounts submitting Dataproc jobs must have the necessary read and write permissions (often `storage.objectUser`) for the staging and temporary buckets. If the job reads input from or writes output to other GCS buckets, ensure that the user or service account has appropriate permissions on those buckets as well (see Table 2-2).

Table 2-2. Example storage-level access roles

Role name	Description	Use case
roles/ storage.objectViewer	Provides read-only access to objects in GCS buckets	Dataproc jobs only need to read data from GCS (e.g., reading input files)
roles/storage.objectUser	Includes both read and write permissions on objects in GCS buckets	Dataproc jobs need to read and write data to GCS (e.g., reading input and writing output/logs)

You can find the staging and temporary bucket details in the configuration section of your existing Dataproc cluster in the Google Cloud console, as shown in Figure 2-4.

		Master node	Standard (1 master, N workers)
	Clusters	Machine type	n2-standard-4
	Jobs	Number of GPUs	0
		Primary disk type	pd-standard
	Workflows	Primary disk size	500GB
	Autoscaling policies	Local SSDs	0
		Worker nodes	2
Serverless	^	Machine type	n2-standard-4
	Batches	Number of GPUs	0
		Primary disk type	pd-standard
	Interactive	Primary disk size	500GB
		Local SSDs	0
Metastore Services	^	Secondary worker nodes	0
	Metastore	Secure Boot	Disabled
		VTPM	Disabled
	Federation	Integrity Monitoring	Disabled
		Cloud Storage staging bucket	dataproc-staging-us-east1-1072535324208-lo1vjjhc
Utilities	^	Network	default

Figure 2-4. Dataproc cluster configuration

At the staging and temporary bucket level, grant the Storage Object User role, as shown in Figure 2-5.

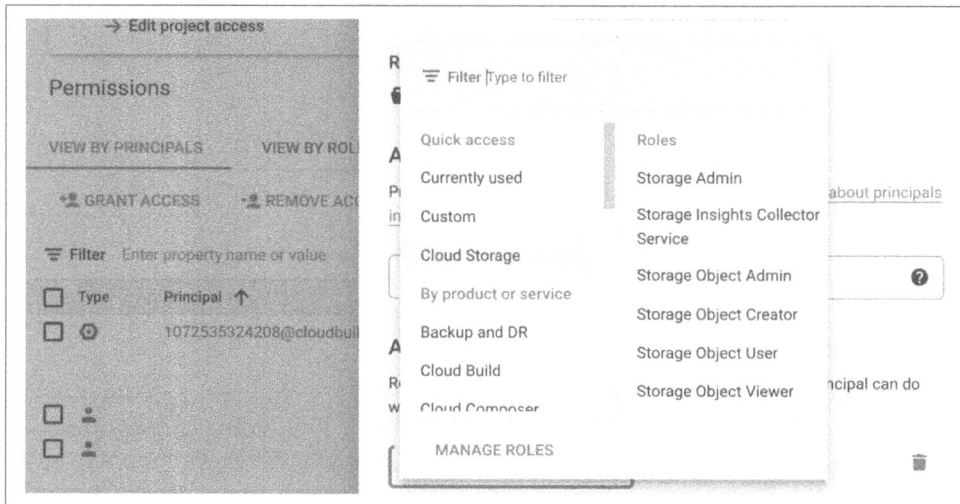

Figure 2-5. GCS bucket permissions

See Also

If you need guidance on granting IAM privileges for creating Dataproc clusters, refer to Recipe 1.2. Also see the official Google Cloud documentation:

- Submitting jobs with granular IAM (*https://oreil.ly/e8MqK*)
- Authenticating to Dataproc (*https://oreil.ly/VYyRL*)
- Configuring temporary and staging buckets (*https://oreil.ly/0yJVl*)

2.2 Generating 1 TB of Data Using a MapReduce Job

Problem

You have an existing MapReduce job (from an on-premises Hadoop setup, another cloud, or a new application) that generates 1 TB of data. You want to execute this job on a Dataproc cluster.

Solution

You can submit a MapReduce job to the Dataproc cluster in various ways, such as gcloud, REST API, or the console (see the ""Discussion" on page 36" section).

Here is the gcloud command to submit a MapReduce job:

```
gcloud dataproc jobs submit hadoop --project={project_name_here} \
  --cluster={cluster_name_here} --region={region_name_here} \
  --jar={application_jar_location_here} -- teragen 1000000 \
  {gcs_teragen_output_file_path_here}
```

{gcs_teragen_output_file_path_here} is a directory in the bucket where you want the TeraGen output to be stored. Ensure that the user submitting this command has write permissions to write output to {gcs_teragen_output_file_path_here}.

Discussion

Here are the detailed steps for submitting a MapReduce job to a Dataproc cluster via the console.

Submitting the job

Stage the MapReduce JAR file and dependencies like data and packages to GCS. To submit your MapReduce job JAR to a Dataproc cluster, choose the cluster to submit to and pass the paths of the JAR and dependency arguments. The example shown in Figure 2-6 runs a TeraGen job that is going to generate terabytes of data.

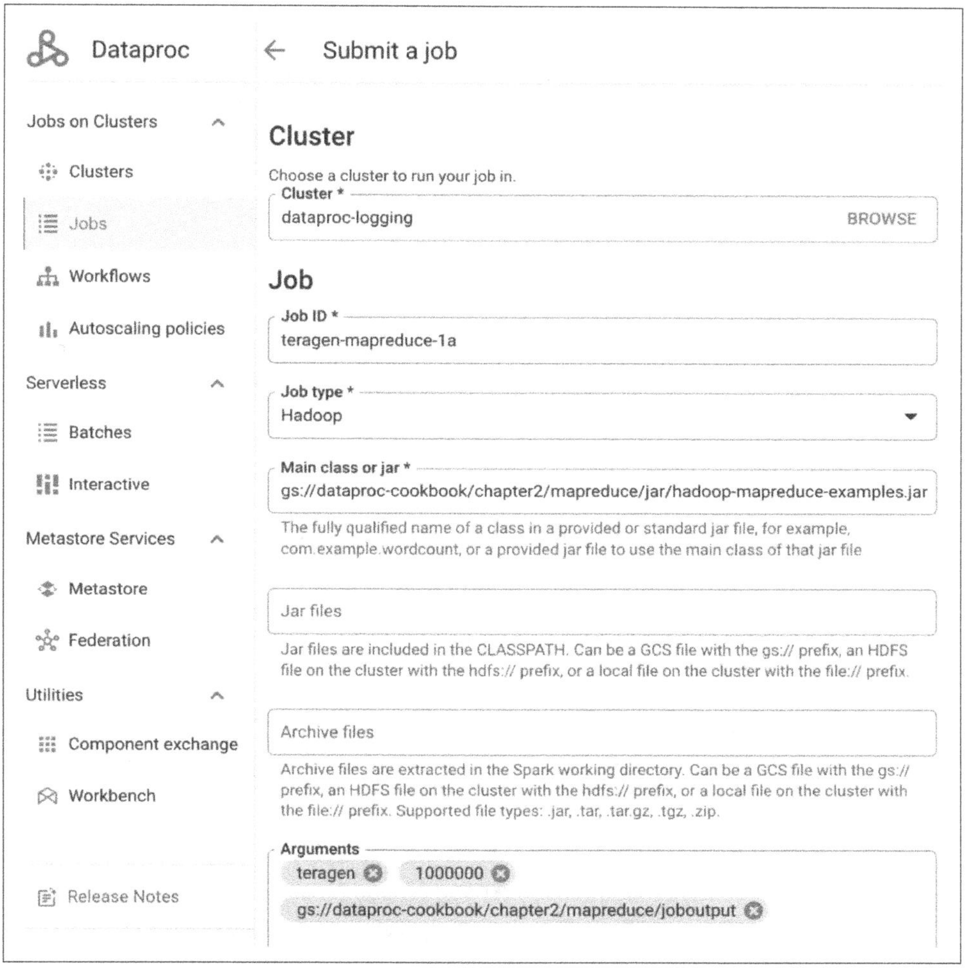

Figure 2-6. Submitting a MapReduce job via the console

Monitoring the job

After submitting the job, you can monitor the progress of the job in the Output section. You'll want to monitor details like YARN memory and CPU utilization in the monitoring UI shown in Figure 2-7 to understand if the job is memory intensive and if it utilizes the resources optimally.

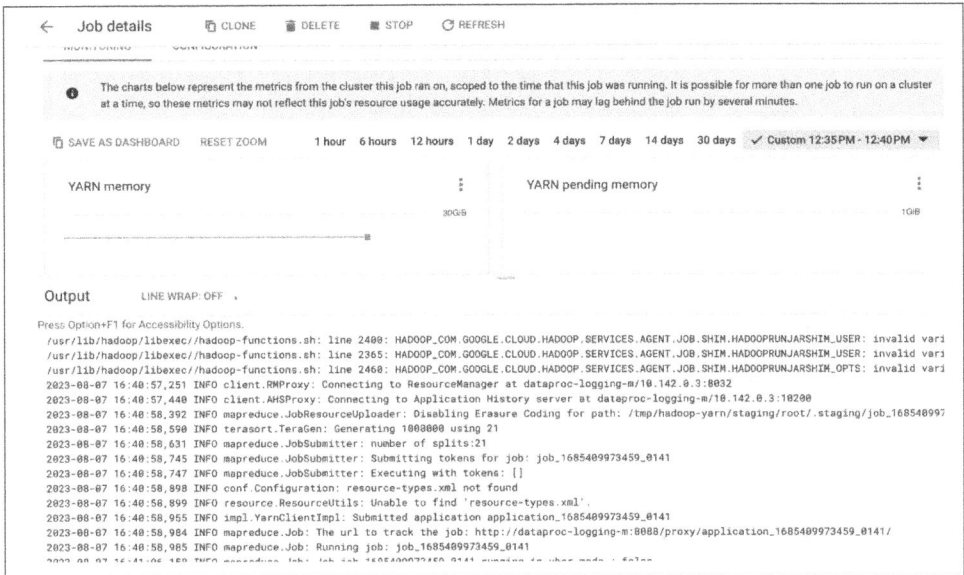

Figure 2-7. Monitoring the job in the monitoring console

In the Dataproc console, go to Web Interfaces for your cluster. Click "YARN ResourceManager" to open the YARN UI, as shown in Figure 2-8. Alternatively, use the `gcloud describe` command as described in Recipe 1.9.

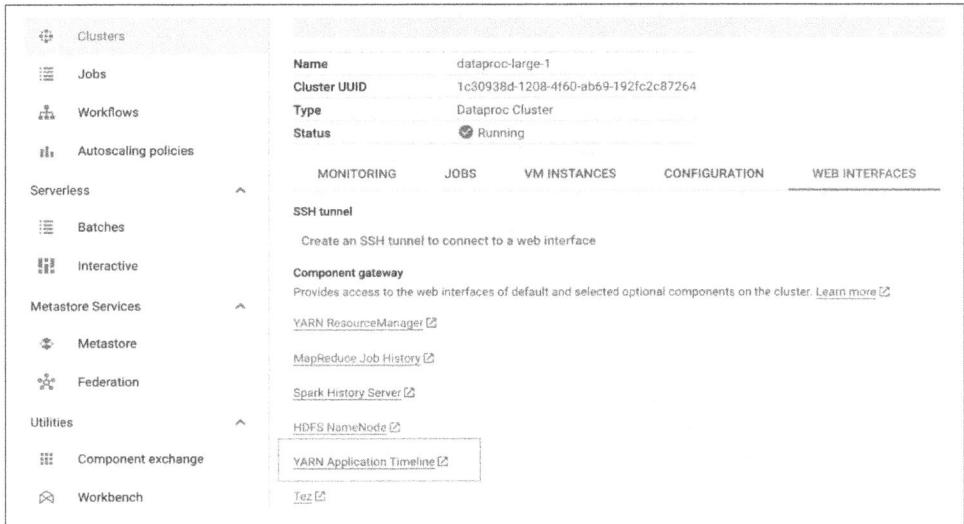

Figure 2-8. Accessing the YARN UI

In the YARN UI, shown in Figure 2-9, you can view the details of running and completed jobs, resource allocation, number of mappers and reducers, and execution time.

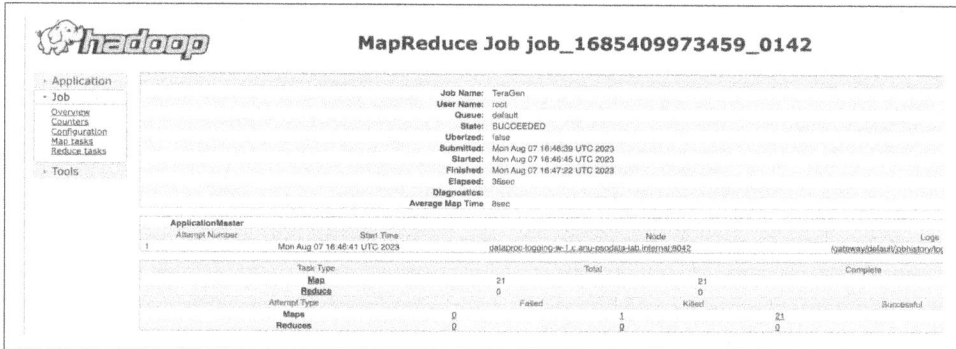

Figure 2-9. Monitoring the job in the YARN UI

2.3 Running a Hive Job to Show Records from an Employee Table

Problem

You have a Hive table on an existing Dataproc cluster and want to execute a HiveQL query against this table.

Solution

Use the `gcloud dataproc jobs submit hive` command.

To submit a HiveQL query from a file, use the following:

```
gcloud dataproc jobs submit hive --project={project_name_here} \
  --region={region_name_here} --cluster={cluster_name_here} \
  --file={hive_query_file_gcs_path_here}
```

To submit a HiveQL query inline, use the following:

```
gcloud dataproc jobs submit hive --cluster {cluster_name_here} \
  --region {region_name_here} --execute {query_here}
```

Discussion

You have an existing Dataproc cluster. Hive provides a familiar SQL-like interface within the Dataproc platform to process structured data. Hive components in Dataproc are HDFS or GCS for storage, execution engines (MapReduce or Tez), and a metastore to manage metadata.

Metastore options

The metastore stores metadata of databases, tables, and partitions. The metastore can be remote—Dataproc Metastore Service (DPMS) or an external metastore—or embedded. DPMS is GCP's serverless offering and supports Apache Thrift and gRPC endpoints.

The solution provided in this recipe focuses on embedded metastore mode. In embedded mode, the metastore is part of the cluster, installed on the master node.

> The embedded metastore is ephemeral, and the metadata will be lost once the cluster is deleted. DPMS or an external metastore is recommended for persisting metadata objects even if the cluster is deleted or stopped. We will discuss DPMS and metadata management in Chapter 6.

Hive query execution process

The steps of executing a Hive query are illustrated in Figure 2-10.

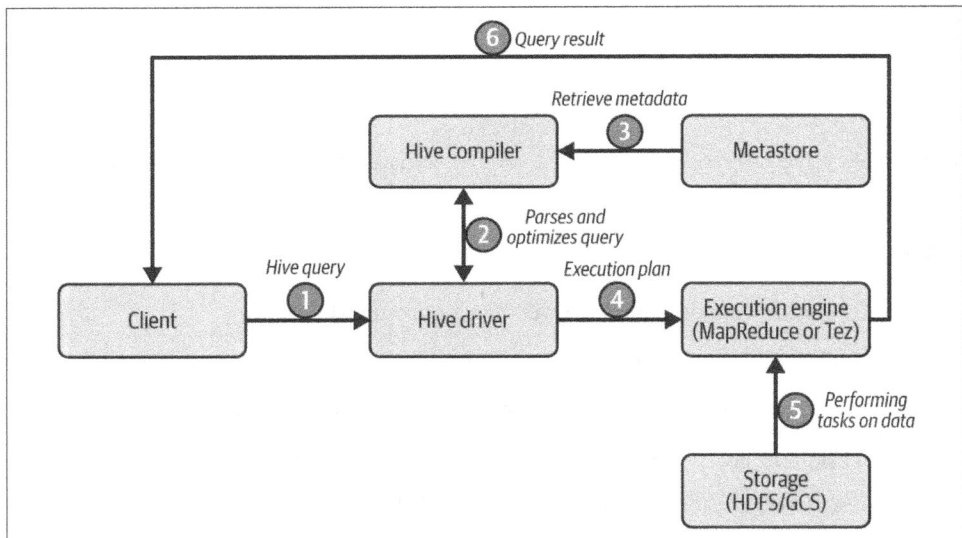

Figure 2-10. Hive query execution process

When you send a Hive query (using a Hive client), the following will occur:

Query submission
> The user sends a Hive query using a Hive client

Query receipt
> The Hive driver component receives the query and forwards it to the compiler.

Parsing and analysis
> The compiler within Hive parses the query to understand its structure and checks it for correctness (semantic analysis).

Metadata retrieval and planning
> The compiler interacts with the metastore (where information about Hive tables and data is stored) to gather the necessary metadata.

Execution plan
> Using this metadata, the compiler creates an optimized plan for executing the query.

Task execution
> The execution engine (such as MapReduce or Tez) takes this plan and coordinates with the resource manager (such as YARN) to get the necessary resources to execute the tasks to process your data based on the query.

Results delivery
> Once the query processing is complete, the execution engine sends the results back to the Hive client that initiated the query.

Let's walk through each of the steps for submitting a HiveQL job to Dataproc.

Step 1: Stage the HiveQL script in the GCS bucket. There are two query source types: query file and query text, as shown in Figure 2-11.

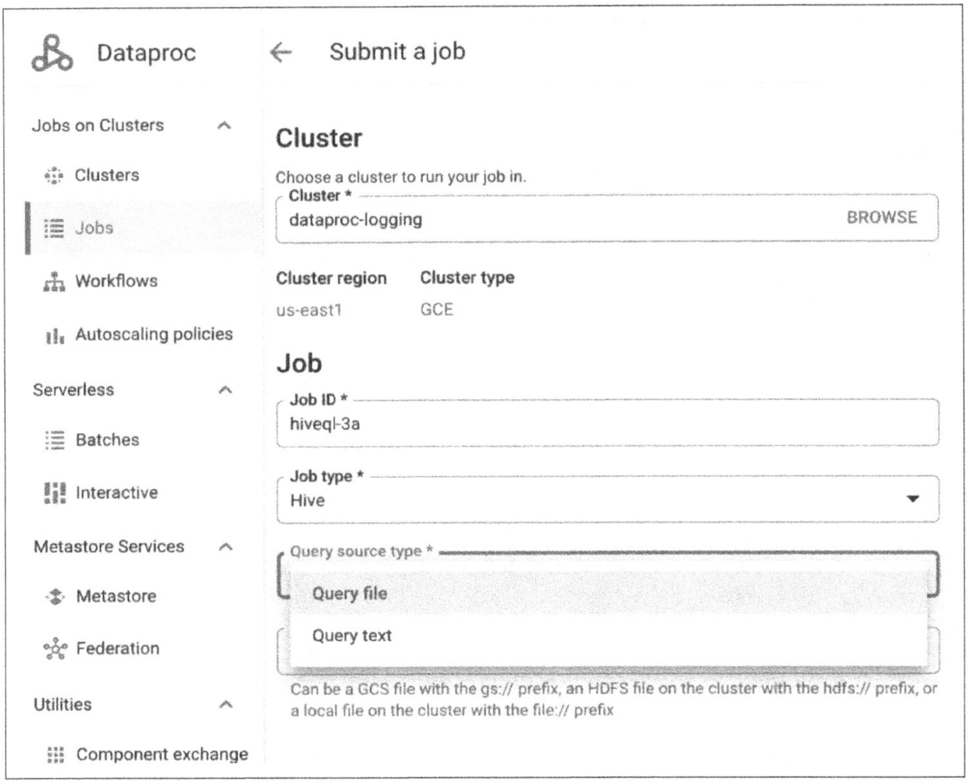

Figure 2-11. Hive query source types can be either a query file or a query text

If you are using a query file, stage the file in the GCS bucket, as shown in Figure 2-12. If you are using a query text, you can choose the "Query text" option from the "Query source type" drop-down menu shown in Figure 2-13.

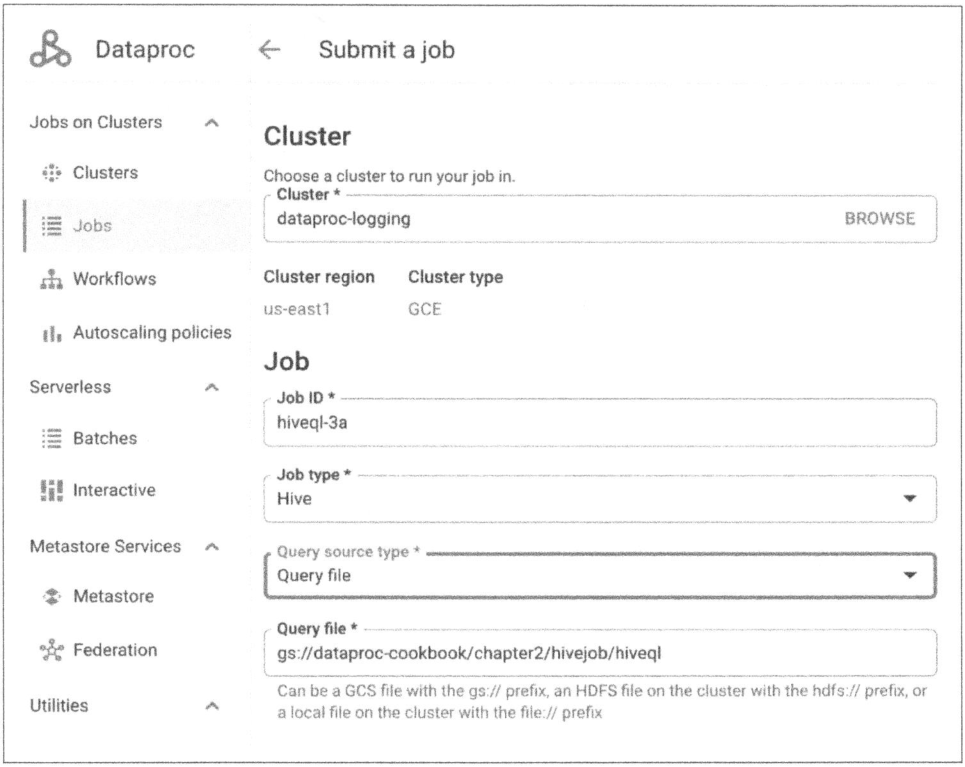

Figure 2-12. Passing the GCS path of the Query file

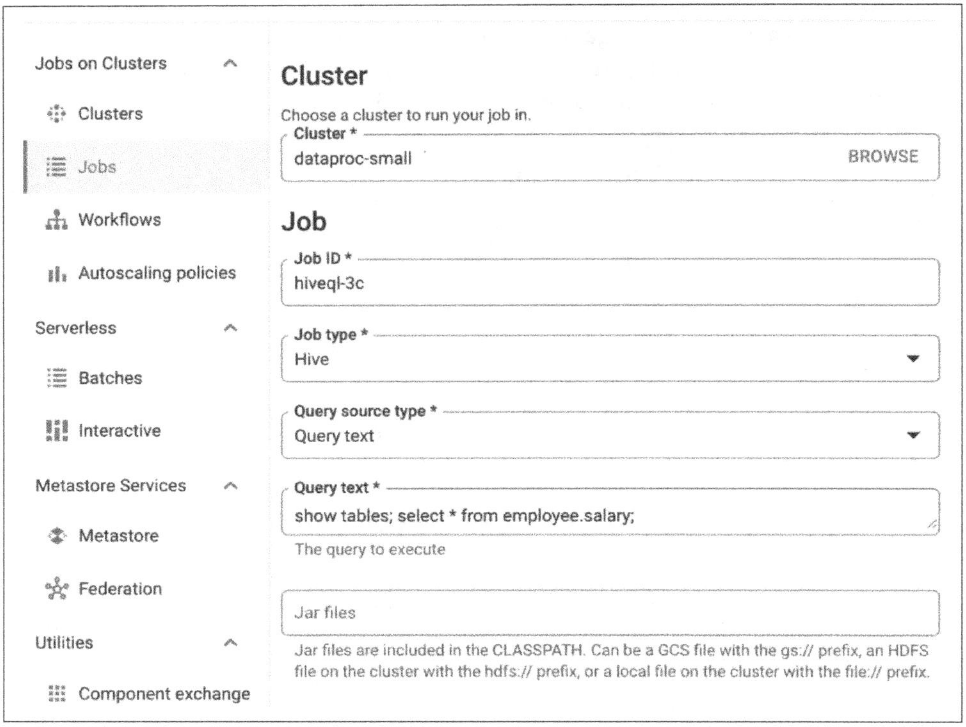

Figure 2-13. Selecting and passing the query text

Step 2: Submit the Hive job to the Dataproc cluster via the console. Use the appropriate submission method (such as the gcloud CLI tool, Dataproc API, Dataproc console, or Cloud Composer) to submit the Hive job to the Dataproc cluster. This will initiate the execution of your job on the cluster.

Once you have successfully submitted the job, you can track its execution and monitor its progress by checking the output section, as shown in Figure 2-14.

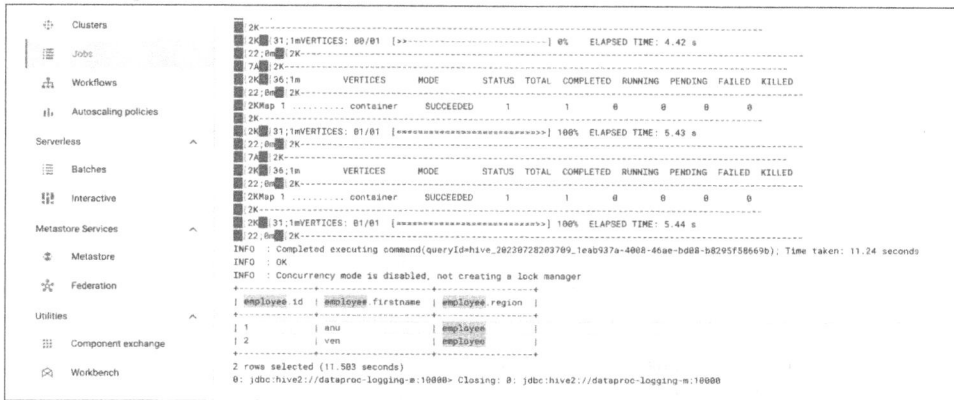

Figure 2-14. Application log in the output section

You can track the job execution and monitor its progress in the cloud shell or terminal, as shown in Figure 2-15.

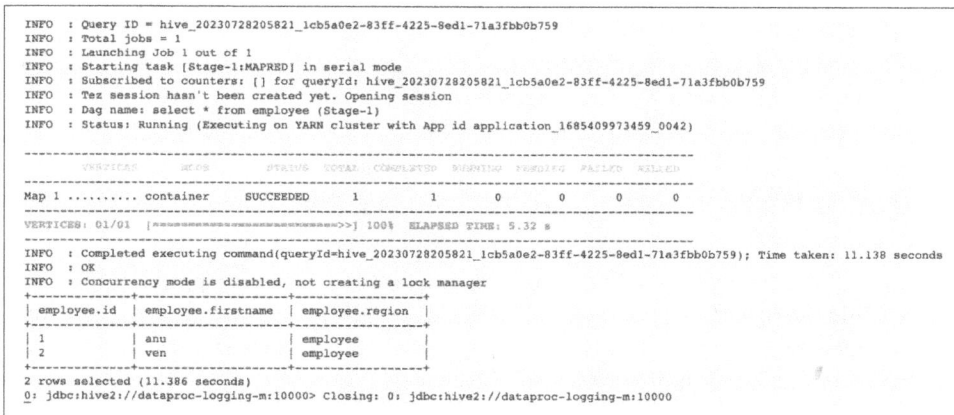

Figure 2-15. Application log in the cloud shell

See Also

See Recipe 2.1 to give appropriate permissions to submit jobs.

2.4 Converting XML Data to Parquet Using Scala Spark on Dataproc

Problem

You have an existing Spark job running on any of the following: on premises, AWS, Microsoft Azure, or a new Spark Scala application, JAR. How do you run and submit a Scala Spark job to a Dataproc cluster?

Solution

The following gcloud command submits Spark Scala jobs to an existing Dataproc cluster:

```
gcloud dataproc jobs submit spark --project={project_name_here} \
  --cluster={cluster_name_here} --region={region_name_here} \
  --class={class_name_here} --jars={jar1_gcs_path_here},{jar2_gcs_path_here} \
  -- {file_input_gcs_path_here} {file_output_gcs_path_here} {arguments}
```

Discussion

Here is a summary of the steps for submitting a Scala Spark job to Dataproc:

- Compile the Scala Spark job into a JAR file. This step will package your code and its dependencies into an executable format that can be run on the cluster. For this example, we are taking a Spark job that reads an XML file and converts the output to Parquet.

- Stage the JAR file to a GCS bucket. This allows the Dataproc cluster to access and execute the job.

- Submit the Spark job to the Dataproc cluster. Use the appropriate submission method (such as the gcloud CLI tool, Dataproc API, Dataproc console, or Cloud Composer) to submit the Scala Spark job to the Dataproc cluster. This will initiate the execution of your job on the cluster.

These are the prerequisite steps before submitting Spark jobs to the Dataproc cluster: ensure that you have created and configured the cluster, compiled your code into a JAR file, staged the JAR file in a GCS bucket, and then submitted the job for execution on the cluster. Create a folder in GCS and stage the JAR in the GCS folder. Use the following `gsutil` commands to create a folder in GCS and copy a JAR file into that folder:

```
gsutil mb gs://{gcs_bucket_name_here}/{folder_name_here}
gsutil cp file.jar gs://{gcs_bucket_name_here}/{folder_name_here}/
```

To submit Spark jobs to the Dataproc cluster, navigate to Jobs in the Dataproc console. Pass the cluster name, indicate the job type as Spark for the Scala Spark application, and add the JAR path or name of the class. If you are adding the name of the class, pass the jar path to "Jar files," as shown in Figure 2-16.

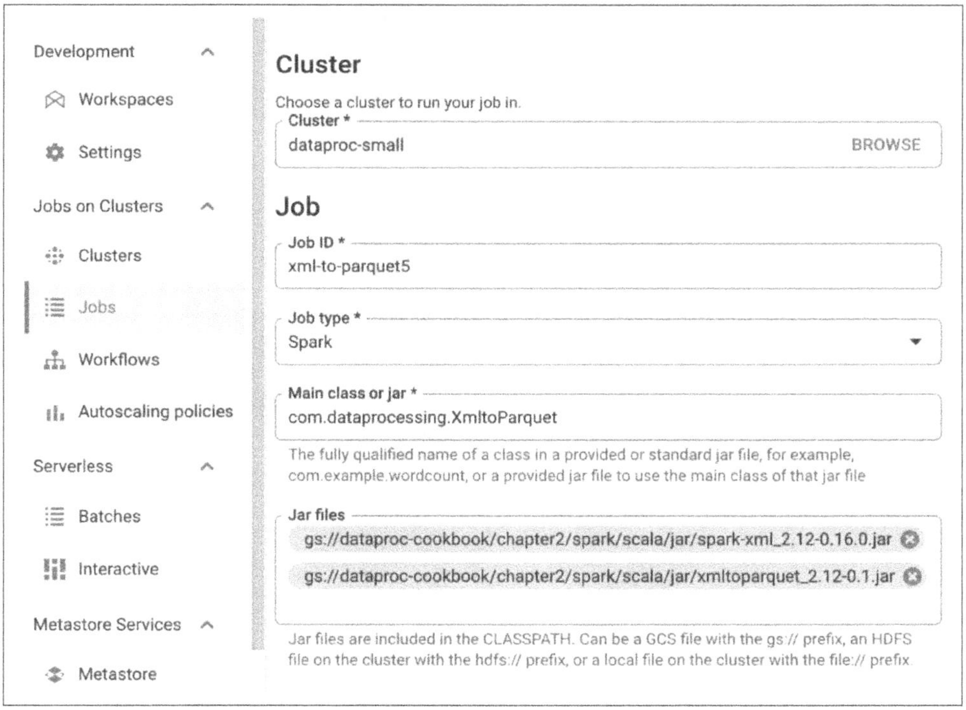

Figure 2-16. Submitting a Spark Scala job via the console

Include any necessary arguments, as shown in Figure 2-17, and indicate if any dependency files are referenced by the Spark code. These dependency files will be extracted into the working directory of each Spark executor.

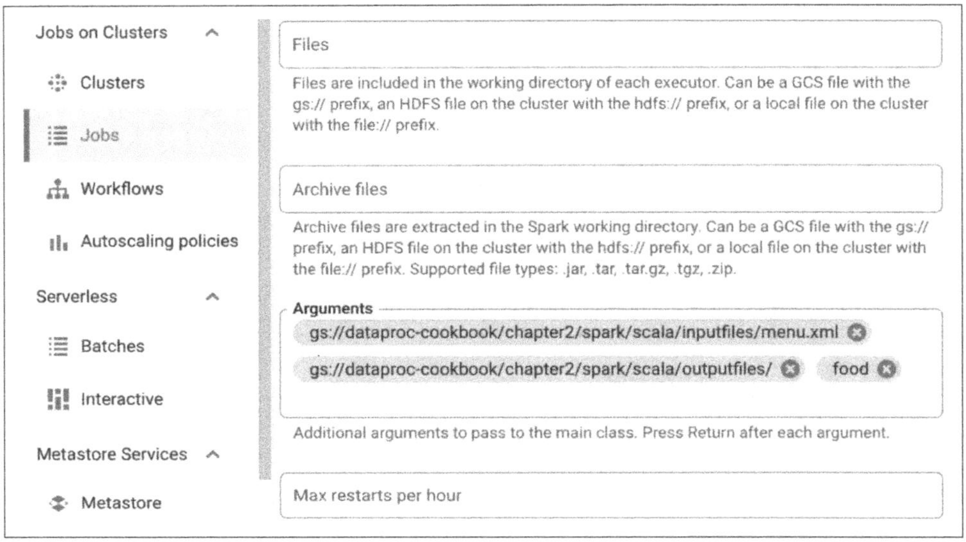

Figure 2-17. Adding arguments for the Spark job

Once you have successfully submitted the job, you can track its execution and monitor its progress by checking the Output section, shown in Figure 2-18. To open the Output section, navigate to the "Dataproc jobs" section of the Google Cloud console and click on the submitted job. Then, locate the section labeled "Output." This section logs all the relevant events and provides information about the job's status and any noteworthy occurrences.

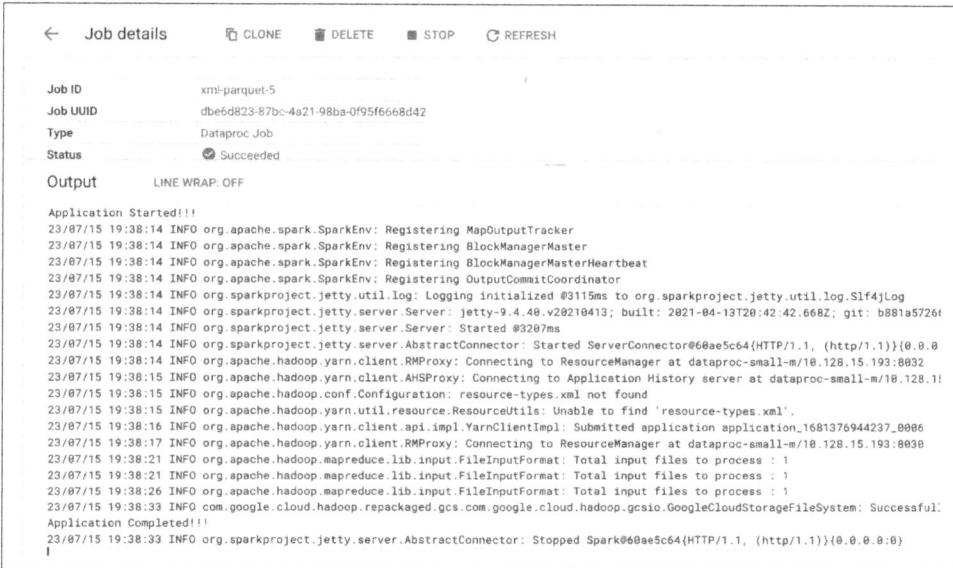

Figure 2-18. Monitoring the job in the output UI

You can configure the Spark parameters, such as driver memory, driver cores, executor memory, executor cores, and shuffle partitions, to tailor the resources to your job's needs. Input the optimal parameters and values under the Properties section, as shown in Figure 2-19.

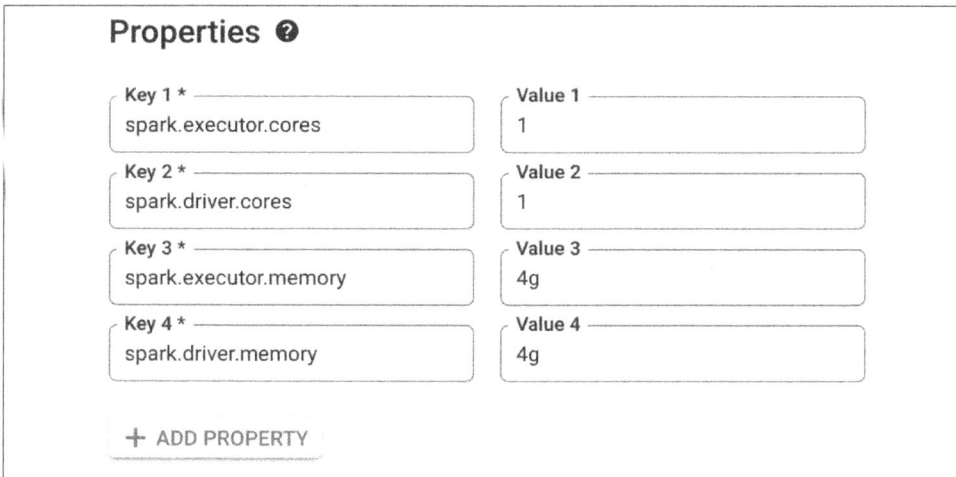

Figure 2-19. Configuring parameters like driver memory, driver cores, executor memory, and executor cores

See Also

Refer to Chapter 9 for more insights into and guidelines on how to achieve the best performance for Spark jobs when running on Dataproc.

2.5 Converting XML Data to Parquet Using PySpark on Dataproc

Problem

You have an existing PySpark 3.x job running on premises, AWS, Azure, or a new PySpark application. How do you run and submit a PySpark job to a Dataproc cluster?

Solution

The following gcloud command shows how to run a PySpark code on an existing Dataproc cluster:

```
gcloud dataproc jobs submit pyspark --project={project_name_here} \
    --cluster={cluster_name_here} --region={region_name_here} \
    {pyspark_file_path_here} -- {file_input_gcs_path_here} \
    {file_output_gcs_path_here} {arguments}
```

Discussion

Here are the detailed steps for submitting a PySpark job to an existing Dataproc cluster via the console.

Staging the PySpark code

Create a GCS storage bucket. Stage the PySpark application code. Include dependency files like data files and JARs in the GCS bucket. Here is example PySpark code that processes an XML input file, infers schema, and writes it as a Parquet output:

```
import sys
from pyspark.sql import SparkSession
from pyspark.sql.functions import *
from pyspark.sql.types import *


spark = SparkSession.builder \
    .appName("XmltoParquet") \
    .master("yarn") \
    .getOrCreate()

#Get the input path, output path and rowTag
inputPath = sys.argv[1]
```

```
outputPath = sys.argv[2]
rowTag = sys.argv[3]

print("Reading the XML input file")

df = spark.read.format('xml').option("rowTag", rowTag).load(inputPath)

# Show the DataFrame to check the data
df.write.mode("overwrite").parquet(outputPath)

print("Application Completed!!!")

# Closing the Spark session
spark.stop()
```

> If you have the PySpark file locally and have the gcloud CLI config-
> ured, you can submit it directly using the local filepath.

Submitting the PySpark job

The PySpark job can be submitted via the console or the gcloud command. It can
also be submitted via orchestrations like Dataproc Workflow Templates or Cloud
Composer directed acyclic graphs (DAGs).

To submit the PySpark job via the console, navigate to the Batches section of the
Dataproc console. Choose the PySpark job type, as shown in Figure 2-20. Pass the
path of the PySpark application code to "Main python file." If the application code has
dependency JARs, pass the paths of the JARs in "Jar files."

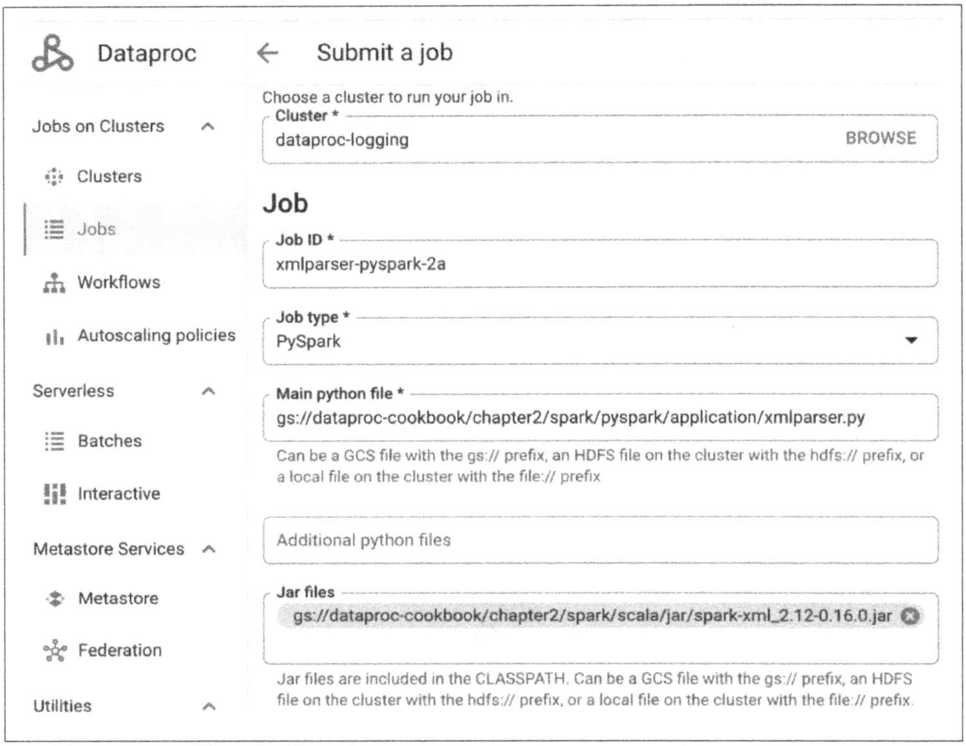

Figure 2-20. Submitting a PySpark job via the console

Pass the arguments, as shown in Figure 2-21, and submit the job to the existing Dataproc cluster.

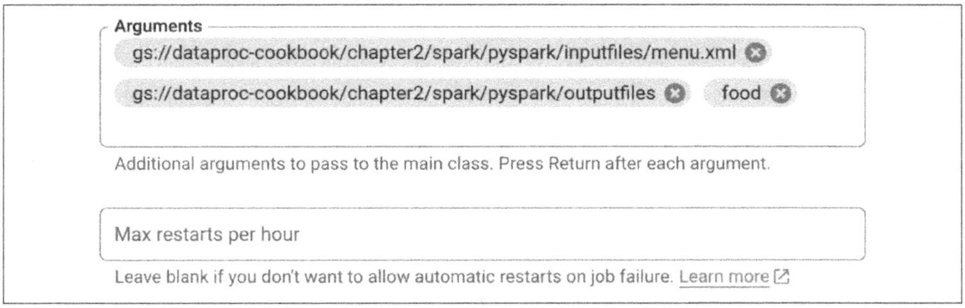

Figure 2-21. Passing arguments required by the PySpark code

See Also

For more guidelines around running Spark jobs on Dataproc, refer to the discussion section in Recipe 2.4.

2.6 Submitting a SparkR Job

Problem

You have an existing SparkR job running on premises, AWS, Azure, or a new SparkR application. How do you run and submit a SparkR job?

Solution

Here is the gcloud command to submit a SparkR job to an existing Dataproc cluster:

```
gcloud dataproc jobs submit spark-r --project={project_name_here} \
  --cluster={cluster_name_here} --region={region_name_here} \
  {R_file_input_gcs_path_here}
```

Discussion

The following are detailed steps for submitting a SparkR job to a Dataproc cluster via the console.

Staging the SparkR code

Create a GCS storage bucket. Stage the SparkR application code. Include dependency files like data files and dependency JARs in the GCS bucket. Here is the sample R script:

```r
# Important: To use SparkR on Dataproc, require(SparkR) line imports the
# necessary library to work with Apache Spark from R.
require(SparkR)

# Initialize a Spark Session
sparkR.session(master = "yarn", appName = "SparkR Example")

data  <- list(
                list(1L, "Anu","India"),
                list(2L, "Ven","India"),
                list(3L, "Gav","Canada")
              )

schema <- structType(
                structField("id",    "integer"),
                structField("name", "string"),
                structField("location", "string")
              )

# Create SparkDataFrame
df      <- createDataFrame(
            data   = data,
            schema = schema
          )

# Print the schema
printSchema(df)

head(df)

#Stop the Spark Session
sparkR.stop()
```

You can use the following `gsutil cp` command to copy the files from a local or source directory to GCS. Stage the application code and sample input file *menu.xml* in GCS:

```
gsutil cp {source_file_path_here} {gcs_destination_path}
```

Submitting the SparkR job

The SparkR job can be submitted via the console, the gcloud CLI, or orchestrations like Dataproc Workflow Templates or Cloud Composer DAGs.

In the console, choose the SparkR job type, as shown in Figure 2-22. Pass the path of the SparkR application code to "Main R file." If the application code has dependency R files, pass the paths of the files in "Additional R files."

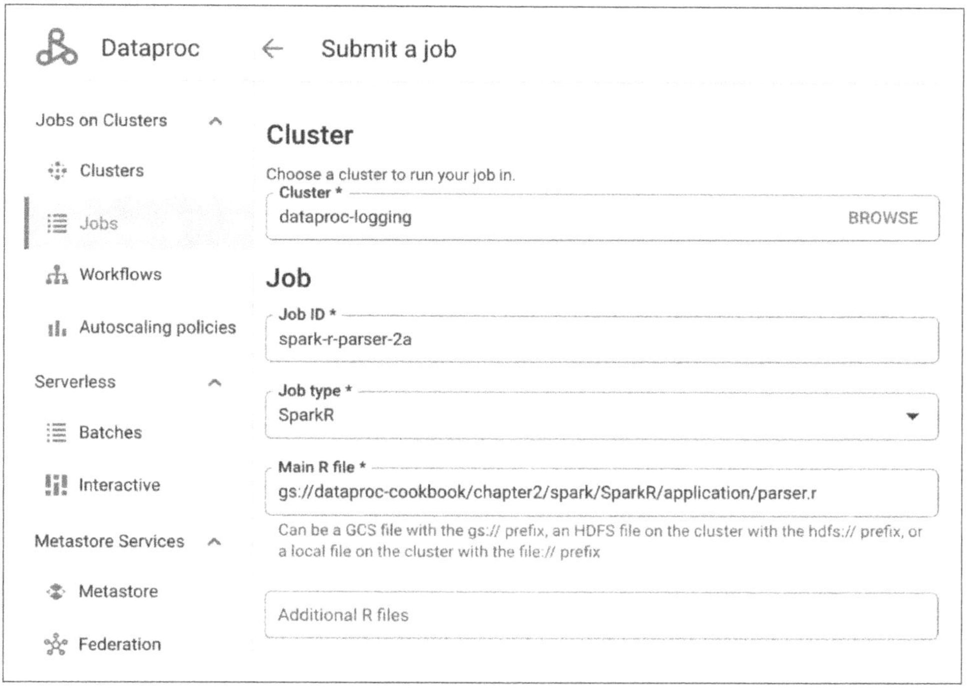

Figure 2-22. Submitting a SparkR job

Monitoring the SparkR job

It is important to monitor the SparkR job after submitting to the cluster. The Jobs section will show the jobs submitted along with each job's status and execution time, as shown in Figure 2-23.

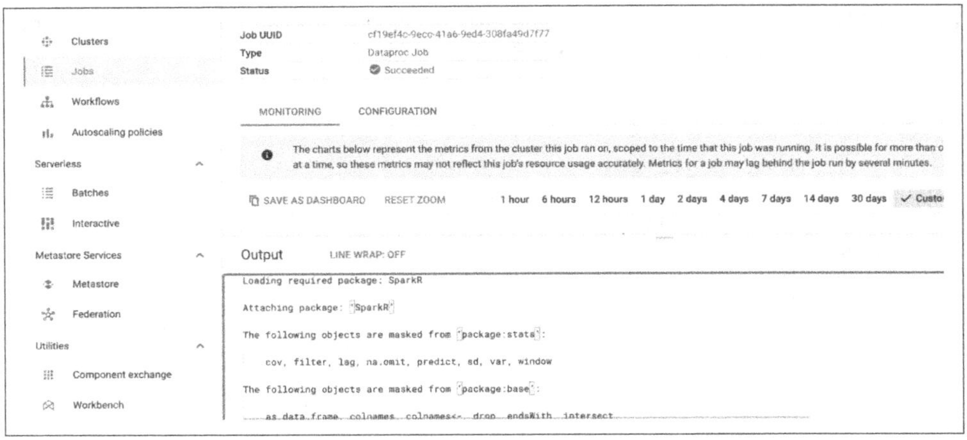

Figure 2-23. Monitoring the UI of the submitted job

The monitoring module will display the results and details of the application, as shown in Figure 2-24.

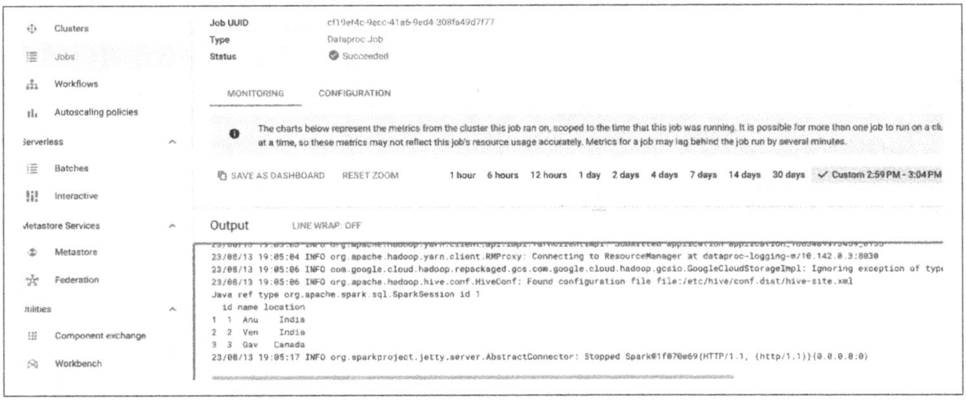

Figure 2-24. Output UI of the submitted job

2.7 Migrating Data from Cloud SQL to Hive Using Sqoop Job

Problem

The source data is in a transactional database like MySQL or another relational database management system (RDBMS). How do you migrate from Cloud SQL to Hive using a Sqoop job?

Solution

The Sqoop job can be submitted via the console or gcloud command. It can also be submitted via orchestrations like Dataproc Workflow Templates or Cloud Composer DAGs.

This can be done with the following code:

```
gcloud dataproc jobs submit hadoop --region={region_name_here} \
  --cluster={cluster_name_here} --class={class_name_here} \
  --jars={jar1_gcs_file_path_here},{jar2_gcs_file_path_here} \
  -- import -Dmapreduce.job.user.classpath.first=true \
  --connect=jdbc:mysql://{mysql_host_ip_or_name_here}:
3306/{database_name_here} \
  --username={username_here} --password={root_password_here} \
  --table={table_name_here}
```

Here is the sample gcloud command that submits a Sqoop job to import a table named "employee":

```
gcloud dataproc jobs submit hadoop --region=us-central1 --cluster=sqooptest \
  --class=org.apache.sqoop.Sqoop --jars=gs://[gcs-bucket-name]/
sqoop-1.5.0-SNAPSHOT.jar,file:///usr/lib/hive/lib/commons-lang-2.6.jar,
file:///usr/share/java/mysql-connector-java.jar \
  -- import -Dmapreduce.job.user.classpath.first=true \
  --connect=jdbc:mysql://[mysql-host-ip-or-name]:3306/[database-name] \
  --username=root --password=rootpassword --table=employee
```

To secure the Cloud SQL instance, configure the instance with a private IP (*https://oreil.ly/OeBoF*). This limits direct exposure of the Cloud SQL instance to the public internet to enhance security.

When connecting from Dataproc to the Cloud SQL instance, use Cloud SQL Auth Proxy (*https://oreil.ly/Ic-r-*). This provides enhanced features to secure the connectivity to the source database.

Discussion

Running a Sqoop job requires that the source and target be present. In this example, we are reading from a Google Cloud MySQL instance and loading it to the Dataproc cluster Hive table.

Setting up a MySQL source database (optional)

If you do not already have a source database, here are the steps to create one in Google Cloud SQL. You can skip this step if you already have a database instance that you can use for testing data migration using Sqoop.

Create a new Cloud SQL instance using the following code:

```
gcloud sql instances create testsqldb --region us-central1 \
  --tier db-n1-standard-2 --assign-ip --root-password rootpassword
```

> Any MySQL instance that is created should be accessible to the Dataproc cluster nodes. To enable this, use the same network and enable private IP allocation to the SQL instance.

Creating a sample table and data in MySQL

Log in to the Cloud SQL instance. Then, create a table and insert sample data for testing purposes, such as the following:

```
CREATE DATABASE testdb;
USE testdb;

CREATE TABLE employee (
  id INT NOT NULL AUTO_INCREMENT,
  name VARCHAR(255) NOT NULL,
  department VARCHAR(255) NOT NULL,
  salary INT NOT NULL,
  PRIMARY KEY (id)
);

INSERT INTO employee (name, department, salary)
VALUES
  ('John Doe', 'Sales', 100000),
  ('Jane Doe', 'Marketing', 80000),
  ('Peter Smith', 'Engineering', 90000),
  ('Mary Johnson', 'Finance', 70000),
  ('David Williams', 'HR', 60000);
```

Triggering the Sqoop data transfer job

Run the following command for triggering a Sqoop job that transfers data from MySQL to Hive:

```
gcloud dataproc jobs submit hadoop --region=us-central1 --cluster=sqooptest \
   --class=org.apache.sqoop.Sqoop --jars=gs://nstestb/sqoop-1.5.0-SNAPSHOT.jar,
file:///usr/lib/hive/lib/commons-lang-2.6.jar,file:///usr/share/java/
mysql-connector-java.jar  -- import -Dmapreduce.job.user.classpath.first=true \
   --connect=jdbc:mysql://10.2.0.3:3306/testdb --username=root \
   --password-file=gs://gcsfolder/passwordfile --table=employee
```

The arguments passed to the job submission include:

region
> Region of the Dataproc cluster.

cluster
> Name of Dataproc cluster where the Sqoop job will run.

class
> Main class name of the Sqoop application.

jars
> JAR files to be included in the class path, which are listed in Table 2-3. Files can be present in GCS, or they can be in your local filesystem of all nodes in the cluster.

Table 2-3. Types of JAR files to include when submitting a Sqoop job

JAR type	Description
Sqoop JARs	Dataproc cluster doesn't come with Sqoop installed by default. You can manually download and add Sqoop JARs to the application classpath at runtime.
Source JARs	JARs required to interact with source databases. In this example, we added a MySQL JAR to be able to communicate with the source MySQL database.
Target JARs	JARs needed to interact with the target database of the filesystem. If writing to Hive, the jars are by default added in the classpath. For those not available by default, add the JAR manually.
Miscellaneous	Add any additional JARs needed, such as Avro framework (for creating Avro files).

import
> Parameters to specify data being imported from the source to the target.

connect
> Connection string for connecting to the source database.

username
> Username to connect to the source database.

`password-file`
Location of the file containing passwords.

`table`
Table to be migrated from the source database.

Upon successful completion, this Sqoop job will migrate the employee table from the source database to Hive.

2.8 Choosing Deployment Modes When Submitting a Spark Job to Dataproc

Problem

You want to know the different deployment modes available when submitting a Spark job to a Dataproc cluster and how they affect where the Spark driver runs.

Solution

Dataproc supports two primary deployment modes: client mode and cluster mode. The Spark job has driver and executor processes running. The deployment mode option controls where the Spark driver is going to run.

Client mode

The Spark driver JVM runs on the master node of the Dataproc cluster. Here is the gcloud command to deploy in client mode:

```
gcloud dataproc jobs submit <SparkType> \
  --project=<gcp-project-name> \
  --cluster=<dataproc-cluster-name> \
  --region=<gcp-region>
  --properties="spark.submit.deployMode=client"
```

Cluster mode

The Spark driver process runs on one of the worker nodes in the Dataproc cluster. Here is the gcloud command to deploy in cluster mode:

```
gcloud dataproc jobs submit <SparkType> \
  --project=<gcp-project-name> \
  --cluster=<dataproc-cluster-name> \
  --region=<gcp-region>
  --properties="spark.submit.deployMode=cluster"
```

Discussion

The Spark driver and executor processes require CPU and memory resources during runtime. Client machines, also referred to as *edge* or *gateway nodes*, are where the jobs are submitted. Running in client mode will start the driver JVM on the master node.

> In Dataproc, the default mode is client mode. You can explicitly set the desired deployment mode using the configuration option `spark.submit.deployMode`.

Here are the key considerations for choosing a deployment mode:

Resource utilization
 Client mode uses resources on the master node. This can potentially consume resources on master, especially when submitting many jobs. Cluster mode distributes the workload better.

 When you submit multiple jobs in parallel from orchestration tools like Cloud Composer, they can quickly occupy resources on the master if you run on client mode. Cluster mode is optimal in this scenario as it distributes resource usage and is generally preferred for production workloads and when running multiple jobs in parallel.

Interactive sessions
 Client mode is often preferred for running Spark Shell (REPL) for interactive exploration or debugging or for interactive workloads (like notebooks) where you need direct access to the driver output.

Logs
 If you want the driver logs to be streamed to the client, client mode is great. In cluster mode, you might need to access driver output from GCS.

See Also

- Client versus cluster deploy modes tutorial (*https://oreil.ly/UkErK*)
- Dataproc logging client versus cluster mode documentation (*https://oreil.ly/ZR07i*)

## CHAPTER 3

Advanced Dataproc Cluster Configuration

So far, we've covered several methods for creating and running a basic Dataproc cluster. This chapter dives into customizing your cluster beyond the basics.

Dataproc offers extensive customization options for underlying resources, including compute machine types, disk types, network configurations, OS image, and scalability settings. The recipes in this chapter will help you:

- Automate resource allocation by implementing dynamic scaling policies to ensure that your cluster has the resources it needs, optimizing costs and performance

- Harness the power of spot instances by leveraging mixed on-demand and spot instance autoscaling policies for significant cost savings without compromising reliability

- Boost worker node performance by attaching local SSDs for optimizing input/output-intensive workloads

- Deploy preconfigured environments by creating custom images with preinstalled software and configurations for a streamlined and repeatable cluster setup

- Build specialized clusters by using custom machine types to match the specific compute, memory, and storage needs of your applications

- Automate cluster initialization by running custom initialization scripts at cluster creation to automate configurations and software installation

- Prevent resource waste by scheduling automatic deletion of unused clusters to optimize resource utilization and cost efficiency

- Fine-tune performance by overriding Hadoop configurations to precisely control your cluster's behavior and maximize performance for your specific needs

3.1 Creating an Autoscaling Policy

Problem

Given your cluster's unpredictable workload, you want to create an autoscaling policy that dynamically adds or removes worker nodes from the Dataproc cluster based on the workload's needs.

Solution

The first step is to create a YAML configuration file with an autoscaling configuration. To do this, use the following *scaling-policy.cfg* file:

```
workerConfig:
  minInstances: 2
  maxInstances: 2
secondaryWorkerConfig:
  maxInstances: 10
basicAlgorithm:
  cooldownPeriod: 2m
  yarnConfig:
    scaleUpFactor: 1
    scaleDownFactor: 0.5
    gracefulDecommissionTimeout: 1h
```

Then, run a command to add the autoscaling policy to your project:

```
gcloud dataproc autoscaling-policies import <scaling_policy_name> \
--source=<config_file_path> \
--region=region
```

Discussion

A typical Dataproc cluster handles variable workloads, where some jobs demand large resources and others require fewer. Using a static cluster size is inefficient and costly because you are billed for the resources at all times. Scaling the cluster nodes up or down according to the workload demand offers a more cost-effective solution.

The autoscaling feature in Dataproc scales the cluster dynamically based on workload demand. Workload demand is calculated by pending memory or pending cores in the cluster. When you submit jobs to the cluster, they request resources (CPU and memory) from the YARN ResourceManager (RM). YARN RM keeps track of the total resources available, allocated, and pending.

By default, Dataproc autoscaling is based solely on pending memory. To enable core-based autoscaling, configure the YARN resource calculator to use the dominant resource calculator. This configuration is achieved using the following property:

```
capacity-scheduler:yarn.scheduler.capacity.resource-calculator=org.apache.hadoop.
yarn.util.resource.DominantResourceCalculator
```

This property must be enabled at the time of cluster creation. The following is an example gcloud command to create a cluster with the dominant resource calculator configured:

```
gcloud dataproc clusters create <CLUSTER_NAME>
  --region <REGION>
  --properties "yarn:yarn.scheduler.capacity.resource-calculator=org.apache.
hadoop.yarn.util.resource.DominantResourceCalculator"
```

How Do I See If There Is a Pending Demand in the Cluster?

All the resources for running jobs are granted in the form of YARN containers. The chart in the Dataproc monitoring dashboard UI provides information about the amount of pending memory in the cluster. To view the chart, shown in Figure 3-1, click Cluster and then the Monitoring tab.

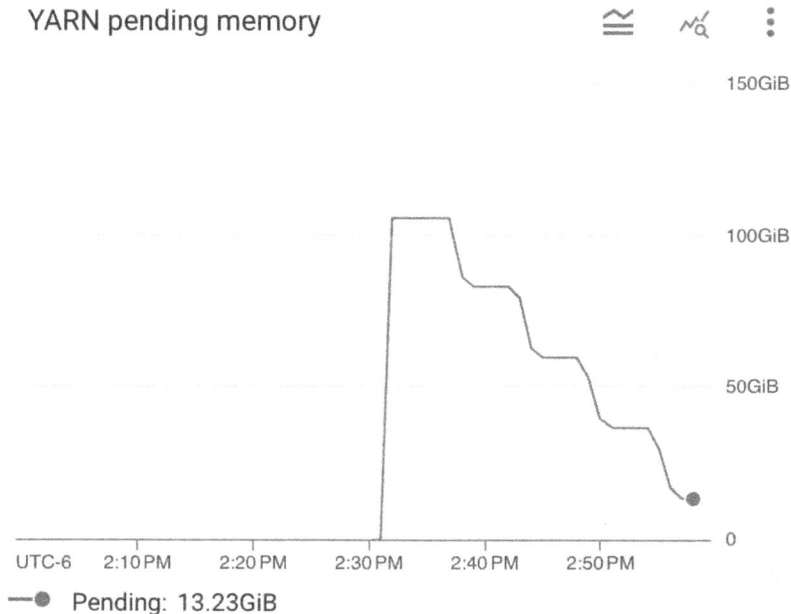

Figure 3-1. Chart showing YARN pending memory in the cluster

For the pending cores metric, you can build a custom chart using Metrics Explorer. Refer to Recipe 9.3 for more details.

The autoscaling configuration involves the following key configurations:

Autoscaling policy name
> The name of the autoscaling policy.

Region
> The regional location where the configuration is saved. A configuration can only be used by the clusters in the same region.

Cooldown duration
> How often Dataproc should check for scaling the cluster. For example, if the configuration is set to five minutes, Dataproc will check once every five minutes and make a decision. The decision result can be to:
>
> - Scale up the cluster, or add more nodes
> - Scale down the cluster, or remove nodes from cluster
> - Make no change, or leave as is
>
> The default cooldown duration time is two minutes.

YARN configuration

The YARN configuration controls how aggressively the cluster can scale up or scale down with the following settings:

Scale-up factor
> Ranges from 0 to 1, where 1 is more aggressive scaling and 0 is least aggressive. For example, if there is a pending memory of 200 GB with 1 set as the scale-up factor, the configuration would add more nodes that can add 200 GB to the cluster. If the scale-up factor is set to 0.5, it would only add nodes that can contribute 100 GB (0.5 × 200 GB).

Scale-up min worker fraction
> Ranges from 0 to 1. This setting decides the percentage of the minimum number of workers to be added as part of the scaling decision.

Scale-down factor
> Ranges from 0 to 1, where 1 indicates the most aggressive scale down, with progressively less aggressive scaling down approaching 0. For example, with 200 GB available memory and a scale-down factor of 1, nodes accounting for 200 GB memory would be removed. With a scale-down factor of 0.5, only nodes accounting for 100 GB memory would be removed. A scale-down factor of 0 signifies that the cluster will never scale down.

Scale-down min worker fraction
> Ranges from 0 to 1. This configuration decides the percentage of the minimum number of workers to be removed as part of the scaling decision.

For the most aggressive scale-up behavior to rapidly meet workload demands with full resources, opting for a scale-up factor of 1 is recommended. Conversely, to promote a more gradual and balanced approach to scaling up (and, similarly, for a less drastic scale down), a factor around 0.5 often strikes a good balance between responsiveness and resource stability.

Worker configuration

The worker configuration configures the type of workers and scaling minimum and maximum limits. Scaling both primary and secondary workers is supported. Dataproc primary workers function as both storage (HDFS) and compute nodes (YARN, Hive, Spark, etc.), while secondary workers are compute-only nodes and do not include HDFS. Most workloads will use GCS for persistent storage and disks for temporary storage. HDFS is only used in limited scenarios, such as storing temporary data or results of Hive jobs with multiple stages storing intermediate stage output in HDFS.

Worker configuration involves the following key settings:

Primary workers
In the UI, this option is disabled by default. Enable it by clicking the option to scale primary workers with the following settings:

Min instances
Minimum number of primary workers to be present in the cluster

Max instances
Maximum number of primary workers to be added to the cluster

 Scaling primary worker nodes involves the additional step of adding the HDFS service to the new node during scale up. Conversely, during scale down, all HDFS data residing on the node being removed must be migrated to other active primary workers. If the services running on the cluster do not rely on HDFS for storage, this is less critical. However, if HDFS is actively used, it is generally advisable to exclude primary workers from autoscaling configurations to avoid potential data movement overhead and complexity.

Secondary workers
> In the UI, this option is disabled by default. Enable it by clicking the option to scale primary workers with the following settings:

Min instances
> Minimum number of primary workers that must be present in the cluster

Max instances
> Maximum number of primary workers to be added to the cluster

> When both primary and secondary workers are configured to scale in each decision configuration, you need to determine what percentage should be primary versus secondary workers. Refer to Chapter 9 to learn more about choosing the ratio of primary to secondary workers in a cluster.

Primary weight
> Weight assigned for calculating the number of primary workers shared in nodes to be scaled up.

Secondary weight
> Weight assigned for calculating the number of secondary workers shared in nodes to be scaled up.

> For example, if primary workers have a weight of 2 and secondary workers have a weight of 1, the cluster will have approximately two primary workers for each secondary worker.

Here are the steps for creating an autoscaling policy. First, create a YAML configuration file *scaling-policy.cfg* with the following autoscaling configuration:

```
workerConfig:
  minInstances: 2
  maxInstances: 2
secondaryWorkerConfig:
  maxInstances: 10
basicAlgorithm:
  cooldownPeriod: 2m
  yarnConfig:
    scaleUpFactor: 1
    scaleDownFactor: 0.5
    gracefulDecommissionTimeout: 1h
```

Then, run gcloud command to add the autoscaling policy to your project:

```
gcloud dataproc autoscaling-policies import test-cluster-auto-scaling \
  --source=scaling-policy.cfg \
  --region=us-central1
```

3.2 Attaching an Autoscaling Policy to a Dataproc Cluster

Problem

You want to attach an existing autoscaling policy to a newly created cluster.

Solution

To create a new cluster with an autoscaling policy, use the gcloud command:

```
gcloud dataproc clusters create {new-cluster-name} \
  --autoscaling-policy={auto-scaling-policy-id} \
  --region={region}
```

To attach an autoscaling policy to the existing cluster, use the gcloud command:

```
gcloud dataproc clusters update {existing-cluster-name} \
  --autoscaling-policy={auto-scaling-policy-id} \
  --region={gcp-region}
```

Discussion

Autoscaling policies can be attached to both new and existing clusters within the same region. Using them across regions isn't supported. For example, a policy created in us-central1 can't be applied to clusters in us-east2. However, a single policy can be attached to multiple clusters.

To view the list of existing autoscaling policies, use the gcloud command:

```
gcloud dataproc autoscaling-policies list \
  --region {gcp-region}
```

To view the details of a scaling policy, use the gcloud command:

```
gcloud dataproc autoscaling-policies describe {auto-scaling-policy-name} \
  --region {gcp-region}
```

To attach a scaling policy to a newly created cluster, add the flag `--autoscaling-policy` and provide the value of the policy name:

```
gcloud dataproc clusters create {new-cluster-name} \
  --autoscaling-policy={auto-scaling-policy-id} \
  --region={region}
```

To attach a scaling policy to an already existing cluster, use this:

```
gcloud dataproc clusters update {existing-cluster-name} \
  --autoscaling-policy={auto-scaling-policy-id} \
  --region={gcp-region}
```

To remove or disable an autoscaling policy from the running cluster, use this:

```
gcloud dataproc clusters update {existing-cluster-name} \
  --disable-autoscaling \
  --region={gcp-region}
```

Can an Autoscaling Policy Be Modified While It's Attached to a Cluster?

Yes, you can modify an autoscaling policy even while it's attached to one or more clusters. A separate process within the Dataproc control API called the *autoscaler* continuously monitors both autoscaling configurations and cluster metrics like pending memory for any changes. Changes to the autoscaling policy will take effect during the next operation, following the cooldown period.

3.3 Optimizing Cluster Costs with a Mixed On-Demand and Spot Instance Autoscaling Policy

Problem

To achieve a balance between cost and stability, you want to design a scaling policy that adds an equal number of on-demand and spot instances when the cluster needs to scale up. You want to ensure sufficient capacity to handle increased demand while minimizing costs.

Solution

To achieve the desired balance between cost and stability, we will use Dataproc's Flexible VMs feature. This feature allows configuring secondary worker groups to include multiple machine types. We will set up the scaling configuration for these secondary workers to target a 50% allocation for on-demand instances and 50% for spot instances during scale-up operations.

Discussion

Dataproc's Flexible VMs feature enables the use of multiple machine types (such as a mix of non-preemptible and spot instances) specifically for the secondary worker role within a cluster. During cluster creation, you can define the target percentages for both on-demand (also referred to as *non-preemptible*) machines and spot instances that should make up this secondary worker group.

You then attach an autoscaling policy to manage these secondary workers, complete with minimum and maximum size limits. When this autoscaling policy triggers a scale-up event, it adds new secondary worker nodes to the cluster according to the percentages you initially specified.

> The Flexible VMs feature is currently not available for primary workers; it applies only to secondary workers.

Here are the flags that will help you create a cluster with mixed VMs:

`--secondary-worker-type`
> Set this as `spot` when you want to create a secondary worker group that includes both spot and non-preemptible instances.

`--secondary-worker-standard-capacity-base`
> Specify the fixed number of secondary worker nodes within the group that you want to guarantee as non-preemptible instances, regardless of the total group size.

`--secondary-worker-standard-capacity-percent-above-base`
> After setting the base non-preemptible nodes, define the percentage of the remaining nodes that should be non-preemptible. The rest will be Spot VMs. For example, if you set 20%, then 80% of the remaining nodes will be Spot VMs:

```
gcloud dataproc clusters create <cluster_name> \
  --project=<project_id> \
  --region=<region> \
  --secondary-worker-type=spot \
  --num-secondary-workers=12 \
  --secondary-worker-standard-capacity-base=2 \
  --secondary-worker-standard-capacity-percent-above-base=20
```

Upon successful creation of the cluster using the previous gcloud command, your cluster secondary workers will look like this:

Total secondary workers
> We specified `--num-secondary-workers=12`, so the cluster will have a total of 12 secondary workers.

Non-preemptible workers
> The `--secondary-worker-standard-capacity-base=2` flag guarantees that two of the secondary workers will always be non-preemptible (on-demand). The `--secondary-worker-standard-capacity-percent-above-base=20` flag tells the system that 20% of the remaining secondary workers (after accounting for the base) should also be non-preemptible. Since the number of remaining workers is 10 (12 total minus the two base workers), 20% of 10 is two additional

non-preemptible workers. In total you see four non-preemptible (standard-type) machines.

Spot workers

The remaining workers will be Spot VMs. Since there are 12 total secondary workers and four are non-preemptible, the remaining eight workers will be Spot VMs.

To create 50% non-preemptible and 50% spot instance, choose the combination of `--num-secondary-workers` and `--secondary-worker-standard-capacity-base` to match 50% of the required capacity. In this example, if your total is 12 and the base is two nodes, then you can configure the percentage above base to be 40 to get 50%, which sets six nodes as non-preemptible. Here is the corresponding gcloud command:

```
gcloud dataproc clusters create <cluster_name> \
  --project=<project_id> \
  --region=<region> \
  --secondary-worker-type=spot \
  --num-secondary-workers=12 \
  --secondary-worker-standard-capacity-base=2 \
  --secondary-worker-standard-capacity-percent-above-base=40
```

Another key benefit of Dataproc Flexible VMs is the ability to define multiple machine types for a secondary worker group. This helps clusters acquire capacity more reliably, which is especially useful for handling "stock-out" scenarios where a specific machine type isn't immediately available.

This flexibility is particularly relevant when using Spot VMs. Spot instance availability can vary considerably across different machine types; some types may be readily available, while others are out of availability in that region. By providing Dataproc with a list of acceptable machine types for your spot secondary workers, you substantially increase the probability of the service successfully provisioning the required compute capacity, even when your preferred machine type is unavailable.

You can configure these multiple machine types using the `--secondary-worker-machine-types` flag. The value provided to this flag should include both the machine type and its rank. You can specify the flag multiple times to define different machine types with their corresponding rank values. These ranks determine the preference order, where zero (0) represents the highest rank (most preferred) and larger numbers indicate lower ranks (less preferred).

Here is a sample gcloud command for specifying n1-standard-4 and n1-standard-8 as rank 0 and n1-highmem-4 and n1-highmem-8 as rank 1 machine types for secondary workers:

```
gcloud dataproc clusters create <cluster_name> \
  --project=<project_id> \
```

```
--region=<region> \
--secondary-worker-type=spot \
--num-secondary-workers=12 \
--secondary-worker-standard-capacity-base=2 \
--secondary-worker-standard-capacity-percent-above-base=40 \
--secondary-worker-machine-types="type=n1-standard-4,n1-standard-8,rank=0" \
--secondary-worker-machine-types="type=n1-highmem-4,type=n1-highmem-8,rank=1"
```

> ## Are Spot Instances Recommended?
>
> Spot instances offer the reward of low cost but come with the risk of being interrupted at any time. This makes them unsuitable for applications with strict service-level agreements (SLAs) that cannot tolerate even brief delays. Instead, consider using spot instances for flexible workloads that can handle losing a few nodes during job execution or that can run during noncritical periods when demand for spot instances is low. Spark and MapReduce jobs are generally resilient and can automatically recover from node loss. However, there may be exceptions, depending on the specific application logic. By default, Spark retries a failed task up to four times, which can be increased by setting the property `spark.task.maxFailures`.
>
> The ratio of on-demand to spot instances you choose depends on your desired balance between reliability and cost savings. Increasing the number of on-demand instances enhances job stability, while increasing the number of spot instances reduces costs. Also increase the probability of securing spot machines by specifying multiple machine types.

3.4 Adding Local SSDs to Dataproc Worker Nodes

Problem

An application running on a Dataproc cluster and using a persistent disk for intermediate data storage is presently running slow. Since it is input/output (I/O) intensive and requires reading and writing large amounts of data, how can you address this issue without significantly affecting costs?

Solution

Local SSDs offer a cost-effective solution with improved read/write performance. Re-create the cluster and attach local SSDs to the primary and secondary worker nodes:

```
gcloud dataproc clusters create <cluster-name> \
  --region=<region> \
  --num-worker-local-ssds=<num-of-local-ssds> \
  --num-secondary-worker-local-ssds=<num-of-local-ssds> \
```

```
--worker-local-ssd-interface=<NVME|SCSI> \
--secondary-worker-local-ssd-interface=<NVME|SCSI>
```

Discussion

When you create a Dataproc cluster, you can choose the type of disk to attach to the Compute Engine machines. Local SSDs are storage devices directly attached to the Compute Engine, offering superior performance. Standard persistent disks, on the other hand, are not directly connected to the VM instance, resulting in some latency during disk read and write operations.

What Goes to the Local Disk?

It depends on what services we configure in the Dataproc cluster. Here are a few examples of data that go to the local disk:

- HDFS data
- Hive job intermediate outputs
- Spark job disk persists cache
- Shuffle data spills

Each local SSD comes with a standard size of 375 GB, and you can customize how many local SSDs to attach to the VM instance. Unlike persistent disks, local SSDs have to be configured in the supported bundles (number of disks). For example, the N2D machine type with 32 cores only supports configuring 2, 4, 8, 16, or 24 local SSDs. Refer to the Google documentation (*https://oreil.ly/VzIMC*) on choosing a valid number of local SSDs based on machine type.

Local SSDs provide greater performance than persistent disk but lower performance than PD SSDs. Typically, price versus performance is the deciding factor when choosing a specific storage type. Based on application I/O requirements, choose the option of PD standard, PD balanced, PD SSD, or local SSDs. Local SSDs are available in two protocol modes: Non-Volatile Memory Express (NVMe) and Small Computer System Interface (SCSI). NVMe is generally recommended over SCSI for low-latency and high-throughput workloads.

Here is the gcloud command to create a Dataproc cluster with primary and secondary workers that has two local SSDs attached:

```
gcloud dataproc clusters create basic-cluster\
  --region=us-central1 \
  --worker-machine-type n2d-standard-16 \
  --secondary-worker-machine-type n2d-standard-16 \
  --num-worker-local-ssds=2 \
  --num-secondary-worker-local-ssds=2 \
  --worker-local-ssd-interface=NVME \
  --secondary-worker-local-ssd-interface=NVME
```

For more details on the supported disk types and their respective pros and cons, refer to Recipe 11.2.

3.5 Creating a Cluster with a Custom Image

Problem

Spark applications running on ephemeral clusters require the installation of a custom Python package. However, installing these packages can be time-consuming, which can delay the deployment of the application and potentially affect critical SLAs.

Solution

Create a custom OS image by cloning the Dataproc base image. Add the dependencies to the custom image and create a new cluster using the custom OS image.

Discussion

Custom images are created by cloning existing Dataproc images and applying your custom changes. Google Cloud offers a tool that can automate the process of custom image creation.

First, clone the GitHub repository (*https://oreil.ly/CRPbX*). Then, create a GCS bucket to store the custom image or use an existing one:

```
gsutil mb gs://custom-image-storage-bucket
```

Create the script with the code or commands to be applied on top of the base image. Then, navigate to the cloned Git project:

```
cd custom-images
```

Run the command to create a custom image:

```
python3 generate_custom_image.py \
  --image-name=custom-image-test \
  --family=custom-2-0-debian10 \
  --dataproc-version=2.0.69-debian10 \
  --customization-script=../../../Downloads/custom-image.sh \
  --zone=us-central1-a \
  --disk-size 30 \
  --gcs-bucket='gs://nstestb/customimages' \
  --no-smoke-test
```

Successful custom image creation will produce a result like the following:

```
INFO:custom_image_utils.image_labeller:Successfully set label on custom image...
INFO:custom_image_utils.expiration_notifier:Successfully built Dataproc custom
image: ns-test
INFO:custom_image_utils.expiration_notifier:
###################################################################
  WARNING: DATAPROC CUSTOM IMAGE 'ns-test'
          WILL EXPIRE ON 2024-02-03 23:13:10.678000.
###################################################################
```

Now, create a Dataproc cluster using the custom image:

```
gcloud dataproc clusters create custom-image-test \
  --image=https://www.googleapis.com/compute/v1/projects/<project-name>/global/
images/custom-image-test \
  --region=us-central1
```

3.6 Building a Cluster with Custom Machine Types

Problem

Your existing workloads require a specific CPU-to-memory ratio that is not available in the predefined machine types offered by your cloud provider. You need to create a cluster with custom machine types that match your workload requirements while minimizing costs compared to using predefined machine types.

Solution

Dataproc allows the creation of clusters with custom machine types, enabling you to select the specific number of vCPU cores and memory for master and worker machines. Use gcloud command to create a Dataproc cluster with each master having 8 cores and 64 GB memory and worker machines having 12 cores and 96 GB memory:

```
gcloud dataproc clusters create custom-machine-cluster  \
  --region us-central1 \
  --master-machine-type custom-8-65536-ext \
  --worker-machine-type custom-12-98304-ext \
  --enable-component-gateway
```

Discussion

Predefined machine types are categorized as:

General purpose
    Designed to give the best combination of price and performance and suit a variety of workloads in a balanced way

Memory optimized

Higher memory limits per core

Compute optimized

Higher CPU limits per machine while also an optimized CPU type for getting higher compute performance

While creating a cluster, choose a machine type that fits your workload. For running workloads that are more CPU intensive, you can choose compute-optimized instances (C2, C2D, and H3 series). For memory-intensive workloads, you can choose memory-optimized instances (M1, M2, and M3 series).

Sometimes, you might have workloads that don't fit any of the predefined machine types (general purpose, memory optimized, or compute optimized). Choosing a custom machine type comes with some limitations. As of the time of writing, Dataproc only supports creating custom machine types for N1 machine families. The maximum memory per core you can configure is 6.5 GB. However, with the extended memory option (which comes with additional charges), you can configure up to 624 GB memory (the maximum allowed total memory for the N1 machine type).

To create a custom machine from the UI at the time of cluster creation, navigate to "Configure nodes." Then, select the machine series as N series, as shown in Figure 3-2.

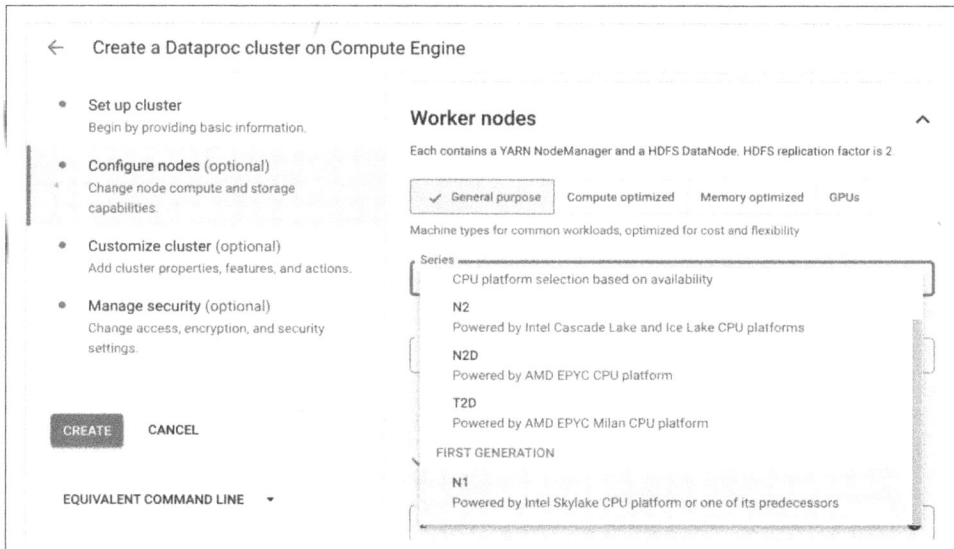

Figure 3-2. Choosing N series worker nodes

Next, select the machine type as Custom, as shown in Figure 3-3.

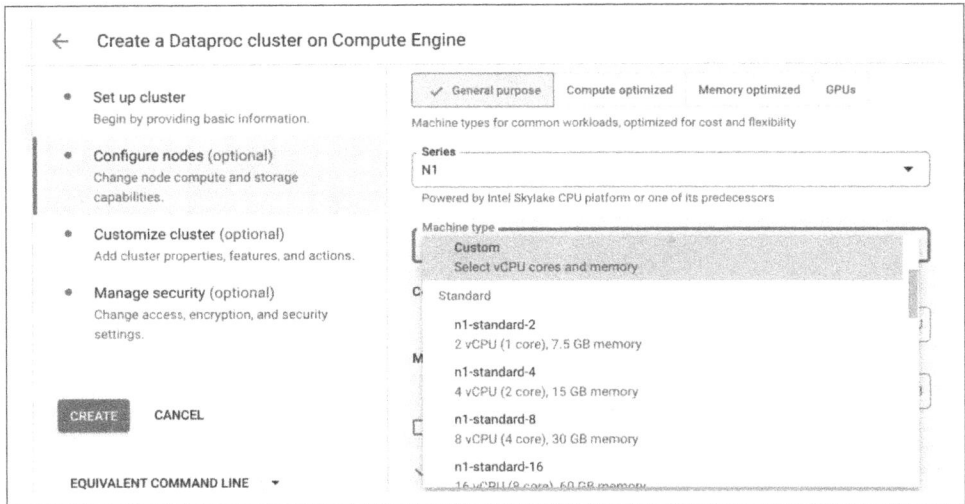

Figure 3-3. Selecting the machine type as Custom

Select the required number of cores and memory for the custom machine type. Check the box for Extend Memory, circled in Figure 3-4. This is necessary because we are requesting more than 6.5 GB per core.

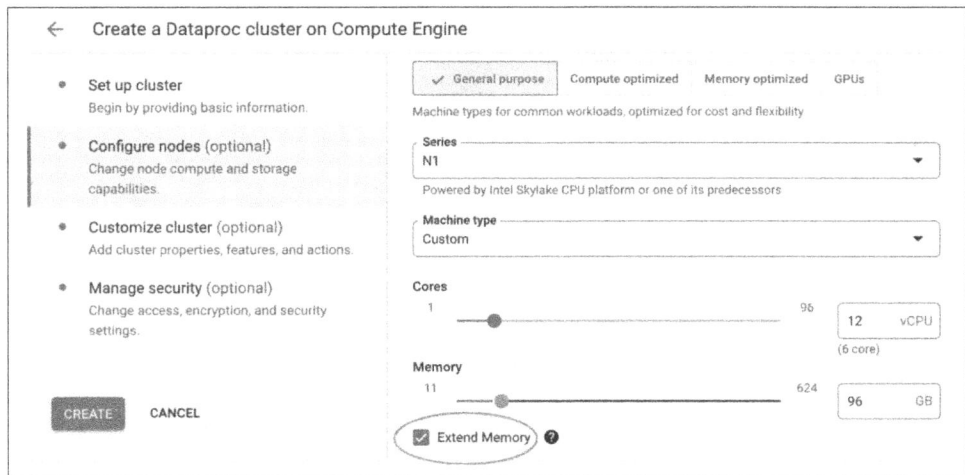

Figure 3-4. Enable the checkbox for extended memory

The equivalent gcloud command for creating a Dataproc cluster with each master having 8 cores and 64 GB memory and worker machines having 12 cores and 96 GB memory is as follows:

```
gcloud dataproc clusters create custom-machine-cluster  \
   --region us-central1 \
   --master-machine-type custom-8-65536-ext \
   --master-boot-disk-size 500 \
   --num-workers 2 \
   --worker-machine-type custom-12-98304-ext \
   --worker-boot-disk-size 500 \
   --image-version 2.1-debian11 \
   --enable-component-gateway
```

The option `--master-machine-type custom-8-65536-ext` specifies that the master machine is a custom machine type with 8 cores and 65,536 MB (64 GB) memory. The ext suffix tells it to use the extended memory option.

The option `--worker-machine-type custom-12-98304-ext` specifies that the worker machine is a custom machine type with 12 cores and 98,304 MB (96 GB) memory. The ext suffix tells it to use the extended memory option.

Custom machines come with a premium price compared to standard machines, which are charged per vCPU and gigabyte of memory. Up to 8 GB of memory per core is priced at the custom machine memory pricing. Each additional gigabyte above that will have separate pricing.

For instance, the N1 custom machine type having 12 cores and 96 GB will be charged as follows:

- Per-hour vCPU price: $0.033174
- Per-hour 1 GB memory price: $0.004446

The machine price for 12 cores and 96 GB = $(12 \times 0.033174) + (96 \times 0.004446)$ = 0.824904.

The nearby predefined machine type, n1-highmem-16, with 16 cores and 104 GB memory, is priced at $0.9464. In this instance, if your workload requires only 12 cores and 96 GB memory, choosing a custom machine type will save costs.

> Cost calculation, as shown in the previous example, is only for VM pricing (compute and memory). There are other costs, such as storage, network, OS (for premium images like Red Hat Enterprise Linux), and so on. Refer to Recipe 11.9 for more details on calculating cluster cost.

3.7 Bootstrapping Dataproc Clusters with Initialization Scripts

Problem

Applications running in the cluster require specific JAR files to be present on worker machines. These files only need to be copied to worker machines at the time of cluster creation.

Solution

First, prepare the initialization script file and copy it to a GCS location. Then, add a flag to the cluster creation command and provide the command's file location:

```
gcloud dataproc clusters create {CLUSTER_NAME} \
  --region={REGION} \
  --initialization-actions={initialization-script-gcs-path}
```

Discussion

When running a set of commands on all cluster nodes, you can use the initialization action feature provided by the Dataproc service. These commands can be used for installing software, copying or modifying files, calling an API, and the like.

The first step is to create a file with the list of commands to be executed. Consider, for example, copying JAR files from GCS to specific folders. Create a file with *initialization-actions.sh* and copy it to a GCS location:

```
ROLE=$(/usr/share/google/get_metadata_value attributes/dataproc-role)
if [[ "${ROLE}" == "Worker" ]]; then
  #code to copy jar files from gcs to local file system
  gsutil cp gs://<bucket-name>/<location-of-jar-file> \
  <target-local-file-system-path>
fi
```

The Unix/Linux version of GCE comes with a built-in script /usr/share/google/get_metadata_value to provide metadata at the project, machine, or service (Dataproc) level. The input is the name of the attribute you want, and the output is its value. For instance, /usr/share/google/get_metadata_value attributes/dataproc-role provides the Dataproc role (master or worker) of the machine, and /usr/share/google/get_metadata_value image provides an OS image of the machine. Run /usr/share/google/get_metadata_value to view all the available options.

Copy the initialization actions script to the GCS location:

```
gsutil cp dataproc-initialization-action.sh gs://<gcs-location-path>
```

Then, run the gcloud command to create a Dataproc cluster with the initialization script:

```
gcloud dataproc clusters create custom-machine-cluster  \
  --region us-central1 \
  --initialization-actions=gs://nstestb/dataproc-initialization-action.sh
```

Successful cluster creation executes the configured initialization actions script. However, any errors in the script will result in a failed cluster state.

At What Phase of Cluster Creation Will the Custom Initialization Script Be Executed?

Initialization actions are run after the cluster creation process is completed. If your initialization action has any requirement to change Hadoop configuration files, you might need to restart the service.

To run the initialization script before the services get started, add the property `dataproc.worker.custom.init.actions.mode` as RUN_BEFORE_SERVICES:

```
gcloud dataproc clusters create custom-machine-cluster \
  --region us-central1 \
  --initialization-actions=gs://nstestb/dataproc-initialization-action.sh \
  --properties 'dataproc:dataproc.worker.custom.init.actions.
mode=RUN_BEFORE_SERVICES'
```

3.8 Scheduling Automatic Deletion of Unused Clusters

Problem

The development team is authorized to create and delete Dataproc clusters. However, they often neglect to delete clusters after use, resulting in unnecessary costs for the company. This issue needs to be addressed.

Solution

The Dataproc service offers scheduled deletions to be triggered by:

- Specific date and time
- Maximum run duration
- Cluster inactive state duration

Use the gcloud command to delete the cluster if there is no activity after an hour:

```
gcloud dataproc clusters create {cluster-name} \
  --region {region-name} \
  --max-idle=60m
```

Discussion

In the cloud environment, it is a common occurrence to create resources and forget to delete them afterward. This oversight can significantly affect your cloud costs since you are still billed for unused resources. The primary advantage of using cloud services is the ability to pay only for what you consume. Failing to delete resources after use defeats this purpose and leads to unnecessary expenses.

Dataproc provides a solution for this problem by offering scheduled deletion with time-based and activity-based triggers. Depending on your use case, you can configure the cluster to be scheduled for deletion based on a specific date and time, the maximum run duration, or the maximum inactive duration of the cluster.

For scheduling a cluster to be deleted following inactivity, use the flag --max-idle. You can set the --max-idle flag at the time of cluster creation. For example, delete a cluster if it's idle for 60 minutes:

```
gcloud dataproc clusters create basic-cluster --region us-central1 --max-idle=60m
```

You can also add a --max-idle flag to an existing cluster—for example, to increase the idle time for deleting a cluster from 60 to 120 minutes:

```
gcloud dataproc clusters update basic-cluster --region us-central1 \
  --max-idle=120m
```

And you can cancel a max-idle time setting configured for an existing cluster:

```
gcloud dataproc clusters update basic-cluster --region us-central1  --no-max-idle
```

> When is a cluster considered to be inactive? Dataproc monitors YARN resource usage in the cluster. When there are no YARN jobs running, the cluster is considered to be idle.

To schedule a cluster to be deleted after a certain time, use the flag --max-age. To create a cluster with a maximum time to run of seven days, use the following:

```
gcloud dataproc clusters create basic-cluster --region us-central1 --max-age=7d
```

To update an existing cluster to run for a maximum of nine days, use the following:

```
gcloud dataproc clusters update basic-cluster --region us-central1 --max-age=9d
```

Updating `max-age` has two prerequisites:

- The cluster must have an existing `max-age` specified.
- The new `max-age` must be greater than the existing `max-age` of the cluster.

To cancel the maximum age configuration for the cluster, use the following:

```
gcloud dataproc clusters update basic-cluster --region us-central1  --no-max-age
```

To schedule a cluster to be deleted on a specific date and time, use the flag `--expiration-time`. For example, create a cluster to expire on November 10, 2023, at 7 A.M. GMT:

```
gcloud dataproc clusters create basic-cluster --region us-central1 \
  --expiration-time=2023-11-1T07:00:00+0000
```

The value of the timestamp has to be in the ISO 8601 datetime format. The maximum expiration time allowed is within 14 days or 336 hours from the current time.

Use the following to update an existing cluster's expiration datetime:

```
gcloud dataproc clusters update basic-cluster --region us-central1 \
  --expiration-time=2023-11-11T07:00:00+0000
```

Use the following to remove a scheduled deletion of an existing cluster's expiration datetime:

```
gcloud dataproc clusters update basic-cluster --region us-central1 --no-max-age
```

Can All Three Settings Be Specified Simultaneously?

No, configuring all three settings at once is not supported. You can configure `--max-idle` along with either `--max-age` or `--expiration-time` but not all three together.

When multiple settings are configured as triggers for scheduled deletion, the earliest trigger takes precedence. For instance, clusters configured with a maximum age of one day and a maximum idle time of one hour will be deleted upon reaching either of these thresholds, whichever occurs first.

3.9 Overriding Hadoop Configurations

Problem

You want to customize the following properties in Hadoop and Spark while creating the cluster:

- Set the HDFS replication factor to 3 for a higher level of data redundancy
- Set the YARN minimum or default container size as 4 GB to ensure that the majority of jobs run fine on a default configuration
- Set Spark dynamic allocation as false at the cluster level to give control to users to enable it on necessary jobs

Solution

Add the following properties to the Dataproc cluster creation command:

```
--properties 'hdfs:dfs.replication=3,yarn:yarn.scheduler.minimum-allocation-mb
=4096,spark:spark.dynamicAllocation.enabled=false'
```

Discussion

The HDFS configuration property to set the replication factor in the cluster to 3 can be modified in *hdfs-site.xml*:

```
hdfs:dfs.replication=3
```

The YARN configuration property for setting a minimum container size in the cluster can be modified in *yarn-site.xml*:

```
yarn:yarn.scheduler.minimum-allocation-mb=4096
```

The Spark configuration to disable Spark dynamic allocation at the cluster level has to be modified in the *spark-defaults.conf* file:

```
spark:spark.dynamicAllocation.enabled=false
```

Modifying cluster configurations involves changing configuration files and ensuring that the services pick the modified changes. The configurations can be modified in two different ways:

- Through the initialization actions file
- By passing as properties at the time of cluster creation

The initialization actions file approach requires you to write code that will execute the property changes in the specific required files and restart the services if needed.

Dataproc allows you to pass properties for specific services during cluster creation. It will automatically apply these changes to specific files before the services in the cluster start up. When specifying a property, mention the name of the service or file first, followed by the property itself and its value to override. Multiple properties can be specified using a comma delimiter. Here's the gcloud command for creating a cluster by overriding the previously mentioned configuration:

```
gcloud dataproc clusters create custom-machine-cluster  \
  --properties 'hdfs:dfs.replication=4,yarn:yarn.scheduler.minimum-allocation-
mb=4096,spark:spark.dynamicAllocation.enabled=false' \
  --region us-central1
```

Alternatively, you can add all of these configurations to the properties file and pass it as a property. Create the properties file *cluster-overrides.properties*:

```
hdfs:dfs.replication=4
yarn:yarn.scheduler.minimum-allocation-mb=4096
spark:spark.dynamicAllocation.enabled=false
```

Pass the properties file while creating a cluster using gcloud command:

```
gcloud dataproc clusters create custom-machine-cluster  \
  --properties-file <property-file-location> \
  --region us-central1
```

How Do We Change These Settings for a Cluster That's Already Running?

Cloud users often adopt a strategy of using transient Dataproc clusters with no direct dependencies. This allows for easy re-creation of the cluster whenever configuration changes are necessary. It's important to note that properties cannot be updated using the --properties flag after the cluster has been created.

In exceptional cases where updating settings on a running cluster is desired without re-creation, it's crucial to understand the nature of the modifications. For example, altering the HDFS replication factor necessitates updating the *hdfs-site.xml* file on all master and worker nodes and subsequently restarting the cluster or relevant HDFS services for the changes to take effect. Similarly, modifying the memory allocation for the Spark history server requires a change on the master node only, followed by a service or cluster restart.

## CHAPTER 4

Serverless Spark and Ephemeral Dataproc Clusters

Dataproc Serverless Spark is an autoscaling serverless product for Spark that simplifies the execution of Spark applications because the user doesn't have to think about infrastructure in order to run Spark jobs.

Ephemeral Dataproc clusters, also known as *transient clusters*, are temporary clusters that run until specific jobs are completed or terminated. Throughout this book, we'll refer to them as *ephemeral clusters*. Both serverless and ephemeral Dataproc clusters support running jobs within a VPC or default network.

In this chapter, you will gain a clear understanding of when to use Dataproc Serverless Spark and ephemeral Dataproc clusters. You will also learn:

- How to submit Spark jobs to Dataproc Serverless
- How to create and run jobs on ephemeral clusters
- How to configure a Spark history server
- How to leverage Spark RAPIDS Accelerator
- How to price and monitor serverless Spark jobs

Let's dive in and explore how to scale your Spark workloads efficiently!

4.1 Running on Dataproc: Serverless Versus Ephemeral Clusters

Problem

Scenario 1: you're tasked with running a Spark job on Dataproc, with the following three criteria:

- You want to avoid managing or customizing the cluster, including hardware selection.
- There's no requirement to sequence Spark jobs on the same cluster.
- The objective is to execute a Spark job, not Hadoop MapReduce jobs or HiveQL scripts.

Scenario 2: you have a Spark job, a Hadoop job, or a sequence involving both, and you intend to run them on a tailored Dataproc cluster, deleting the clusters upon job completion.

Solution

For Scenario 1, use the Dataproc Serverless service to run the Spark job.

For Scenario 2, use an ephemeral Dataproc cluster approach: create a Dataproc cluster, run the jobs on the Dataproc cluster, and delete the cluster upon completion of the jobs.

Discussion

Let's consider the serverless and ephemeral approaches separately.

Dataproc Serverless

Dataproc Serverless for Spark is tailored for Spark jobs. These jobs can be executed on either a private VPC or the default network. To understand how to run a Dataproc Serverless job, refer to Recipe 4.3.

Although you won't be able to change the hardware (such as machine type), you can select between standard or premium tiers. The premium tier offers upgraded compute and storage capabilities, which you can configure for drivers and executors.

> For Spark machine learning, SparkR, or highly computational Spark jobs, opt for the premium tier Dataproc Serverless service for optimal performance.

Ephemeral Dataproc clusters

Customers managing a data lake often require sequential jobs, spanning tasks from data engineering to data science. For example, there might be a Spark job 1 that converts raw data from the bronze zone into a standardized format like Parquet in the silver zone, followed by a Spark job 2, which generates facts, dimensions, or data marts in the gold zone.

These jobs can be executed sequentially on a Dataproc cluster, orchestrated via Cloud Composer or Dataproc Workflow Templates, and subsequently deleted using the cluster postjob completion, as shown in Recipe 4.2. This cluster is called an *ephemeral Dataproc cluster* because it's set up to run on demand and to delete upon job completion.

In this context, leveraging an ephemeral Dataproc cluster offers several benefits:

- Customized, job-scoped clusters that allow benchmarking and optimal selection of machine types and disks to achieve target throughput
- The ability to install necessary packages using initialization scripts during startup
- Cost savings compared to static Dataproc clusters

The differences in the features of Dataproc Serverless and ephemeral Dataproc are compared in Table 4-1.

Table 4-1. Comparison of Dataproc ephemeral and Serverless

Feature	Dataproc Serverless	Ephemeral Dataproc
Autoscaling	Handled by Dataproc Serverless	Configurable
Machine type selection	Cannot be changed	Customizable
Tier selection	Standard or premium	Not applicable, as it is customizable at granular levels, such as machine type, disks, etc.
Startup time	60 seconds	90 seconds
Resource management	Spark based	YARN framework
Use cases	Suitable for Spark workloads	Suitable for Hive, MapReduce, Pig, Spark, and other Hadoop workloads

4.2 Running a Sequence of Jobs on an Ephemeral Cluster

Problem

You're orchestrating an end-to-end analytics pipeline. Your Spark job 1 converts CSV to Parquet format and lands in a GCS bucket. Then, job 2 takes the output of the first Spark job, performs aggregations, and prepares a facts table Parquet output. You want to run these on an ephemeral cluster.

Solution

Use Dataproc Workflow Templates to orchestrate this process following these steps:

- Create a workflow template using the following gcloud command:

```
gcloud dataproc workflow-templates create {template_name_here} \
    --region=us-central1
```

- Import the job details and ephemeral cluster configuration into the workflow template:

```
gcloud dataproc workflow-templates import {template_name_here} \
    --region us-central1 --source {config_file_name.yaml}
```

- Instantiate the workflow template using the following gcloud command:

```
gcloud dataproc workflow-templates instantiate my_template \
    --region us-central1
```

Discussion

There are different ways to orchestrate an end-to-end analytics pipeline on GCP. For a deeper dive into this topic, refer to Chapter 13.

Dataproc Workflow Templates provide a simple way to manage and orchestrate one or more Hadoop or Spark jobs on an existing or ephemeral Dataproc cluster. They are especially useful for big data workloads involving Hive, Spark, and MapReduce. You can also define dependencies between jobs to ensure that they run sequentially.

To run jobs sequentially on an ephemeral Dataproc cluster, specify the job details and ephemeral cluster configuration in a YAML file as follows:

```
jobs:
- pysparkJob:
    args:
    - gcs://lakehouse/input.xml
    mainPythonFileUri: gs://lakehouse/process.py
  stepId: process-large-pyspark

- pysparkJob:
    args:
```

```
        - gcs://lakehouse/silver/standard.parquet
      mainPythonFileUri: gs://lakehouse/analytics.py
    stepId: analytics-large-pyspark

  placement:
    managedCluster:
      clusterName: three-node-cluster
      config:
        gceClusterConfig:
          zoneUri: us-central1-b
        masterConfig:
          diskConfig:
            bootDiskSizeGb: 500
          machineTypeUri: n1-standard-2
        workerConfig:
          diskConfig:
            bootDiskSizeGb: 500
          machineTypeUri: n1-standard-2
          numInstances: 2
```

Before instantiating the workflow, import the YAML configuration file into the workflow template using this command:

```
gcloud dataproc workflow-templates import my_template --region us-central1 \
  --source template.yaml
```

4.3 Executing a Spark Batch Job to Convert XML Data to Parquet on Dataproc Serverless

Problem

You want to run a Spark Scala batch job that converts XML to Parquet on Dataproc Serverless.

Solution

Here is the gcloud command to submit a Spark Scala batch job to Dataproc Serverless:

```
gcloud dataproc batches submit spark --region=us-central1 \
  --class={class_name_here} \
  --jars={jar1_gcs_path_here},{jar2_gcs_path_here} \
  -- {file_input_gcs_path_here} \
  -- {argument1} {argument2} {argument3}
```

This job requires three arguments and relies on two dependency JARs. Replace the placeholders with the appropriate values as follows:

{class_name_here}
: Replace with the name of the main class for your Spark job.

{jar1_gcs_path_here}
: This is the GCS path for the Spark XML JAR.

{jar2_gcs_path_here}
: This is the GCS path for the Spark application JAR.

{file_input_gcs_path_here}
: This is the GCS path for the input XML file.

{argument1}
: Replace with the appropriate value for the input XML file.

{argument2}
: Replace with the desired output filepath.

{argument3}
: Replace with the root tag for the input XML file.

Discussion

Here's a snippet of the Scala code for the XML-to-Parquet conversion:

```
import org.apache.spark.sql.{SparkSession, SaveMode}

object xmltoparquet {
  def main(args: Array[String]): Unit = {

    val spark = SparkSession.builder()
      .appName("XmlToParquet")
      .master("yarn")
      .getOrCreate()

    // Reading XML and infer schema
    val xmlDataFrame = spark.read
      .format("com.databricks.spark.xml")
      .option("rowTag", args(2)) // Passing Root Tag
      .load(args(0)) // Input XML file GCS path

    xmlDataFrame.write
      .mode(SaveMode.Overwrite)
      .parquet(args(1)) // Output Parquet files GCS Path

    spark.stop()
  }
}
```

Execute a batch Spark job on Dataproc Serverless using the provided gcloud command. Here's an example:

```
gcloud dataproc batches submit spark \
  --region=us-central1 \
  --jars="gs://dataproc-cookbook/chapter2/spark/scala/jar/spark-xml_2.12-0.16.0.
jar","gs://dataproc-cookbook/chapter2/spark/scala/jar/
xmltoparquet_2.12-0.1.jar" \
  --class=com.dataprocessing.XmltoParquet \
  -- "gs://dataproc-cookbook/chapter2/spark/scala/inputfiles/menu.xml" \
  "gs://dataproc-cookbook/chapter2/spark/scala/outputfiles/" "food"
```

To run a batch Spark job on Dataproc Serverless via the console, navigate to Dataproc and select Batches under the Serverless section. Then, create a batch job by clicking Create, as shown in Figure 4-1.

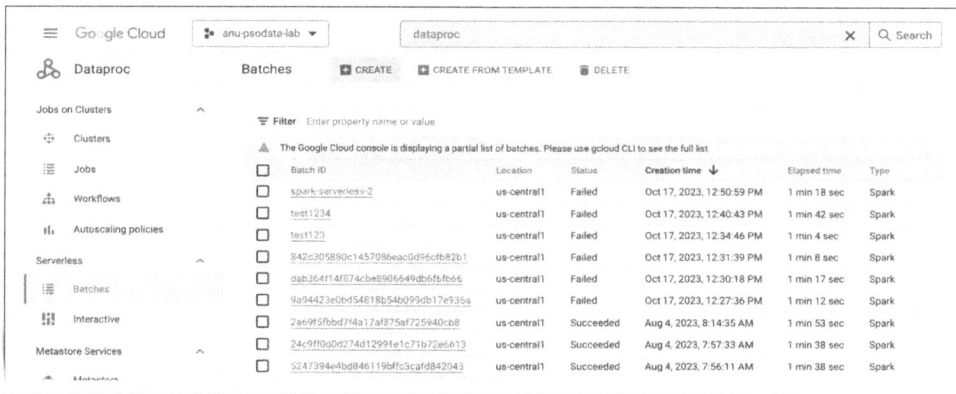

Figure 4-1. Creating a batch Spark job

Choose the appropriate runtime environment for your Spark job by selecting from the "Runtime version" drop-down menu and include any necessary arguments or dependency files that are referenced by the Spark code, as shown in Figure 4-2 and Figure 4-3. These dependency files will be extracted into the working directory of each Spark executor.

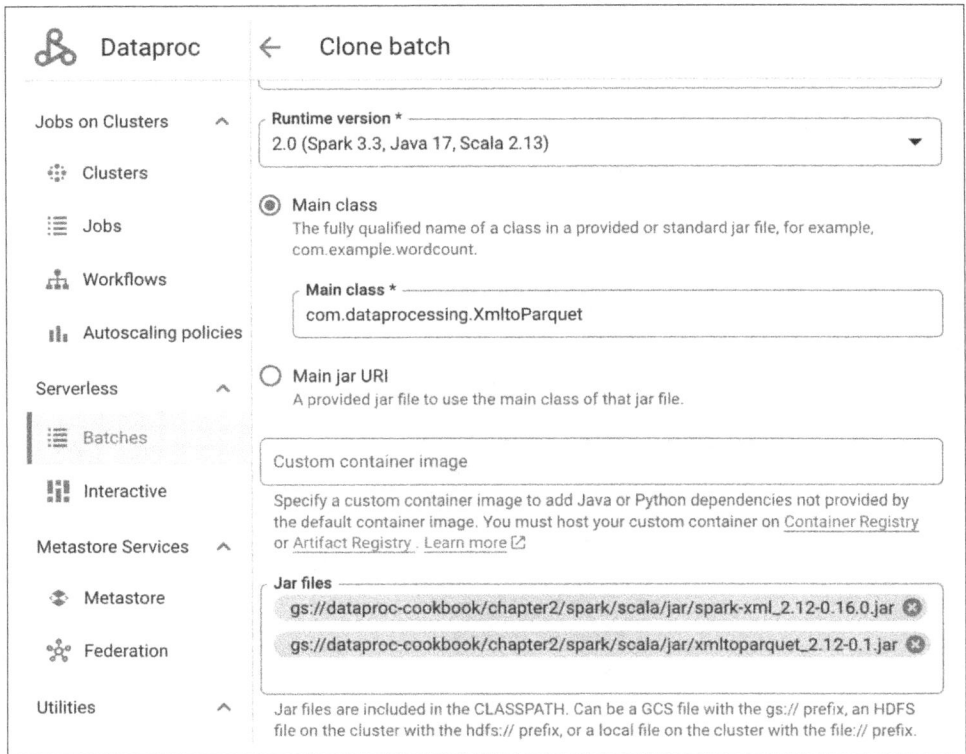

Figure 4-2. Adding application and dependency JARs

Figure 4-3. Adding arguments for the Spark job

Next, choose appropriate tiers for the driver and executor and then submit the job, as shown in Figure 4-4.

Figure 4-4. Choosing the executor and driver compute and storage tiers

> A best practice is to configure `--ttl` to terminate the application if it keeps running indefinitely.

See Also

Submitting a Spark workload public documentation (*https://oreil.ly/fgB-3*)

4.4 Running a Serverless Job Using the Premium Tier Configuration

Problem

You're dealing with a serverless Spark application that processes large volumes of data and is currently experiencing extended runtime. Monitoring shows that the job involves compute-intensive and disk-write-intensive operations. How can you optimize this job?

Solution

Leverage premium tier. Use the following gcloud command template to submit a Spark Scala batch job to Dataproc Serverless with a premium driver, executor compute, and disk tier configuration:

```
gcloud dataproc batches submit --project {project_name_here} \
  --region {region_name_here} pyspark --batch {unique_job_name_here} \
  {pyspark_file_path_here} --version {version_number_here} \
  --properties spark.dataproc.driver.compute.tier=premium,
spark.dataproc.executor.compute.tier=premium,spark.dataproc.driver.disk.
tier=premium,spark.dataproc.executor.disk.tier=premium,spark.dataproc.
driver.disk.size=750g,spark.dataproc.executor.disk.size=1500g
```

Discussion

The choice between premium and standard tiers in Dataproc Serverless depends on your task's demands and cost considerations.

Standard tier offers a comparatively lower per-core performance but is a lower-cost option than premium tier. Choose standard tier for:

- Cost-sensitive tasks, when performance requirements are minimal and cost optimization is a primary concern
- Tasks with minimal memory or shuffle requirements, when memory usage and shuffle operations are not significant bottlenecks

Premium tier offers higher per-core performance, but the pricing is also high. Premium tier allows for a higher limit of driver and executor memory per core compared to standard tier. For instance, `spark.driver.memory` must fall within the range of 1,024–7,424 MB for the standard compute tier, whereas it can go up to 24,576 MB for the premium tier. Premium disk tiers upgrade local and shuffle storage on both the driver and executors for better IOPS and throughput. Choose premium tier for memory-intensive tasks and shuffle-heavy tasks.

Table 4-2 provides a breakdown of what each tier is and when you should consider using each one.

Table 4-2. Feature comparison between premium and standard tiers

Feature	Premium tier	Standard tier
Compute tier	Higher per-core performance	Lower per-core performance
Cost	Higher billing rate	Comparatively lower
Driver and executor disk tier	Better IOPS and throughput	Standard performance
Driver memory overhead	Higher (larger memory allocations)	Lower

The following is a sample PySpark code snippet for reading a Parquet file, aggregating it, and writing it as Parquet output:

```
import sys
import pyspark
from pyspark.sql import SparkSession
from pyspark.sql.functions import *

spark=SparkSession.builder.appName("dailyinsights").getOrCreate()

#Get the input path, output path
inputPath = sys.argv[1]
outputPath = sys.argv[2]

# Read parquet file using read.parquet()
parquetDF=spark.read.parquet(inputPath)

aggregateDF=parquetDF.groupBy("day").agg(sum("views"))

aggregateDF.show(10)

aggregateDF.printSchema()

aggregateDF.write.mode("Overwrite").parquet(outputPath)

print("Application Completed!!!")

# Closing the Spark session
spark.stop()
```

Here's a sample gcloud command for running a PySpark job that reads a Parquet file and runs on premium storage and compute:

```
gcloud dataproc batches submit --project anu-psodata-lab --region us-central1 \
    pyspark --batch test-1d gs://dataproc-cookbook/chapter4/gcstobiqguery.py \
    --version 1.1.37 --subnet default --service-account \
    1072535324208-compute@developer.gserviceaccount.com --properties spark.dataproc
.driver.compute.tier=premium,spark.dataproc.executor.compute.tier=premium,spark.
dataproc.driver.disk.tier=premium,spark.dataproc.executor.disk.tier=premium,spark
.dataproc.driver.disk.size=750g,spark.dataproc.executor.disk.size=1500g,spark.
```

```
executor.instances=2,spark.driver.cores=4,spark.executor.cores=4,spark.
dynamicAllocation.executorAllocationRatio=0.3
```

When this PySpark job starts, autoscaling will adjust the number of active executors.

4.5 Giving a Unique Custom Name to a Dataproc Serverless Spark Job

Problem

You want to run a batch Spark job on Dataproc Serverless and give a name to the batch job. How do you give a custom name to a Dataproc Serverless batch Spark job?

Solution

By default, the gcloud command in Recipe 4.3 generates a unique name. You can use the `--batch` property in the gcloud command to customize the name of the batch Spark job. The following gcloud command submits a Spark Scala batch job to Dataproc Serverless with the custom name you passed:

```
gcloud dataproc batches submit spark --batch={job_name_here} \
    --region=us-central1 --class={class_name_here} \
    --jars={jar1_gcs_path_here},{jar2_gcs_path_here} \
        -- {file_input_gcs_path_here} -- {argument1} {argument2} {argument3}
```

Discussion

The following is a sample gcloud command that creates and runs a batch serverless Spark job named "serverless-1h" to convert XML input *menu.xml* to Parquet:

```
gcloud dataproc batches submit spark --batch=serverless-1h --region=us-central1 \
    --jars="gs://dataproc-cookbook/chapter2/spark/scala/jar/spark-xml_2.12-0.16.0.
jar","gs://dataproc-cookbook/chapter2/spark/scala/jar/xmltoparquet_
2.12-0.1.jar" \
    --class=com.dataprocessing.XmltoParquet --version=1.1 \
    -- "gs://dataproc-cookbook/chapter2/spark/scala/inputfiles/menu.xml" \
    "gs://dataproc-cookbook/chapter2/spark/scala/outputfiles/" "food"
```

The gcloud command creates the serverless Spark job, which you can see in the Batches console. Navigate to the job's output section to see the status and output of the Spark application, as shown in Figure 4-5.

Figure 4-5. Validating the name of the serverless Spark job in the Batches console

> The batch ID has to be unique for every run, add a logic to apply a unique string for every run, and adhere to appropriate naming conventions for the Dataproc Serverless Spark jobs.

See Also

Gcloud Dataproc batch submit command public documentation (*https://oreil.ly/yg4Yh*)

4.6 Cloning a Dataproc Serverless Spark Job

Problem

You have successfully run a Dataproc Serverless Spark job with specific arguments and Spark properties. How do you clone and submit this job again without manually updating the arguments and properties?

Solution

Navigate to the Jobs list in the Batches console, identify the successful Spark run, and click Clone to replicate the job, as shown in Figure 4-6.

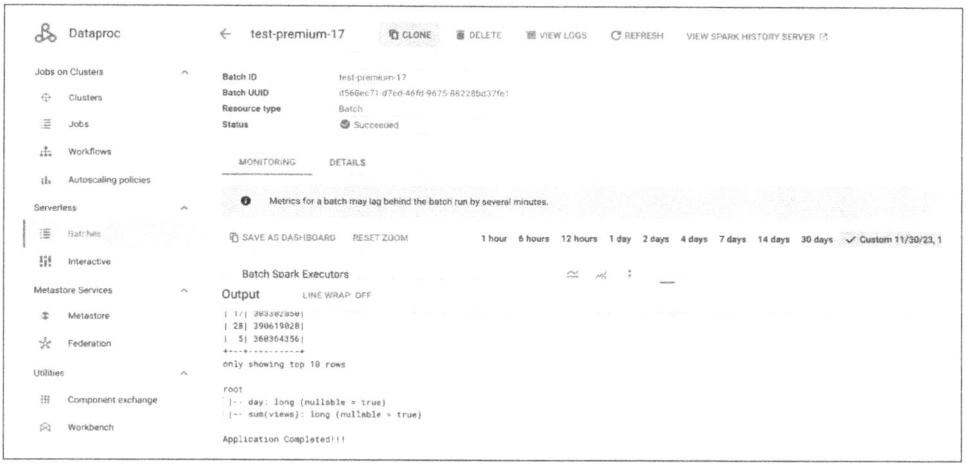

Figure 4-6. Job details of the job "test-premium-17" in the Batches section

Discussion

Once you clone the job, provide a unique name for the Spark batch ID. Then, validate all of the configurations and hit Submit, as shown in Figure 4-7.

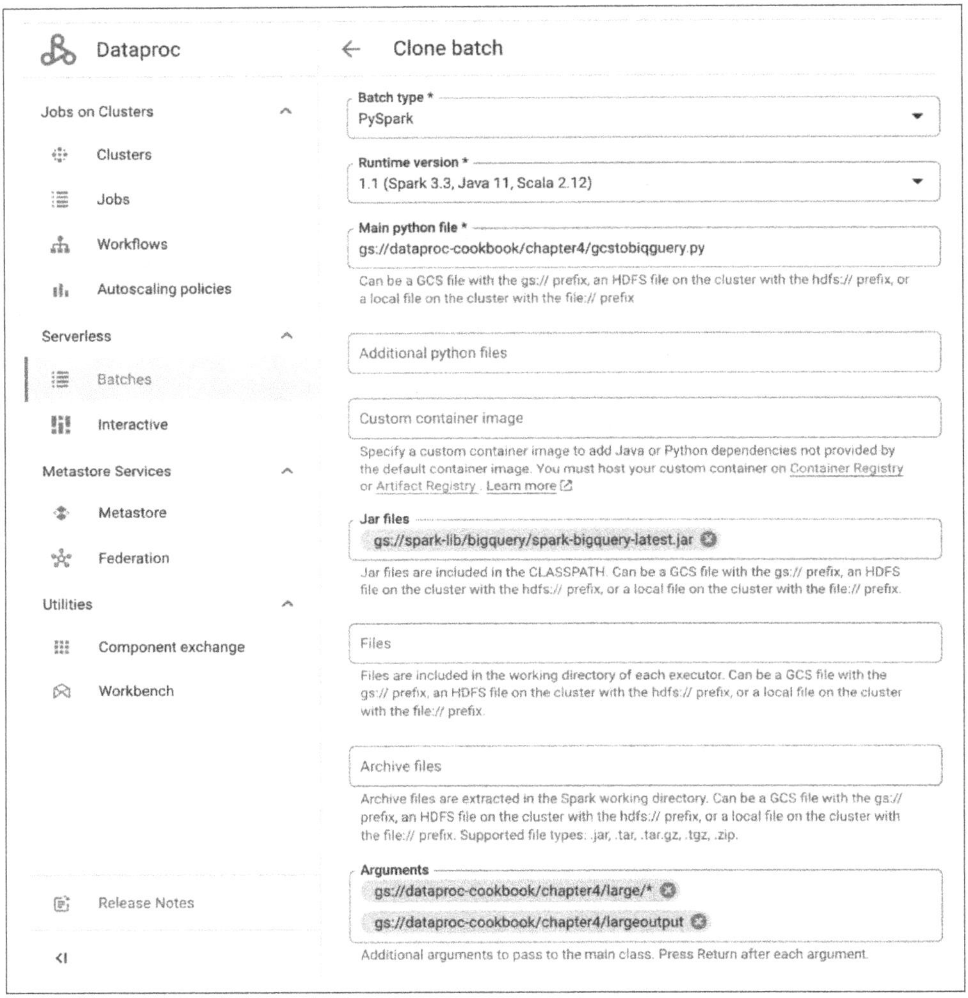

Figure 4-7. Cloning the job in the Batches section

Alternatively, you can clone the batch job by copying the gcloud command using the "equivalent command line" feature, as shown in Figure 4-8.

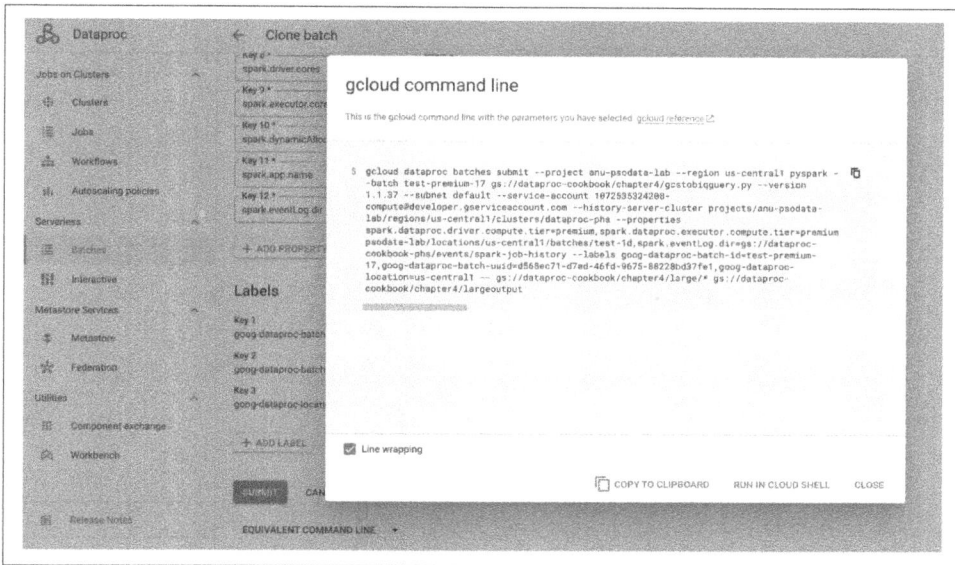

Figure 4-8. Equivalent command line of the job

In a production environment, developers might not have access to create clusters or run jobs from the console. To ensure best practices, orchestrate and standardize the cloning process through services like Cloud Composer.

4.7 Running a Serverless Job on Spark RAPIDS Accelerator

Problem

You have a serverless Spark job that runs large-scale data analytics workloads. How do you accelerate Spark workload performance using Spark RAPIDS?

Solution

You can configure GPU accelerators when submitting batch jobs to Dataproc Serverless to boost the performance significantly. Here is the gcloud command to configure GPUs when submitting the batch job:

```
gcloud dataproc batches submit --project {project_name_here} \
  --region {region_here} pyspark --batch {batch_job_name} \
  {pyspark_gcs_file_path} --version 1.1 --subnet default \
  --service-account {service_account_email} --properties \
  spark.dataproc.executor.resource.accelerator.type=l4,spark.
dataproc.driver.compute.tier=premium,spark.dataproc.executor.
compute.tier=premium,spark.dataproc.driver.disk.tier=premium,
spark.dataproc.executor.disk.tier=premium --labels org-domain=finance
```

Discussion

Spark RAPIDS accelerates Spark workloads by enabling the power of GPUs in Spark DataFrame and Spark SQL. GPU offers higher throughput. Spark features like resilient distributed datasets (RDDs) are not GPU enabled, and they will directly run on CPUs.

Here is the sample gcloud batch job that runs a PySpark job to read and process large input data on Spark RAPIDS:

```
gcloud dataproc batches submit --project anu-psodata-lab --region us-central1 \
    pyspark --batch test-accelerator-1 gs://dataproc-cookbook/chapter4/
gcstobiqguery.py \
    --version 1.1 --jars gs://spark-lib/bigquery/spark-bigquery-latest.jar \
    --subnet default --service-account 1072535324208-compute@developer.
gserviceaccount.com \
    --properties spark.dataproc.executor.resource.accelerator.type=l4,
spark.dataproc.driver.compute.tier=premium,spark.dataproc.executor.compute.
tier=premium,spark.dataproc.driver.disk.tier=premium,spark.dataproc.executor.
disk.tier=premium,spark.dataproc.driver.disk.size=375g,spark.dataproc.
executor.disk.size=1500g \
    --labels org-domain=finance -- gs://dataproc-cookbook/chapter4/large/* \
    gs://dataproc-cookbook/chapter4/largeoutput
```

The property `spark.dataproc.executor.resource.accelerator.type` configures GPUs. The accelerator type could be `l4` or `a100` (NVIDIA L4 or NVIDIA A100).

> When using the accelerator, you must configure the compute and storage tiers to be premium for both the driver and the executor.

When there is a wide dependency transformation, Spark performs a shuffle. Accelerator supports Spark Shuffle plug-in. It leverages Unified Communication X (UCX) as the transport and speeds up shuffle transfers.

You can also configure the accelerators via the console, as shown in Figure 4-9.

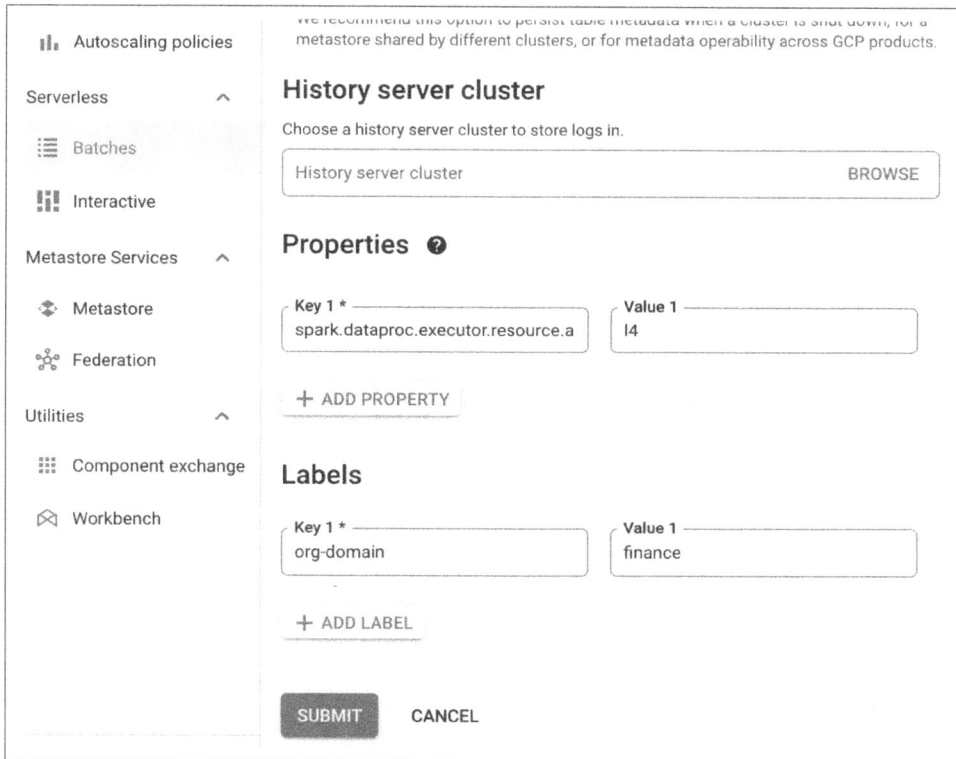

Figure 4-9. Configuring accelerators in the Properties section

The driver and executor compute and storage must be premium, and the value of the driver and executor disk sizes must not be set, as shown in Figure 4-10.

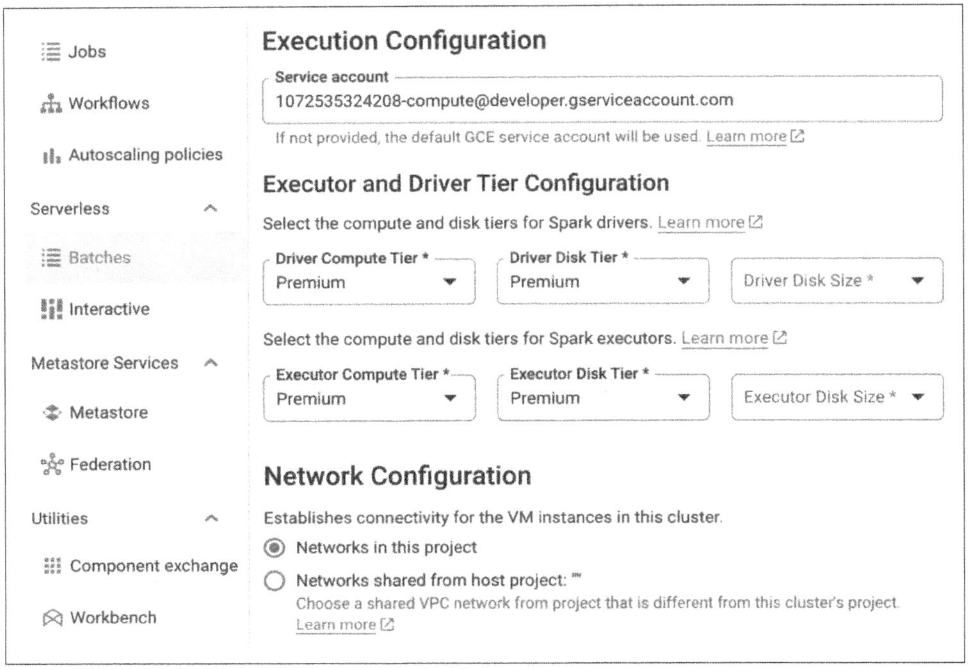

Figure 4-10. Configuring premium disk and compute tiers

When the accelerator is configured, the pricing will include accelerator usage as well as the premium tiers.

See Also

- Official documentation on GPUs with Dataproc Serverless (*https://oreil.ly/XVs6-*)
- NVIDIA RAPIDS Shuffle Manager documentation (*https://oreil.ly/QWZBr*)

4.8 Configuring a Spark History Server

Problem

You are running Spark jobs on Dataproc Serverless or ephemeral Dataproc clusters, and you want to set up a Spark history server to monitor the completed Spark applications. How do you configure a Spark history server?

Solution

Here is the sample gcloud command to set up a Spark history server:

```
gcloud dataproc clusters create {phs_cluster_name_here} \
  --enable-component-gateway --region {region_name_here} --zone {zone_here} \
  --single-node --master-machine-type n2-highmem-8 --master-boot-disk-size 500 \
  --image-version 2.0-debian10 --properties yarn:yarn.nodemanager.
remote-app-log-dir=gs://$GCS_BUCKET/yarn-logs,mapred:mapreduce.jobhistory.
done-dir=gs://$GCS_BUCKET/events/mapreduce-job-history/done,mapred:mapreduce.
jobhistory.intermediate-done-dir=gs://$GCS_BUCKET/events/mapreduce-job-history/
intermediate-done,spark:spark.eventLog.dir=gs://$GCS_BUCKET/events/
spark-job-history,spark:spark.history.fs.logDirectory=gs://$GCS_BUCKET/events/
spark-job-history,spark:SPARK_DAEMON_MEMORY=16000m,spark:spark.history.custom.
executor.log.url.applyIncompleteApplication=false,spark:spark.history.custom.
executor.log.url={{YARN_LOG_SERVER_URL}}/{{NM_HOST}}:{{NM_PORT}}/{{CONTAINER_ID}}
/{{CONTAINER_ID}}/{{USER}}/{{FILE_NAME}} \
  --project $PROJECT_NAME
```

Discussion

Spark history server is stateless; it parses the Spark application event logs persisted in GCS and constructs the Spark UI of the applications. If the event logs are significantly large, increase the history server memory config `SPARK_DAEMON_MEMORY` to render the large logs. As a start, configure the server on n2-highmem-8 or n2-highmem-8+ machines.

If you want to configure an ephemeral Dataproc cluster to write logs to the Spark history server, here is the gcloud command:

```
gcloud dataproc clusters create {job_cluster_name_here} \
  --enable-component-gateway --region {region_name_here} --zone {zone} \
  --master-machine-type n1-standard-4 --master-boot-disk-size 100 \
  --single-node --worker-machine-type n1-standard-4 --image-version \
  2.0-debian10 --properties yarn:yarn.nodemanager.remote-app-log-dir=gs://
$GCS_BUCKET/yarn-logs,spark:spark.eventLog.dir=gs://$GCS_BUCKET/events/
spark-job-history,spark:spark.history.fs.logDirectory=gs://$GCS_BUCKET/
events/spark-job-history,spark:spark.history.fs.gs.outputstream.type=
FLUSHABLE_COMPOSITE,spark:spark.history.fs.gs.outputstream.sync.min.
interval.ms=1000ms --project $PROJECT_NAME
```

If you want to configure Dataproc Serverless to write logs to the Spark history server, refer to Recipe 4.9.

See Also

Google Cloud blog post "Best Practices of Dataproc Persistent History Server" (*https://oreil.ly/xEdQL*)

4.9 Writing Spark Events to the Spark History Server from Dataproc Serverless

Problem

You have a Spark history server set up, and you are submitting Spark jobs to Dataproc Serverless. You want to read the logs in the Spark history server after a Spark application is successfully completed.

Solution

If you haven't configured a Spark history server yet, refer to Recipe 4.8 for setup instructions. Once the history server is configured, use the `--history-server-cluster` property to set up Spark UI for serverless Spark jobs.

The following gcloud command submits a Spark Scala batch job to Dataproc Serverless and writes logs to the specified Spark history server:

```
gcloud dataproc batches submit spark --batch={job_name_here} \
  --region={region_here} --class={class_name_here} \
  --jars={jar1_gcs_path_here},{jar2_gcs_path_here} \
  -- {file_input_gcs_path_here} -- {argument1} {argument2} {argument3} \
  --history-server-cluster projects/{project_name_here}/regions/
{region_here}/clusters/{cluster_name_here}
```

Discussion

The following is a sample gcloud command that submits the Scala Spark job converting XML files to Parquet to Dataproc Serverless and logs the Spark events to the "dataproc-phs" history server:

```
gcloud dataproc batches submit --project anu-psodata-lab --region us-central1 \
  spark --batch serverless-1s --class com.dataprocessing.XmltoParquet \
  --version 1.1 --jars gs://dataproc-cookbook/chapter2/spark/scala/jar/spark-xml
_2.12-0.16.0.jar,gs://dataproc-cookbook/chapter2/spark/scala/jar/xmltoparquet_
2.12-0.1.jar --subnet default --history-server-cluster projects/anu-psodata-lab/
regions/us-central1/clusters/dataproc-phs \
  -- gs://dataproc-cookbook/chapter2/spark/scala/inputfiles/menu.xml \
  gs://dataproc-cookbook/chapter2/spark/scala/outputfiles/ food
```

Another way to configure this is via the console, as shown in Figure 4-11.

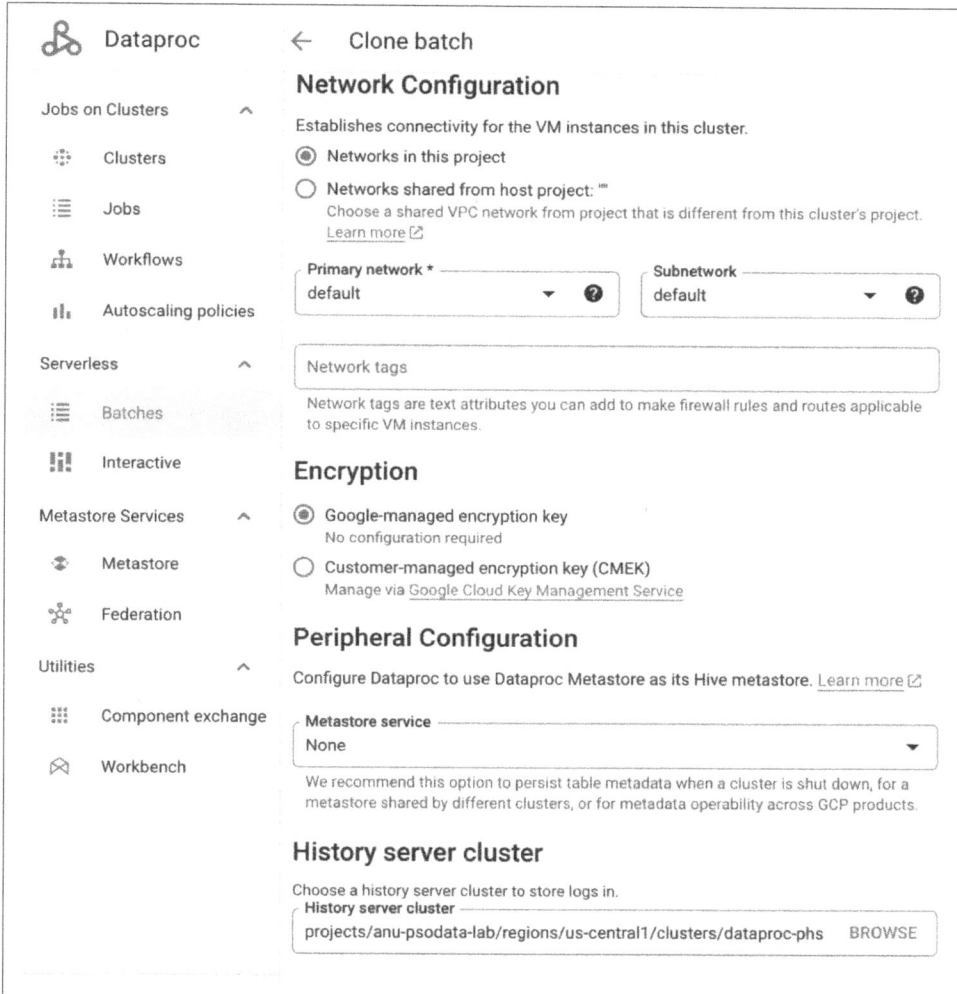

Figure 4-11. Configuring a history server cluster when submitting a Dataproc Serverless job via the console

There are two ways to monitor completed batch serverless Spark jobs in the Spark web UI:

- Navigate to the corresponding batch job. "View Spark History Server" will be active if it's configured, as shown in Figure 4-12. If it's not active, configure the history server when submitting a job.

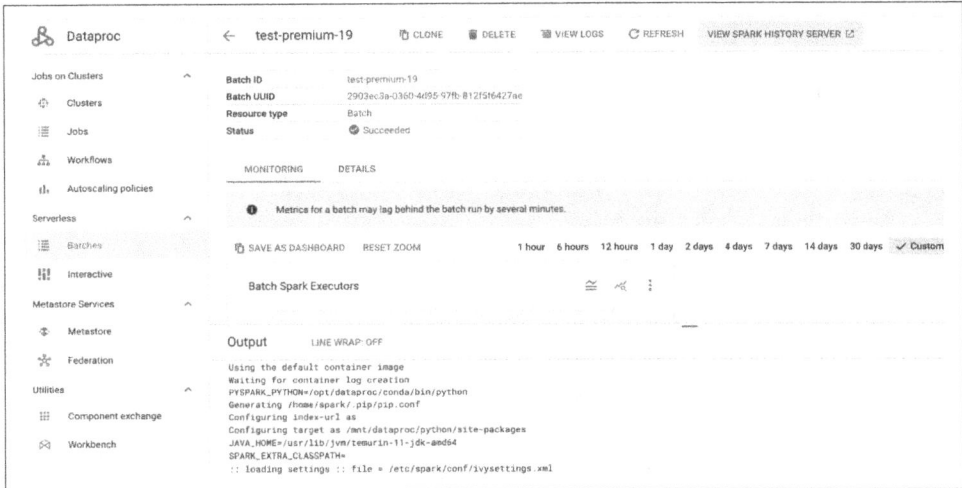

Figure 4-12. The batch job has "View Spark History Server" active to show the completed applications

- Navigate to the Spark history server cluster and access Web Interfaces, as shown in Figure 4-13. The component gateway will show the details of the Spark applications.

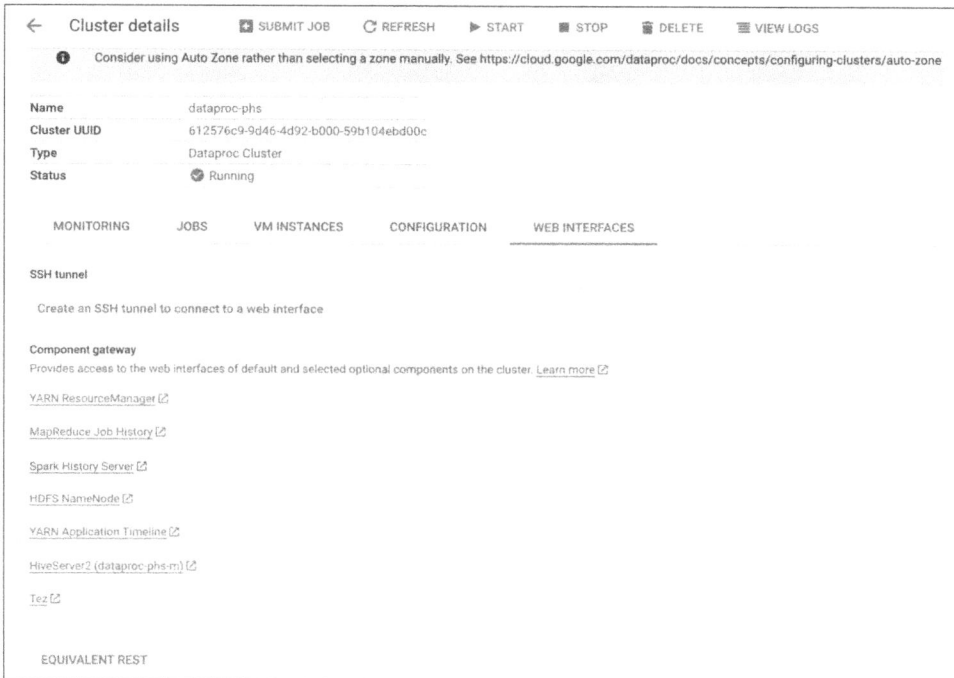

Figure 4-13. The Web Interfaces section in the Spark history server cluster

Here is the gcloud command to configure an ephemeral Dataproc cluster to write event logs to the Spark history server:

```
gcloud dataproc clusters create {job_cluster_name_here} \
    --enable-component-gateway --region {region_here} --zone {zone_here} \
    --master-machine-type {machine_type_here} --master-boot-disk-size 500 \
    --num-workers 2 --worker-machine-type {machine_type_here} \
    --worker-boot-disk-size 500 --image-version 2.0-debian10 \
    --properties yarn:yarn.nodemanager.remote-app-log-dir=
gs://{gcs_bucket_here}/yarn-logs,mapred:mapreduce.jobhistory.done-dir=
gs://{gcs_bucket_here}/events/mapreduce-job-history/done,mapred:mapreduce.
jobhistory.intermediate-done-dir=gs://{gcs_bucket_here}/events/
mapreduce-job-history/intermediate-done,spark:spark.eventLog.dir=
gs://{gcs_bucket_here}/events/spark-job-history,spark:spark.history.fs.
logDirectory=gs://{gcs_bucket_here}/events/spark-job-history,spark:spark.
history.fs.gs.outputstream.type=FLUSHABLE_COMPOSITE,spark:spark.history.fs.
gs.outputstream.sync.min.interval.ms=1000ms --project {project_name_here}
```

If your organization has a Spark history server configured and you want to monitor all jobs in the same server, follow the steps described previously in this section. However, if a Spark history server is not configured, and you need to access the Spark UI for a specific Dataproc Serverless job, you can enable the Spark UI, which will launch an on-demand history server, as shown in Figure 4-14.

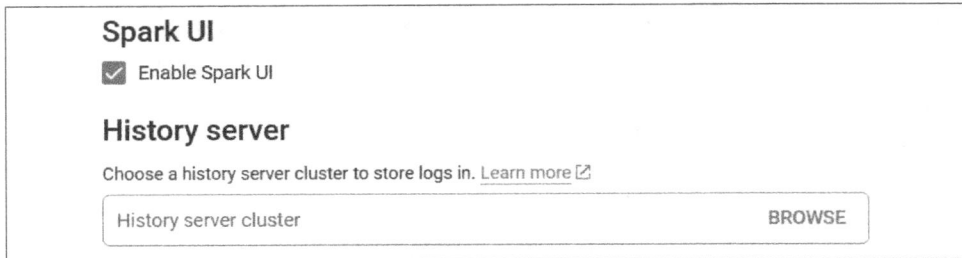

Figure 4-14. Enabling Spark UI to launch an on-demand history server

4.10 Monitoring Serverless Spark Jobs

Problem

You are running a Spark job on Dataproc Serverless, and you want to monitor the status and performance of the job.

Solution

To monitor your serverless Spark jobs on Dataproc from the UI, navigate to the job in the Dataproc Batches console. Here, you will see details like the number of running Spark executors and the job status (running, failed, or succeeded). Navigate to the corresponding job in the Dataproc Batches console, as shown in Figure 4-15.

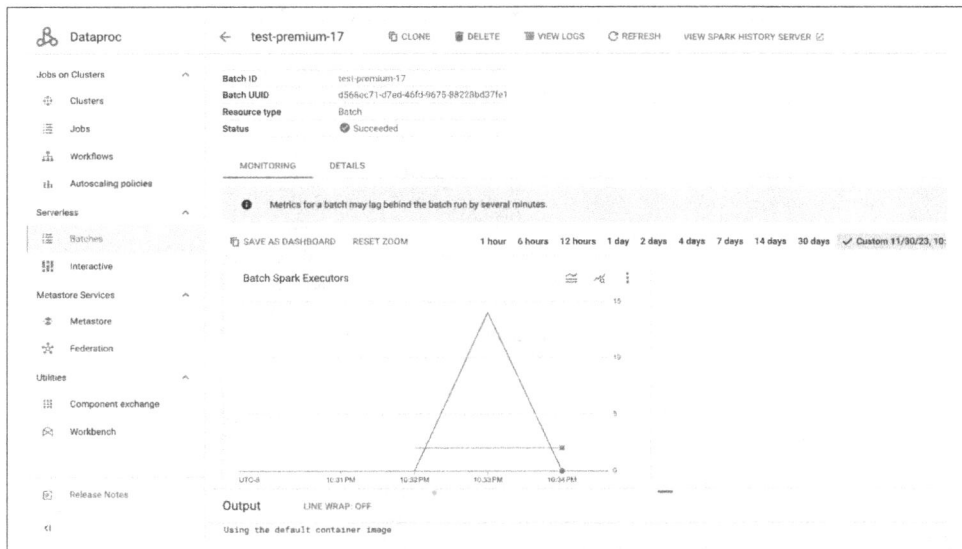

Figure 4-15. The Job monitoring UI in the Batches console

To monitor using gcloud, use the gcloud command to submit the job (refer to Recipe 4.3) and observe the streamed details:

- Job status
- Driver output

For deeper insights, use the Spark web UI (Spark history server) outlined in Recipe 11.4.

Discussion

Here is the sample gcloud command that creates a serverless Spark job named "serverless-1j," which converts the XML to Parquet:

```
gcloud dataproc batches submit spark --batch=serverless-1j --region=us-central1 \
    --jars="gs://dataproc-cookbook/chapter2/spark/scala/jar/spark-xml_2.12-0.16.0.
jar","gs://dataproc-cookbook/chapter2/spark/scala/jar/xmltoparquet
_2.12-0.1.jar" \
    --class=com.dataprocessing.XmltoParquet \
    -- "gs://dataproc-cookbook/chapter2/spark/scala/inputfiles/menu.xml" \
    "gs://dataproc-cookbook/chapter2/spark/scala/outputfiles/" "food"
```

The gcloud command in the terminal will stream details like job status and driver output, as shown in the Figure 4-16.

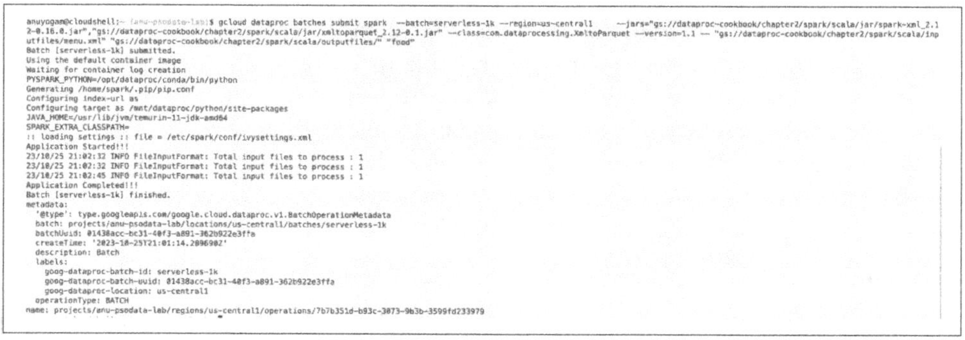

Figure 4-16. Job status and driver output in the terminal

Let's take an example PySpark job that reads Parquet input that has daily Wikipedia views. The Spark code snippet shown here groups by day, sums the views, and writes the output as Parquet:

```
import sys
import pyspark
from pyspark.sql import SparkSession
from pyspark.sql.functions import *

spark=SparkSession.builder.appName("dailyinsights").getOrCreate()
```

```
#Get the input path, output path
inputPath = sys.argv[1]
outputPath = sys.argv[2]

# Read parquet file using read.parquet()
parquetDF=spark.read.parquet(inputPath)

aggregateDF=parquetDF.groupBy("day").agg(sum("views"))

aggregateDF.write.mode("Overwrite").parquet(outputPath)

print("Application Completed!!!")

# Closing the Spark session
spark.stop()
```

To determine what aspects to monitor and where, let's first understand the Spark execution model at a broad level. When the sample PySpark job is submitted, Spark Resource Manager will create an application master (AM) container. The AM container launches the driver thread and runs the main method. Spark is written in Scala and always runs on JVM. PySpark will start a JVM application using the Py4J connection. The driver doesn't perform any processing; it requests the resource managers for executor containers. Both the Spark driver and the executors are JVM applications.

There are two actions (a read action and a write action) in the previous application code snippet. Each of these actions triggers one or more Spark jobs. There is one transformation in this code: a wide dependency transformation (group by and aggregate). The driver prepares a logical query plan for each of the jobs; then, the driver breaks the job into stages after each wide dependency transformation. Each stage consists of a collection of tasks that are run concurrently on executors, and all the tasks in a stage perform the same type of operation on different data.

To monitor your Spark application, there are four important places to look:

- Monitoring UI
- Output UI
- Details UI
- Spark web UI (Spark history server)

The monitoring UI will display the number of executors that run the tasks at each moment, as was shown in Figure 4-16.

The Spark driver log is streamed to the Output console, which contains the application output and status, as shown in Figure 4-17.

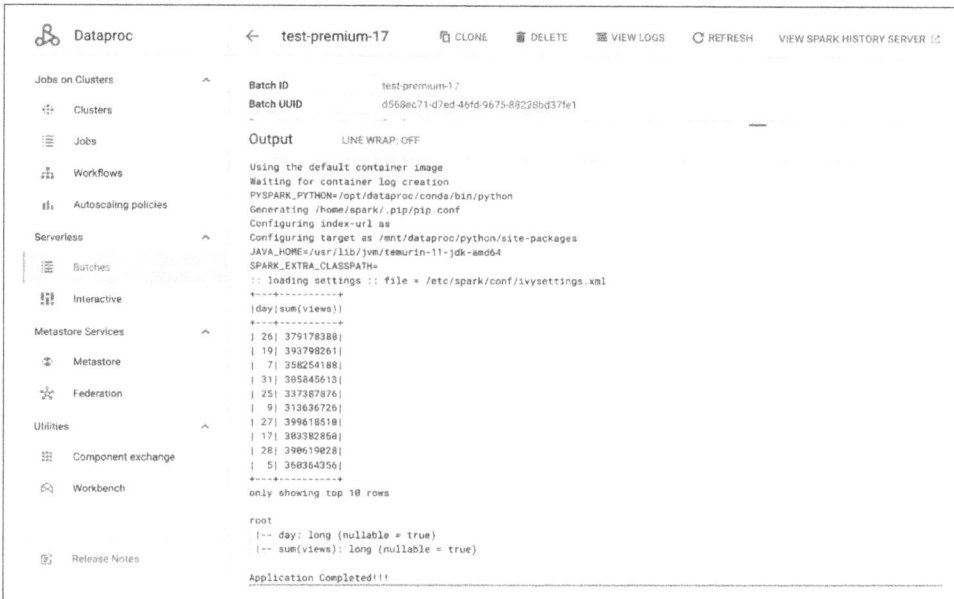

Figure 4-17. Application driver output in Output console

The details section will list the executor and driver compute tiers, the size of the memory, and the cores, as shown in Figure 4-18.

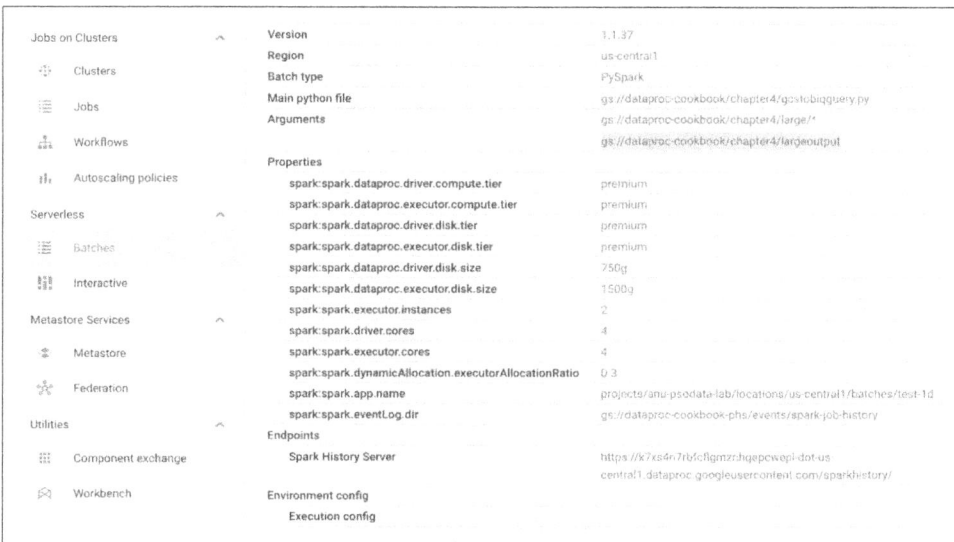

Figure 4-18. The details section displaying the Spark properties of the job

The Spark UI will help you monitor granular levels like jobs, stages, and tasks. You can look for stragglers and data skew. Stragglers are the tasks within a stage that take longer to execute than the other tasks in that stage. Refer to Recipe 4.3 to understand how to monitor using the Spark UI.

See Also

"The Anatomy of a Spark Job" section in *High Performance Spark* by Holden Karau and Rachel Warren (O'Reilly)

4.11 Calculating the Price of a Serverless Batch

Problem

You're running Spark jobs on Dataproc Serverless, but you're unsure how to figure out the cost of a batch job. How do you calculate the price of a single batch job?

Solution

The most straightforward way is using the Billing console. Navigate to the Billing Reports section in the GCP console and then choose the correct project and Dataproc under the Services drop-down menu. Select the label "goog-dataproc-batch-id" and the corresponding batch job ID under Value, as shown in Figure 4-19.

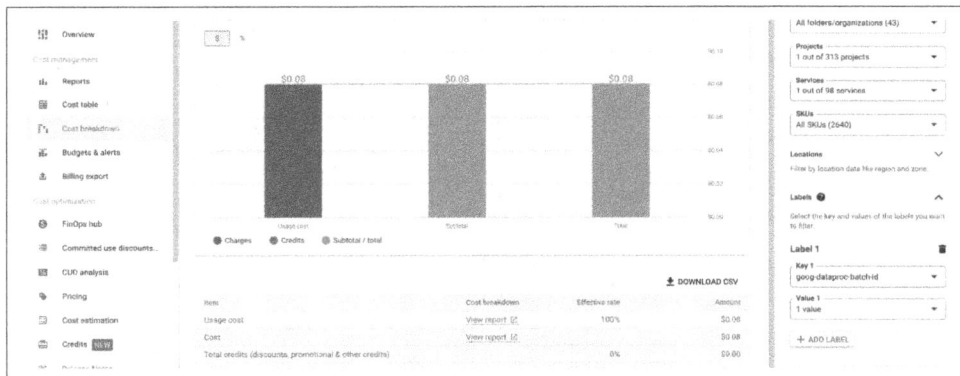

Figure 4-19. Dataproc Serverless cost breakdown

Discussion

When you submit a Spark job to Dataproc Serverless, it generates default labels like "goog-dataproc-batch-id," "goog-dataproc-batch-uuid," and "goog-dataproc-location." You can filter using these labels to find the cost of the job. To calculate the cost grouped by domain or teams, add custom labels like "org-domain" when running the Spark jobs, as shown in Figure 4-20.

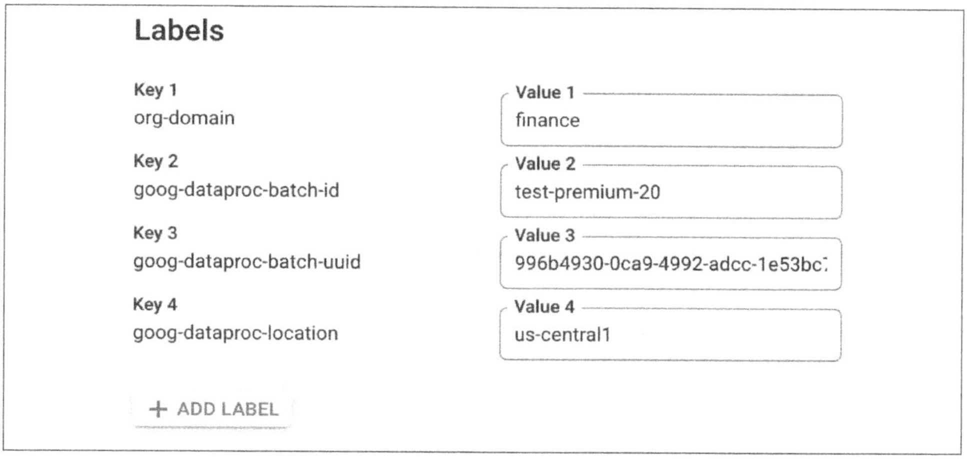

Labels

Key 1	Value 1
org-domain	finance

Key 2	Value 2
goog-dataproc-batch-id	test-premium-20

Key 3	Value 3
goog-dataproc-batch-uuid	996b4930-0ca9-4992-adcc-1e53bc:

Key 4	Value 4
goog-dataproc-location	us-central1

+ ADD LABEL

Figure 4-20. Adding custom labels to the Dataproc Serverless batch job

Remember, Dataproc Serverless for Spark pricing is based on the amount of compute, storage, and accelerators used. The total compute cost is the number of data compute units (DCUs), the number of accelerators used, and the amount of shuffle storage used. The Details section will display the approximate DCU, accelerator, and shuffle storage usage metrics, as shown in Figure 4-21.

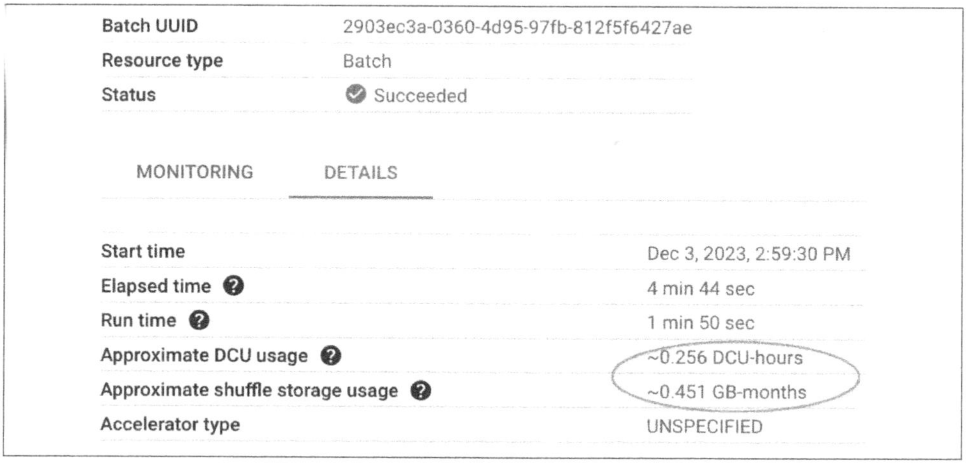

Batch UUID	2903ec3a-0360-4d95-97fb-812f5f6427ae
Resource type	Batch
Status	✓ Succeeded

MONITORING   DETAILS

Start time	Dec 3, 2023, 2:59:30 PM
Elapsed time ❓	4 min 44 sec
Run time ❓	1 min 50 sec
Approximate DCU usage ❓	~0.256 DCU-hours
Approximate shuffle storage usage ❓	~0.451 GB-months
Accelerator type	UNSPECIFIED

Figure 4-21. Approximate DCU, accelerator, and shuffle storage usage metrics of the batch job

To view billing information in a Google Cloud project, the user must have the following privileges:

Billing account viewer role
Allows the user to view billing information for the entire billing account, including all projects in the account

Project viewer role
Allows the user to view billing information for a specific project.

This job ran on the premium compute and storage tiers, as indicated in Figure 4-22. The cost breaks down as follows:

Total compute cost: ~0.256 DCU-hours × $0.100232 per hour = $0.0256

Total storage cost: ~0.451 GB-months × $0.11 per GB = $0.04961

Total cost: ~$0.0256 + ~$0.04961 = ~$0.07521

Properties	
spark:spark.dataproc.driver.compute.tier	premium
spark:spark.dataproc.executor.compute.tier	premium
spark:spark.dataproc.driver.disk.tier	premium
spark:spark.dataproc.executor.disk.tier	premium
spark:spark.dataproc.driver.disk.size	750g
spark:spark.dataproc.executor.disk.size	1500g
spark:spark.executor.instances	2
spark:spark.driver.cores	4
spark:spark.executor.cores	4
spark:spark.dynamicAllocation.executorAllocationRatio	0.3

Figure 4-22. Batch job configured to run on premium tier

See Also

Dataproc Serverless pricing documentation (*https://oreil.ly/RPoYm*)

## CHAPTER 5

Dataproc on Google Kubernetes Engine

Kubernetes is an open source platform that is designed to automate deploying, scaling, and operating containerized applications and has widespread adoption. With Kubernetes, developers can deploy complex applications more quickly and ensure high availability, fault tolerance, and better resource utilization across their infrastructure.

In standard Dataproc clusters running on GCE, the YARN framework handles resource management. Like YARN, Kubernetes functions as a resource manager, capable of allocating resources to meet framework needs. While YARN is specifically designed for Hadoop ecosystems, though, Kubernetes offers a more general-purpose approach to container orchestration, extending its capabilities beyond big data workloads.

Deploying Dataproc on GCE requires managing the underlying hardware infrastructure, including VMs and networking. In contrast, Dataproc on GKE leverages the existing infrastructure of a GKE cluster, simplifying deployment and management. Dataproc on GKE clusters are essentially virtual clusters, allowing you to define which node pools to utilize without directly managing the underlying hardware.

It's important to note that Dataproc on GKE currently focuses primarily on Spark workloads. For use cases involving Hive, HDFS, or other Hadoop components, you'll need to opt for Dataproc on GCE. Table 5-1 compares Dataproc on GCE to Dataproc on GKE.

Table 5-1. Dataproc on GCE versus Dataproc on GKE

Feature	Dataproc on GCE	Dataproc on GKE
Underlying infrastructure	Google Compute Engine (VMs)	Google Kubernetes Engine (containers within Pods)
Supported frameworks	Core Hadoop services HDFS, YARN, Hive, and Spark and optional services like JupyterHub, Flink, etc.	Primarily supports Spark workloads with future plans to extend to more services
Resource management	VM based; scaling involves adding and removing VMs	Container based
Scalability	Autoscaling is based on resource usage (YARN metrics)	Autoscaling is more granular (scaling both nodes and Pods using Kubernetes horizontal Pod autoscaling and cluster autoscaler)
Cost efficiency	Cost optimizations though preemptible VMs and autoscaling	More control over costs with a mix of preemptible VMs, custom node pools, and autoscaling policies
Operational complexity	Lower complexity (optimized for big data workloads)	Higher complexity (due to Kubernetes management and multiworkload integration)
Common use cases	Traditional Hadoop/Spark workloads; lift-and-shift migrations	Organizations heavily invested in Kubernetes or microservices architectures and those seeking enhanced resource utilization and integration

This chapter provides a comprehensive guide to deploying and managing Dataproc on GKE. While Kubernetes itself is a vast ecosystem with extensive customization options, we will focus specifically on the areas relevant to creating and configuring GKE clusters for use as a Dataproc backend. This includes the foundational steps of establishing a Kubernetes cluster, deploying Dataproc as a virtual cluster within that environment, and executing Spark workloads on the Dataproc cluster.

5.1 Creating a Kubernetes Cluster

Problem

You want to create a Kubernetes cluster that will be used as the underlying compute platform for a Dataproc cluster.

Solution

You can use the web UI or gcloud command to create a Kubernetes cluster. While the GKE cluster can be created in autopilot or standard mode, Dataproc requires a standard GKE cluster and does not support autopilot clusters. Here is the gcloud command to create a standard Kubernetes cluster for Dataproc:

```
gcloud container clusters create <CLUSTER_NAME> \
    --zone <ZONE> \
    --num-nodes <NUMBER_OF_NODES> \
    --machine-type <MACHINE_TYPE>
```

Replace the cluster name, zone, and machine type as needed.

Discussion

Creating a Kubernetes cluster involves several key considerations to ensure scalability, security, and proper resource allocation. In this section, we provide a list of the main factors you need to address when setting up a cluster for Dataproc. These considerations were limited to the context of creating GKE clusters for Dataproc clusters. You'll encounter additional configurations when customizing your work.

Application requirements

Application factors to consider include the following:

Data volume and processing
> How much data will you process, and what kind of processing will you perform? This helps determine storage and compute requirements.

Performance goals
> Define your performance expectations (latency and throughput). Consider factors like data locality, network performance, and the need for specialized hardware (e.g., GPUs for accelerated processing).

Infrastructure

Select machine types that provide sufficient CPU, memory, and disk space for your Dataproc workloads. Consider preemptible VMs for cost-effective solutions when appropriate.

Node pool strategy

A *node pool* is a group of machines within the cluster that share the same configuration. Cluster configurations, such as machine type or disk type, can be overridden at the node-pool level to provide flexibility. You can have the following strategies while creating node pools:

Default node pool
> Create a general-purpose node pool for system components and potentially smaller Dataproc jobs.

Dedicated Dataproc node pools
> Create separate node pools specifically for Dataproc workloads. This allows customization (e.g., increased memory, high-performance disks) and independent scaling based on workload demands.

Network

Network factors to consider include the following:

VPC
> Choose the VPC where your cluster will be deployed.

Subnets
> Allocate subnets within the VPC for cluster communication. For better isolation, use separate subnets for your Dataproc node pools to enhance security and network performance.

Number of Pods
> A *Pod* is a basic unit for running containers in Kubernetes. Estimate the number of Pods to be run, which helps with allocating IPs and planning the cluster's scale.

IP ranges
> The *IP range* is a sequence of IP addresses in a network used to assign unique addresses to Pods and services in your GKE cluster. Define IP ranges to ensure that Pods and services get the required number of IP addresses.

Kubernetes distribution

These considerations include the following:

GKE (standard mode)
> Use standard GKE clusters since Dataproc on GKE does not support autopilot mode. This gives you greater control over node configurations, which is crucial for optimizing Dataproc performance.

GKE version compatibility
> Ensure that the GKE version you choose is compatible with your desired Dataproc version. Refer to the Dataproc documentation for compatibility information.

Security

Security factors to consider include the following:

Service accounts
> Create service accounts with the necessary permissions for Dataproc to access other Google Cloud services like GCS.

Network policies
> Implement network policies to control and secure traffic flow between your Dataproc Pods and other components within your cluster.

Kubernetes Autopilot Versus Standard Clusters

Autopilot clusters are fully managed by Google Cloud. Google Cloud handles the provisioning, scaling, security, and lifecycle management of the underlying infrastructure. This means you can focus on deploying and managing your applications without worrying about the underlying infrastructure.

Standard GKE clusters are clusters where you manage the control plane and the worker nodes. This gives you more control over the configuration of your cluster, but it also means you are responsible for provisioning, scaling, and maintaining the underlying infrastructure. Dataproc currently supports only the standard GKE clusters.

Prerequisites

You will need a higher privilege role editor at the project or service level and admin roles like Kubernetes engine admin (`roles/container.admin`) to create GKE clusters.

Run the following gcloud command to create a standard Kubernetes cluster:

```
gcloud container clusters create "gke-cluster-dataproc" \
    --region "us-central1" \            ❶
    --machine-type "n2d-standard-8" \       ❷
    --disk-type "pd-balanced"  \        ❸
    --disk-size "250" \          ❹
    --num-nodes "3" \       ❺
    --workload-pool=<YOUR_PROJECT_ID>.svc.id.goog        ❻
```

This initiates a new GKE cluster named "gke-cluster-dataproc."

Let's look at the significance of the flags specified in the command:

❶ Specifies the region where the cluster will be created. In this case, it's the us-central1 region. Note that this creates a regional cluster (recommended for production). If you wanted a zonal cluster, you would use the `--zone` flag instead.

❷ Sets the machine type for the nodes in the cluster's default node pool. Machines of type n2d-standard-8 have eight vCPUs and 32 GB memory. This is a good starting point for many workloads, including Dataproc.

❸ Specifies the type of persistent disk (PD) to be attached to the nodes. The disk type `pd-balanced` offers a good balance of price and performance.

❹ Sets the size of the PD for each node to 250 GB.

⑤ Creates a default node pool with three nodes initially.

⑥ This is a crucial flag that enables workload identity on your GKE cluster. *Workload identity* is a secure way for workloads running in your cluster (like Dataproc jobs) to access Google Cloud services without needing to manage service account keys. Replace *<YOUR_PROJECT_ID>* with the actual ID of your Google Cloud project.

There are a few more things to note about this command. This command creates a cluster with a single, default node pool. In this example, the three nodes of the cluster will be part of this default node pool.

This command uses the default network settings for the cluster. This means it will use the default VPC network and subnetwork in your project. To customize the network configuration, you can use the following flags:

`--network`
Specifies the name of the VPC network to use for the cluster

`--subnetwork`
Specifies the name of the subnetwork within the VPC to use for the cluster

Here's an example gcloud command for a custom network:

```
gcloud container clusters create "gke-cluster-dataproc" \
  --region "us-central1" \
  --machine-type "n2d-standard-8" \
  --disk-type "pd-balanced" \
  --disk-size "250" \
  --num-nodes "3" \
  --network "my-custom-network" \
  --subnetwork "my-custom-subnet" \
  --workload-pool=<YOUR_PROJECT_ID>.svc.id.goog
```

As you can see, the network and subnetwork flags have been added to the command.

5.2 Creating a Dataproc Cluster on a GKE Cluster

Problem

You would like to create a Dataproc cluster on top of an existing Kubernetes cluster.

Solution

Here is a gcloud command to create a cluster on GKE.

```
gcloud dataproc clusters gke create ${DP_CLUSTER} \
  --region=${REGION} \
  --gke-cluster=${GKE_CLUSTER} \
  --staging-bucket=${BUCKET} \
  --pools="name=${DP_POOLNAME},roles=default" \
  --setup-workload-identity \
  --history-server-cluster=${PHS_CLUSTER} \
  --spark-engine-version=latest
```

Discussion

Before you can spin up a Dataproc cluster on GKE, you'll need a GKE cluster with workload identity enabled. This is essential for secure authentication and interaction with Google Cloud services. Once you have your GKE cluster ready, you can create your Dataproc cluster by specifying a few key parameters:

Staging bucket
Configure a GCS bucket for temporary files.

Spark version
Use the latest Spark version or specify your desired version.

History server
Optionally designate a separate cluster to act as the Spark history server for tracking job history.

Workload identity
Include the flag to set up workload identity for secure access to Google Cloud services.

Region
Select the region where your Dataproc cluster will reside (this must match your GKE cluster's region).

Cluster name
Choose a unique name for your Dataproc cluster.

Use the following gcloud command to create a Dataproc cluster on an existing GKE cluster:

```
gcloud dataproc clusters gke create dataproc-cluster-on-gke-1 \
  --region="us-central1" \          ❶
  --gke-cluster="gke-cluster" \        ❷
  --staging-bucket=<gcs_bucket_name_here> \       ❸
  --pools="name=default,roles=default" \       ❹
  --setup-workload-identity  \       ❺
  --history-server-cluster=spark-phs \        ❻
  --spark-engine-version=latest        ❼
```

Let's go through the fields specified in the command:

❶ Specifies the region where the Dataproc cluster will be created. This should match the region of your GKE cluster.

❷ This is a key parameter that specifies the name of your existing GKE cluster where the Dataproc cluster will be deployed. Make sure you replace gke-cluster with the actual name of your GKE cluster.

❸ Specifies a GCS bucket that Dataproc will use for staging temporary files and data. Replace <gcs_bucket_name_here> with the name of your GCS bucket.

❹ This defines a node pool for the Dataproc cluster. In this case, it's using the default node pool with the default role (which typically includes both master and worker nodes). You can customize this to use different node pools with specific configurations if needed.

❺ This is an important flag that instructs Dataproc to set up the workload identity for your cluster. Workload identity allows your Dataproc jobs to securely access other Google Cloud services without needing to manage service account keys.

❻ This configures the Dataproc cluster to use a separate Dataproc cluster named "spark-phs" as the Spark history server. The history server stores information about completed Spark jobs, allowing you to analyze their performance and troubleshoot issues. See Recipe 4.8 for more details on Spark history server creation.

❼ Specifies that the Dataproc cluster should use the latest available Spark version. You can also specify a specific Spark version if needed.

5.3 Running Spark Jobs on a Dataproc GKE Cluster

Problem

You want to run a Spark job on a Dataproc GKE cluster.

Solution

To run a Spark job on a Dataproc GKE cluster, use the `gcloud dataproc jobs submit spark` command with the correct values for the cluster name, region, and job parameters:

```
gcloud dataproc jobs submit spark \
  --cluster=<DATAPROC_CLUSTER_NAME> \
  --region=<REGION> \
  --class=org.apache.spark.examples.SparkPi \
  --jars=file:///usr/lib/spark/examples/jars/spark-examples*.jar \
  -- 1000
```

Discussion

Submitting a Spark job to a Dataproc cluster on GKE, as opposed to Dataproc on GCE, involves no difference for the end user; the process remains the same. You use the same gcloud command or API for both environments. The distinction lies in how the job is executed at runtime. In GKE, jobs run in a containerized environment orchestrated by Kubernetes, allowing for better resource allocation, flexibility, and integration with Kubernetes-native services. In contrast, Dataproc on GCE runs on traditional VMs, offering simplicity but with less flexibility compared to GKE's containerized approach. Here is an example:

```
gcloud dataproc jobs submit spark \
  --cluster=dataproc-cluster-on-gke-1 \
  --region=us-central1 \
  --class=org.apache.spark.examples.SparkPi \
  --jars=file:///usr/lib/spark/examples/jars/spark-examples*.jar \
  -- 1000
```

This command submits a SparkPi job to the Dataproc cluster dataproc-cluster-on-gke-1. The SparkPi job calculates an approximation of the value of pi using Spark's distributed computation framework. The `1000` argument specifies the number of iterations the job will run to improve the accuracy of the pi calculation.

When running your custom code, you should modify the JAR filepath and arguments accordingly. For example, replace the `--jars` argument with the path to your own JAR file and provide any additional arguments required by your job.

To submit a PySpark job to a Dataproc cluster, use a command like this:

```
gcloud dataproc jobs submit pyspark \
  --cluster=<DATAPROC_CLUSTER_NAME> \
  --region=<REGION> \
  <YOUR_PYSPARK_SCRIPT>.py \
  -- <ARGUMENTS>
```

Replace *<YOUR_PYSPARK_SCRIPT>.py* with the path to your PySpark script and *<ARGU MENTS>* with any parameters your script needs.

5.4 Customizing Node Pools

Problem

How can you satisfy the differing compute capacity requirements of two teams—one needing compute-optimized machines and the other requiring memory-optimized machines—for their Dataproc workloads on GKE?

Solution

This can be achieved using a single standard GKE cluster with multiple node pools and two Dataproc virtual clusters. First, create a standard GKE cluster. Then, create two Dataproc virtual clusters: one with a memory-optimized node pool and another with a compute-optimized node pool.

Discussion

In a multitenant environment where different application teams have varying compute-capacity needs, using different node pools is an ideal solution. Node pools in GKE allow you to group nodes (VM instances) with the same configuration (such as machine type, disk size, etc.) to cater to the specific needs of each team or workload.

Here's how you can map these node pools to predefined Dataproc GKE roles:

Default role
> This is the general-purpose role for running system and auxiliary tasks on the cluster. The nodes in the default role handle basic operations and administrative tasks. These machines don't typically handle Spark jobs but support cluster-level operations.

Controller role
> Nodes assigned to this role are responsible for managing and orchestrating workloads. They are involved in scheduling jobs, managing resources, and coordinating between other nodes in the cluster. The controller role ensures that jobs submitted to the cluster are handled efficiently.

Spark-driver role

The driver node is critical in a Spark application as it controls the execution of the job. Nodes in the Spark-driver pool should have enough compute or memory capacity based on the team's requirements to run the Spark job efficiently, particularly when the driver program needs significant resources to manage the tasks.

Spark-executor role

Executors are the workers that actually perform the computations defined in the Spark job. These nodes in the Spark-executor pool can be optimized based on the type of workload. Teams needing compute-intensive jobs can use compute-optimized nodes, while teams running memory-heavy jobs can use memory-optimized nodes. Each executor node will receive tasks from the driver and process them based on the allocated resources.

By assigning node pools to these predefined roles, you can ensure that each team's workload runs on the appropriate hardware, optimizing resource usage for their specific needs. This structure allows for flexibility in running diverse workloads within the same cluster, leveraging the power of GKE.

While it's possible to use a single node pool for all roles, this isn't a good practice in shared environments. If multiple jobs run in parallel, resource contention could occur where executor-heavy tasks block the driver, preventing new applications from being submitted. To avoid this, it's recommended to isolate driver and worker pools, ensuring that driver tasks always have the resources they need to manage job execution without being blocked by worker tasks. This setup improves reliability and performance when running multiple jobs concurrently.

Creating node pools can be done either on the GKE cluster side or while creating a new Dataproc GKE cluster. Let's look at the steps to create a Dataproc cluster with new node pools. Run the following gcloud command for creating a compute-optimized Dataproc GKE cluster:

```
gcloud dataproc clusters gke create compute-cluster \
  --region=us-central1 \
  --staging-bucket=nstestb \
  --gke-cluster=gke-cluster \
      --pools="name=compute-optimized-pool,machineType=c2-standard-8,min=2,
max=5,roles=spark-driver,spark-executor" \
  --pools="name=driver-pool,machineType=c2-standard-8,min=1,max=3,roles=
spark-driver,spark-driver" \
  --pools="name=default,machineType=c2-standard-8,min=2,max=5,roles=default" \
  --setup-workload-identity \
  --history-server-cluster=spark-phs --spark-engine-version=latest
```

Run the following gcloud command for creating a memory-optimized cluster with n2d-highmem machine types:

```
gcloud dataproc clusters gke create memory-cluster \
  --region=us-central1 \
  --staging-bucket=nstestb \
  --gke-cluster=gke-cluster \
  --pools="name=memory-optimized-pool,machineType=n2d-highmem-8,min=2,
max=5,roles=spark-driver,spark-executor" \
  --pools="name=driver-pool,machineType=n2d-standard-8,min=1,max=3,
roles=spark-driver" \
  --pools="name=default,machineType=n2d-standard-8,min=1,max=3,
roles=default" \
  --setup-workload-identity \
  --history-server-cluster=spark-phs \
  --spark-engine-version=latest
```

With the successful creation of two distinct GKE clusters, each configured with its own specialized node pools, application teams now have the flexibility to choose the cluster that best meets their specific resource requirements.

5.5 Autoscaling in a GKE Cluster

Problem

Your application team needs to run diverse workloads on GKE. Some jobs require many Spark executors for large-scale data processing while others process smaller datasets and need fewer executors. How can you ensure optimal performance for all jobs without overspending on resources?

Solution

Configure autoscaling for each node pool to dynamically adjust the number of nodes based on workload needs. This ensures efficient resource allocation, prevents overspending, and optimizes performance for all jobs.

Discussion

Dataproc GKE clusters depend on the underlying GKE node pools for resource provisioning. These node pools can be configured either with a static number of resources or with a dynamic autoscaling configuration. In the dynamic approach, you define the minimum and maximum number of nodes for a pool. When a job demands more resources, such as more Spark executors or other workloads, the autoscaler will trigger a scale up to add more nodes and accommodate the additional Pods. Likewise, if the resource demand decreases, it will trigger a scale down to free up nodes and save costs.

A key consideration when using autoscaling is ensuring that the underlying network can support the required number of IP addresses. Each Pod in a GKE cluster requires its own unique IP address. For instance, if you're running a Spark job with two thousand executors, the subnet must have at least two thousand available IP addresses for the Pods. Without sufficient IPs, scaling operations may fail, leading to job delays or disruptions.

To avoid these issues, it's important to plan your network configuration carefully. You can either ensure a large enough IP range in your subnet or use GKE's VPC-native mode, which makes IP management easier by assigning each Pod an IP address from a secondary CIDR block. This helps in large-scale environments where there might be frequent scaling of resources.

The following gcloud command creates a Dataproc cluster associated with an autoscaling-enabled node pool:

```
gcloud dataproc clusters gke create compute-cluster \
  --region=us-central1 \
  --staging-bucket=nstestb \
  --gke-cluster=gke-cluster \
      --pools="name=auto-scale-pool,machineType=n2d-standard-8,min=1,
max=5,roles=default" \
  --setup-workload-identity \
  --history-server-cluster=spark-phs \
  --spark-engine-version=latest
```

Let's explore the `--pools` flag configuration where we define the node pool with autoscaling:

`name=auto-scale-pool`
    The name of the node pool.

`machineType=n2-standard-8`
    The machine type for the nodes in the pool.

`min=1`
    The minimum number of nodes in the pool (the lower bound for autoscaling).

`max=5`
    The maximum number of nodes in the pool (the upper bound for autoscaling).

`roles=default`
    This pool will be used for all roles of the cluster driver, worker, and controller.

With this configuration, jobs submitted to the Dataproc cluster can now dynamically scale their resource consumption up to the maximum limit set for the node pool. The cluster will intelligently adapt to varying workloads, scaling up resources when demand is high (many pending executors) and scaling down when demand is low, optimizing both performance and cost efficiency.

Dataproc on GKE, relying on GKE's node pool autoscaling, provides a simpler, more reactive approach that quickly scales up or down based on demand. On the other hand, Dataproc on GCE's native autoscaling policies offer fine-grained control with settings like cooldown periods and custom thresholds. You can learn more about Dataproc GCE autoscaling policies in Recipes 3.1 and 3.2.

5.6 Achieving Zonal High Availability for Dataproc Jobs

Problem

Your application has an enterprise requirement for high availability within a specific zone. This means that even if some nodes within that zone experience issues, your Dataproc jobs should continue to run without interruption. How can you achieve this level of zonal high availability with Dataproc GKE clusters, ensuring that your critical Spark and Hadoop workloads remain resilient and operational even in the face of disruptions within the zone?

Solution

To achieve zonal high availability for your Dataproc jobs on GKE, ensure that your GKE cluster is regional, distributing nodes across multiple zones. Then, when creating your Dataproc virtual cluster, use the `--pools` flag to specify a regional node pool with enough nodes to handle your workload even if one zone experiences disruptions. This ensures redundancy and keeps your jobs running despite zonal issues.

Discussion

In enterprise application development, ensuring that applications and services remain accessible and operational at all times is a common strategy. This is where the concept of *high availability (HA)* comes into play. Most enterprises follow these HA concepts:

High availability
    HA ensures that applications remain operational even when facing underlying failures. This means that your system can tolerate disruptions at various points without affecting users. For example, a website hosted on multiple web servers with redundant database instances exhibits HA. Even if one web server or database fails, the website remains accessible. This is achieved by having multiple instances of each component, ensuring redundancy and fault tolerance.

Zonal HA

GCP organizes its infrastructure into regions, which are further divided into zones. Zones are isolated locations within a region, each with independent resources like power and networking. Zonal HA focuses on protecting your applications from disruptions within a specific zone.

Regional HA

Regional HA takes things a step further by protecting your applications from outages that affect an entire zone. Resources are replicated or distributed across multiple zones within a region. If one zone becomes unavailable, your application continues to run using resources in the other zones.

Returning to the problem we are trying to solve here, your application requires high availability within a specific zone. This means that even if some nodes within that zone fail, your Dataproc jobs must continue running uninterrupted. Dataproc on GCE is a zonal resource, so when you create a Dataproc cluster on GCE, it's tied to a single zone. Any jobs submitted to that cluster will run exclusively within that zone. Therefore, if your chosen zone experiences disruptions, your Dataproc jobs on GCE will be affected.

Dataproc on GKE can offer zonal HA by leveraging a regional GKE cluster. When you create an underlying node pool at the regional level, it distributes machines across multiple zones. This ensures that even if one zone is unavailable, applications can continue running in other zones.

Here's the command to create a GKE cluster as a regional resource:

```
gcloud container clusters create <CLUSTER_NAME> \
  --region <region-name> \
  --num-nodes <NUMBER_OF_NODES> \
  --machine-type n2d-standard-4
```

When you create a GKE cluster with the `--region` flag, you're establishing a regional control plane. This control plane is responsible for managing the cluster's resources and is itself distributed across multiple zones for high availability. By default, any node pool you create within a regional cluster inherits the regional property. This means that the node pool will also span multiple zones within that region.

GKE automatically distributes the nodes of a regional node pool across multiple zones. It typically selects three zones within the region to provide a good balance of availability and cost efficiency. While the default behavior is to distribute nodes across multiple zones, you can customize this if needed. You can specify the zones you want to use for a particular node pool using the `--node-locations` flag during node pool creation. Here's the gcloud command for creating a node pool with three custom zonal locations:

```
gcloud container clusters create regional-cluster \
  --region=us-central1 \
  --machine-type=n2d-standard-4 \
  --num-nodes=2 \
  --node-locations=us-central1-a,us-central1-b,us-central1-c
```

This command will create a cluster with a default node pool containing six nodes
(two nodes in each of the three specified zones). Use the `--zone` flag to create a zonal
GKE cluster.

Now, let's create a Dataproc cluster leveraging this GKE cluster:

```
gcloud dataproc clusters gke create regional-dp-cluster-1 \
  --region=us-central1 \
  --gke-cluster=gke-cluster --pools="name=default,roles=default" \
  --setup-workload-identity \
  --history-server-cluster=spark-phs \
  --spark-engine-version=latest
```

> When using an existing GKE node pool for your Dataproc clus-
> ter, remember that you cannot modify the node pool's properties
> (machine type, minimum nodes, maximum nodes) through the
> Dataproc cluster creation command. You can only assign the node
> pool to specific roles (driver, worker) within your Dataproc cluster.
> If you need to change the node pool's configuration, you must do
> so directly through GKE.

## CHAPTER 6

Dataproc Metastore

In the realm of data processing, efficiently handling metadata is essential for effectively maintaining and organizing data. This chapter dives into the practical aspects of working with metadata using Apache Metastore–based services. The focus will be on the Apache Hive Metastore, a crucial component of the Hive framework that leverages an RDBMS database for robust metadata storage.

The Hive Metastore plays a pivotal role in facilitating Spark and Hive jobs that operate on structured data stored in tables. These jobs rely on reading and storing metadata to perform their operations efficiently. By using Apache Metastore–based services, data engineers and analysts gain a powerful tool for managing and utilizing metadata effectively, ensuring the integrity and accessibility of their data. Let's explore some of the key concepts, advantages, and integration methods of Hive Metastore.

The key concepts and components of Hive Metastore are:

Metastore
> A centralized repository that stores and manages metadata about data, including table definitions, column schemas, and partition information

Catalog
> A logical grouping of databases within the metastore

Database
> A container for tables and other data objects within the metastore

Table
> A collection of structured data organized into rows and columns, along with its schema and other properties

Partition

> A logical division of a table based on specific criteria, enabling efficient data management and querying

The advantages of using Hive Metastore include the following:

Centralized metadata management

> Provides a single source of truth for metadata, ensuring consistency and reducing the risk of data inconsistencies

Scalability

> Supports large-scale data environments with high throughput and concurrent access

Flexibility

> Allows for the integration of various data sources and storage systems, making it a versatile solution for diverse data ecosystems

Extensibility

> Offers customization and extensibility through plug-ins, enabling integration with custom data sources and specialized metadata management requirements

Hive Metastore offers these integration methods:

- The Spark SQL API enables direct metadata access from Hive Metastore for Spark integration.
- Hive jobs leverage the Metastore for metadata, such as table definitions and partition info, for Hive integration.

In the Dataproc world, you can handle metastore configuration in three ways:

- Embedded metastore
- DPMS from Google Cloud
- External metastore

We compare these three options in Table 6-1.

Table 6-1. Comparison of metastore options in Google Cloud

Feature	Embedded metastore	DPMS	External metastore
Installation	Build in Dataproc cluster	Create a new DPMS instance	Requires complex manual setup
Persistence	Ephemeral; lost once a cluster is deleted	Persistent	Persistent
Endpoint interface	Thrift	Thrift gRPC (high performance)	Thrift

Feature	Embedded metastore	DPMS	External metastore
Interoperability	Not recommended to share across other clusters	Easy to share across multiple clusters	Possible but requires additional manual configurations
High availability	N/A	Built in	Manual setup
Scalability	Limited to the size of the master node	Offers various sizes to support development and production workloads	Scalable and requires complex self-managed setup
Backup and restore	Manual process	API call to create and restore backup	Manual process
Cost	Included as part of Dataproc cluster	Additional charge for DPMS	Additional; charges vary by setup
Customization	Limited	Moderate	Greater control
Support	Google	Google	Open source

In addition to the Dataproc Metastore, the Google Cloud product team is introducing a new option: BigQuery metastore. The vision is for BigQuery metastore to become a unified platform for metadata storage across GCP data products. However, at the time of writing, BigQuery metastore is still in preview and has limited support for file formats, currently including Iceberg. Given the rapid pace of cloud technology development, you should keep an eye on the evolution of BigQuery metastore as it may become the preferred option as it matures and supports more use cases.

Let's begin by creating a Dataproc Metastore instance and then proceed with hands-on exercises involving the creation and access of tables within Spark jobs.

6.1 Creating a Dataproc Metastore Service Instance

Problem

You want to create a Dataproc Metastore instance for managing and storing metadata.

Solution

To create a DPMS instance, you have several options: web UI, gcloud, REST API, and Terraform.

Here is the gcloud command for creating a DPMS instance:

```
gcloud metastore services create <name_of_dps_instance> \
    --location=<region> \
    --tier=<developer|enterprise> \
    --network=<network-name>
```

Discussion

Assign IAM privileges for creating the DPMS instance:

```
gcloud projects add-iam-policy-binding dataproc-samples \
  --member='serviceAccount:metastore-export-user@dataproc-samples.iam.
gserviceaccount.com' \
  --role='roles/metastore.editor'
```

DPMS offers predefined IAM roles like metastore admin, metastore editor, and so on. Refer to the documentation (*https://oreil.ly/s0jbD*) for a full list of DPMS IAM roles.

To create a metastore instance from the web UI, navigate to the Dataproc home page in the Google Cloud console and click Metastore. Then, click Create, as shown in Figure 6-1.

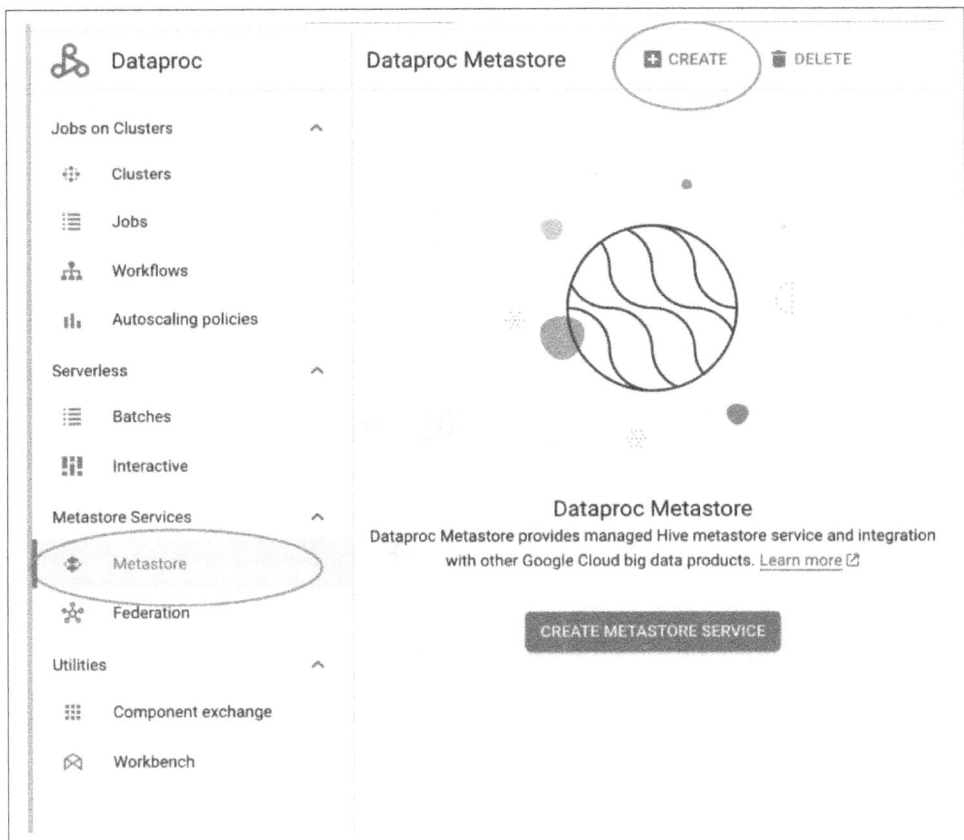

Figure 6-1. The Create button for creating a DPMS instance

Select Dataproc Metastore 1, as shown in Figure 6-2.

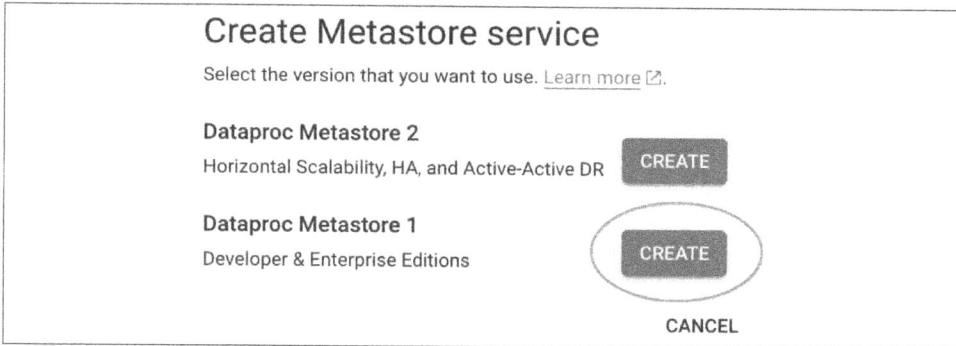

Create Metastore service

Select the version that you want to use. Learn more ⬀.

Dataproc Metastore 2

Horizontal Scalability, HA, and Active-Active DR   `CREATE`

Dataproc Metastore 1

Developer & Enterprise Editions   `CREATE`

CANCEL

Figure 6-2. Selecting the version of the DPMS instance

What Is the Difference Between DPMS Versions 1 and 2?

DPMS 2 is the latest version, offering several enhancements, like the ability to span DPMS across multiple regions for regional failover. It also provides options for choosing the size of your DPMS instance, ranging from extra small to extra large, each with dedicated storage and compute capacity. The pricing structure for this service differs for version 1 and version 2. Refer to the Google Cloud documentation (*https://oreil.ly/zyabo*) for detailed pricing information.

Configure the following fields and click the Create button to create a DPMS instance:

Service name
    Name of the Dataproc Metastore service.

Data location
    Region in which the data gets stored.

Metastore version
    Version of the Dataproc Metastore.

Release channel
    Stable (contains all tested features) or Canary (all new features and may contain some unstable changes).

Port
    The TCP port to be used for connecting to DPMS. The gRPC method defaults to 443. For Thrift, you can use the default 9083 or choose a custom port number.

Service tier

Select between the developer and enterprise editions. The developer edition possesses a finite capacity, making it suitable for development workloads. The enterprise edition is recommended for production workloads.

End protocol

GRPC is a modern high-performance protocol. Thrift protocol is a legacy (native Hadoop) way of connecting to the metastore service. The gRPC protocol is a recommended approach for high-performance workloads.

Metadata integration with Dataplex

Select Metadata integration to be synced to the Data Catalog. The Data Catalog is a Google Cloud product for indexing all objects and enabling search features on all metadata objects.

Maintenance window

Allows configuring a specific time to be reserved for Google Cloud to try applying patches and potential restarts in the background. Not all updates require service downtime, but some may require the service to be restarted and cause unavailability of DPMS. Configuring maintenance windows allows more control over when such updates can happen.

Security

DPMS supports Kerberos-based authentication for clients connecting to the metastore. To enable Kerberos, it is required to configure *krb5.conf* along with keytab for the service account and the name of the principle.

Encryption

This setting defaults to Google-managed encryption. This enables you to choose a customer-managed encryption key for encrypting data.

Metastore config overrides

Properties to be overridden from the default metastore configuration. For instance, override `hive.metastore.connect.retries` from the default of three to five.

Auxiliary version config

DPMS allows exposing the endpoint of the previous version for backward compatibility. Optionally, enable the auxiliary version to expose more than one version (the auxiliary version has to be prior to the selected DPMS version).

Labels

User-provided custom metadata to be added in key-value pairs.

After you've filled in the required information, click the Submit button. Creating the Dataproc Metastore (shown in Figure 6-3) should take around 10 minutes or more depending on the configuration (version, release, service tier, etc.).

Figure 6-3. Successful creation of a DPMS instance showing in the active state

See Also

Refer to the Google Cloud public documentation (*https://oreil.ly/4DIo7*) for additional options for creating a DPMS instance using gcloud command.

6.2 Attaching a DPMS Instance to One or More Clusters

Problem

Multiple teams in your organization have their own Dataproc clusters and want to share the same DPMS instance for reading and writing metadata.

Solution

Create a Dataproc cluster and configure it to utilize a DPMS instance as a metastore.

Discussion

By default, Dataproc clusters use an embedded metastore instance on the master node. This means metadata is lost when the cluster is deleted. To preserve metadata beyond the cluster's lifespan, you can attach a DPMS instance to the cluster. Once the DPMS instance is attached, metadata for newly created tables will be stored in it instead of in the embedded metastore.

> DPMS must be attached to the cluster at the time of creation. It cannot be added after the cluster is created.

To attach DPMS using the web UI, select the external metastore option at the time of cluster creation, as shown in Figure 6-4.

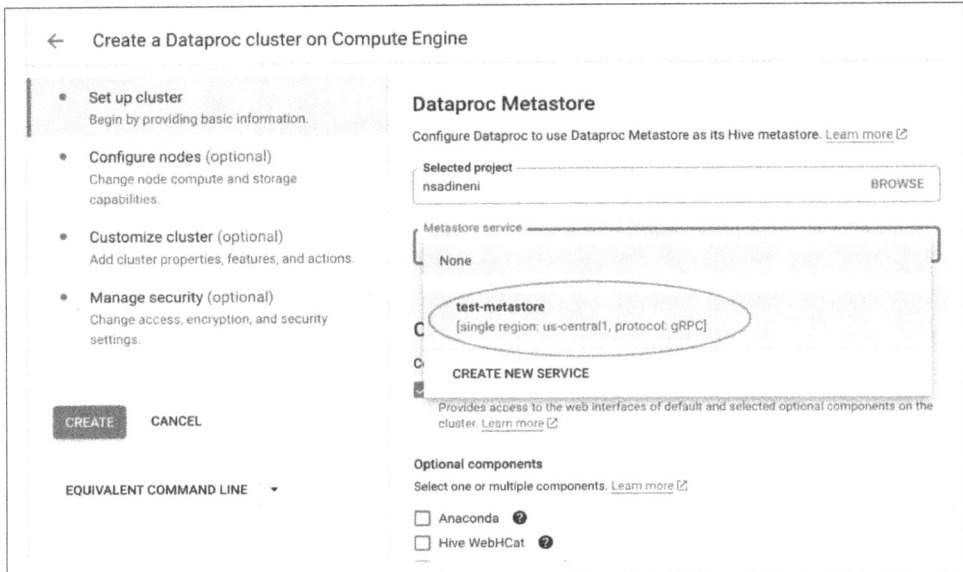

Figure 6-4. Cluster creation screen selecting a DPMS instance as a metastore

To attach DPMS using gcloud, enter the following command:

```
gcloud dataproc clusters create dataproc-ext-metastore1 \
  --enable-component-gateway --region us-central1 \
  --master-machine-type n2-standard-4 --master-boot-disk-size 500 \
  --num-workers 2 --worker-machine-type n2-standard-4 \
  --worker-boot-disk-size 500 --image-version 2.1-debian11 \
  --dataproc-metastore projects/dataproc-samples/locations/us-central1/
services/test-metastore
```

How Many DPMS Instances Should I Create?

DPMS instances can be shared across multiple Dataproc clusters. However, you might need to create a new DPMS instance to control access isolation. Currently, DPMS does not offer fine-grained access control to its stored metadata. IAM roles are granted at the instance level, meaning users with access to one database within an instance could potentially access all other databases within that instance. Creating a new instance helps isolate access. Also consider performance and regional requirements when deciding on the number of DPMS instances.

6.3 Creating Tables and Verifying Metadata in DPMS

Problem

You want to verify that tables created from the Dataproc cluster are present in DPMS and not in the embedded metastore.

Solution

Follow these steps to validate that tables created in Dataproc clusters are being stored in a DPMS instance and not in the embedded metastore (as part of the cluster):

1. Create Cluster 1, attaching the DPMS instance.
2. Create a new table using Spark SQL or Hive from Cluster 1.
3. Create Cluster 2, attaching the DPMS instance.
4. Verify that the table created from Cluster 1 is accessible in Cluster 2.

This process is shown in Figure 6-5.

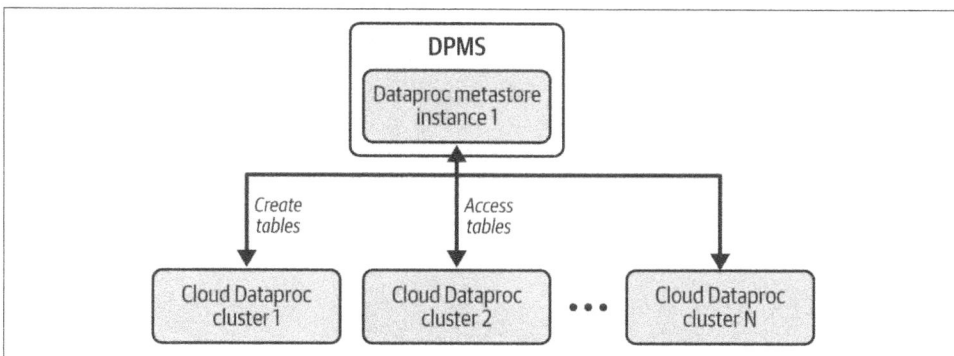

Figure 6-5. Multiple Dataproc clusters accessing the same DPMS instance

Discussion

This discussion assumes that you have already created a Dataproc metastore instance. Create a cluster and attach the DPMS instance per the instructions in Recipe 6.2. Here is a sample gcloud command for creating Cluster 1:

```
gcloud dataproc clusters create dataproc-ext-metastore1 \
  --enable-component-gateway --region us-central1 \
  --master-machine-type n2-standard-4 --master-boot-disk-size 500 \
  --num-workers 2 --worker-machine-type n2-standard-4 \
  --worker-boot-disk-size 500 --image-version 2.1-debian11 \
  --dataproc-metastore projects/dataproc-samples/locations/
us-central1/services/test-metastore
```

Creating new tables can be achieved in multiple ways. Let's use spark-sql shell for testing purposes. Log in to the master node of the Dataproc cluster using the following gcloud command:

```
gcloud compute ssh --zone "us-central1-b" "dataproc-ext-metastore-m" --project \
  "dataproc-samples"
```

Log in to spark-sql shell and run the SQL commands as shown in the following code to create a new table named employee and insert some sample records into the table:

```
testuser@dataproc-ext-metastore-m:~$ spark-sql
Setting default log level to "WARN".
To adjust logging level use sc.setLogLevel(newLevel). For SparkR, use
setLogLevel(newLevel).
ivysettings.xml file not found in HIVE_HOME or HIVE_CONF_DIR,/etc/hive/conf.
dist/ivysettings.xml will be used
24/01/15 21:12:31 INFO SparkEnv: Registering MapOutputTracker
24/01/15 21:12:31 INFO SparkEnv: Registering BlockManagerMaster
24/01/15 21:12:31 INFO SparkEnv: Registering BlockManagerMasterHeartbeat
24/01/15 21:12:31 INFO SparkEnv: Registering OutputCommitCoordinator
Spark master: yarn, Application Id: application_1705352339609_0001
spark-sql> show tables;
Time taken: 3.306 seconds
spark-sql> show databases;
default
Time taken: 0.533 seconds, Fetched 1 row(s)
spark-sql> create table employee(eid int,ename string);
24/01/15 21:13:19 WARN ResolveSessionCatalog: A Hive serde table will be created
as there is no table provider specified. You can set spark.sql.legacy.createHive
TableByDefault to false so that a native data source table will be created
instead.
24/01/15 21:13:20 WARN SessionState: METASTORE_FILTER_HOOK will be ignored,
since hive.security.authorization.manager is set to an instance of
HiveAuthorizerFactory.
Time taken: 3.247 seconds
spark-sql> insert into table employee values(1,"Tom");
Time taken: 8.268 seconds
spark-sql> insert into table employee values(1,"Reilly");
Time taken: 4.092 seconds
spark-sql> select * from employee;
1       Reilly
1       Tom
Time taken: 2.175 seconds, Fetched 2 row(s)
```

Now, let's verify that these tables are actually created in DPMS and not in the embedded metastore that is part of the cluster. Create a new cluster with the name "dataproc-ext-metastore2" using the following command:

```
gcloud dataproc clusters create dataproc-ext-metastore2  \
  --enable-component-gateway --region us-central1 \
  --master-machine-type n2-standard-4 --master-boot-disk-size 500 \
  --num-workers 2 --worker-machine-type n2-standard-4 \
  --worker-boot-disk-size 500 --image-version 2.1-debian11 \
```

```
--dataproc-metastore projects/dataproc-samples/locations/us-central1/
services/test-metastore
```

Log in to the master node of the second cluster we created using the gcloud command. Next, log in to the spark-sql shell and run the commands to view the existing tables. Try selecting data from the table created using Cluster 1.

As shown in the following code, the tables are automatically available for Cluster 2 because it has the same DPMS instance attached as Cluster 1. This test confirms that metadata is getting stored in DPMS and not in the embedded metastore:

```
testuser@dataproc-ext-metastore2-m:~$ spark-sql
Setting default log level to "WARN".
To adjust logging level use sc.setLogLevel(newLevel). For SparkR, use
setLogLevel(newLevel).
ivysettings.xml file not found in HIVE_HOME or HIVE_CONF_DIR,/etc/hive/conf.dist/
ivysettings.xml will be used
24/01/15 21:24:57 INFO SparkEnv: Registering MapOutputTracker
24/01/15 21:24:57 INFO SparkEnv: Registering BlockManagerMaster
24/01/15 21:24:57 INFO SparkEnv: Registering BlockManagerMasterHeartbeat
24/01/15 21:24:57 INFO SparkEnv: Registering OutputCommitCoordinator
Spark master: yarn, Application Id: application_1705353525190_0001
spark-sql> show tables;
employee
Time taken: 3.528 seconds, Fetched 1 row(s)
spark-sql> select * from employee;
24/01/15 21:25:21 WARN SessionState: METASTORE_FILTER_HOOK will be ignored,
since hive.security.authorization.manager is set to an instance of
HiveAuthorizerFactory.
1    Reilly
1    Tom
Time taken: 6.555 seconds, Fetched 2 row(s)
spark-sql>
```

6.4 Installing an External Hive Metastore

Problem

You want to install an open source Apache Hive Metastore in standalone mode.

Solution

Download the Hive Metastore standalone software and install it on a compute engine. Installation of Hive Metastore has the prerequisites of installing the Java and Hadoop dependency libraries.

Discussion

Installing the open source Hive Metastore on a Google Compute Engine VM provides flexibility and adaptability that may be better aligned with your specific needs than Dataproc Metastore or embedded options. Although using an open source Hive Metastore is not a popular option, here are a few use cases that are suitable:

- Greater control over customization of the Hive Metastore
- Custom scalability requirements
- Cost control
- Hybrid cloud usage where the metastore is shared across Hadoop clusters on multiple clouds

Installing an open source Hive Metastore has two prerequisites:

- Java installation
- Hadoop dependency libraries (this will download Hadoop software, but it doesn't run any of its services)

To get started, install Java and add the JAVA_HOME classpath:

```
sudo apt-get install openjdk-11-jre openjdk-11-jdk
```

Configure the JAVA_HOME classpath:

```
export JAVA_HOME="/usr/lib/jvm/java-11-openjdk-amd64"
```

Download the Hadoop software:

```
wget https://archive.apache.org/dist/hadoop/common/hadoop-3.3.1/
hadoop-3.3.1.tar.gz
```

Extract the Hadoop software from the *tar.gz* file:

```
tar -xvzf hadoop-3.3.1.tar.gz
```

Configure the HADOOP_HOME classpath:

```
export HADOOP_HOME="/home/testuser/hadoop-3.3.1"
```

Download the Apache Hive Metastore standalone software:

```
wget https://downloads.apache.org/hive/hive-standalone-metastore-3.0.0/
hive-standalone-metastore-3.0.0-bin.tar.gz
```

Extract it from the *tar.gz* file:

```
tar -xczf hive-standalone-metastore-3.0.0-bin.tar.gz
```

The Hive Metastore uses a database in the backend. You can install the MySQL on the same machine or use Google Cloud's SQL service. Once the database is installed, you need to create a schema and a user and grant privileges to the user. Here are the steps to do this.

First, create your schema and databases inside the metastore:

```
create database metastoredb;
GRANT ALL PRIVILEGES ON metastoredb.* TO 'metastoreuser'@'%';
FLUSH PRIVILEGES;
```

Create your user in Cloud SQL:

```
gcloud sql users create metastoreuser --host=% --instance=nstest --password=hive
```

Now, configure the Hive Metastore to use the SQL database. Modify the *metastore-site.xml* file under the *lib* folder to add the following properties:

```
<property>
    <name>javax.jdo.option.ConnectionURL</name>
    <value>jdbc:mysql://10.2.0.3:3306/enterprise</value>
    <description>the URL of the MySQL database</description>
</property>
<property>
    <name>javax.jdo.option.ConnectionDriverName</name>
    <value>com.mysql.jdbc.Driver</value>
</property>
<property>
    <name>javax.jdo.option.ConnectionUserName</name>
    <value>metastoreuser</value>
</property>
<property>
    <name>javax.jdo.option.ConnectionPassword</name>
    <value>hive</value>
</property>
```

Run the Schema Tool to initialize the metastore for the first time:

```
bin/schematool -initSchema -dbType mysql
```

Here is sample output of a successful Schema Tool setup command:

```
Closing: com.mysql.cj.jdbc.ConnectionImpl
sqlline version 1.3.0
Initialization script completed
schemaTool completed
2024-01-22 23:59:39,861 shutdown-hook-0 INFO Log4j appears to be running in a
Servlet environment, but there's no log4j-web module available. If you want
better web container support, please add the log4j-web JAR to your web archive
or server lib directory.
2024-01-22 23:59:39,864 shutdown-hook-0 INFO Log4j appears to be running in a
Servlet environment, but there's no log4j-web module available. If you want
better web container support, please add the log4j-web JAR to your web archive
or server lib directory.
2024-01-22 23:59:39,874 shutdown-hook-0 WARN Unable to register Log4j shutdown
hook because JVM is shutting down. Using SimpleLogger
```

Now, the standalone Hive Metastore is ready to be attached to Dataproc clusters.

6.5 Attaching an External Apache Hive Metastore to the Cluster

Problem

You want to create a Dataproc cluster by attaching an external Apache Hive Metastore.

Solution

Specify the metastore to be used at the time of cluster creation using the `metastore-service` property.

Discussion

To add an external metastore to the Dataproc cluster, specify the metastore to be used at the time of cluster creation. Use the following gcloud command to attach an external Hive Metastore while creating a cluster:

```
gcloud dataproc clusters create test1 \
  --region us-central1 \
  --zone us-central1-a \
  --metastore-service=thrift://<metastore-service-ip-address>:
<metastore-service-port> \
  --num-workers=2 \
  --master-machine-type n1-standard-2 \
  --worker-machine-type n1-standard-2
```

Adding the –metastore-service property tells the Dataproc cluster not to initialize the metastore and to use the existing schema. Also make sure to check version compatibility between the external metastore and the Hive version of the Dataproc cluster.

> It's essential to enable firewall rules that allow your Dataproc cluster to communicate with the external Hive Metastore. This means permitting traffic on the necessary ports (e.g., 9083 for the Hive Metastore) from the Dataproc cluster's network to the metastore's network. Without this, your Dataproc cluster won't be able to access the metastore.

6.6 Searching for Metadata in a Dataplex Data Catalog

Problem

Users want to browse tables and their schema stored in a Dataproc metastore instance.

Solution

Complete these two steps:

1. Enable Data Catalog sync in your DPMS instance
2. Search for metadata objects from the Dataplex

Discussion

Dataplex is Google Cloud's solution for managing and organizing data across data lakes, data warehouses, and data marts. It automatically extracts technical and business metadata, making your data easily searchable and discoverable. Integrating DPMS with Dataplex lets users search the DPMS metadata in the Dataplex Discover UI.

Create a DPMS instance with Data Catalog sync enabled, as shown in Figure 6-6. Follow the instructions in Recipe 6.1 to create a new DPMS instance.

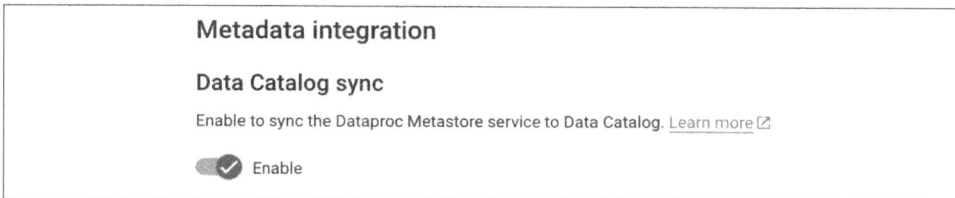

Metadata integration

Data Catalog sync

Enable to sync the Dataproc Metastore service to Data Catalog. Learn more ↗

✅ Enable

Figure 6-6. Option to enable Data Catalog sync while creating a DPMS service

Create a new Dataproc cluster attached with a DPMS instance that has Data Catalog sync enabled with the following command:

```
gcloud dataproc clusters create dataproc-ext-metastore2 \
  --enable-component-gateway --region us-central1 \
  --master-machine-type n2-standard-4 --master-boot-disk-size 500 \
  --num-workers 2 --worker-machine-type n2-standard-4 \
  --worker-boot-disk-size 500 --image-version 2.1-debian11 \
  --dataproc-metastore projects/dataproc-samples/locations/us-central1
/services/test-metastore
```

Log in to the master node of the cluster and launch spark-sql shell:

```
gcloud compute ssh --zone "us-central1-b" "dataproc-ext-metastore3-m" \
  --project "dataproc-samples"
```

Run the following commands to create new tables for employees and departments:

```
testuser@dataproc-ext-metastore3-m:~$ spark-sql
Setting default log level to "WARN".
To adjust logging level use sc.setLogLevel(newLevel). For SparkR, use setLogLevel
(newLevel).
ivysettings.xml file not found in HIVE_HOME or HIVE_CONF_DIR,/etc/hive/conf.dist/
ivysettings.xml will be used
24/01/15 22:03:34 INFO SparkEnv: Registering MapOutputTracker
24/01/15 22:03:34 INFO SparkEnv: Registering BlockManagerMaster
24/01/15 22:03:34 INFO SparkEnv: Registering BlockManagerMasterHeartbeat
24/01/15 22:03:34 INFO SparkEnv: Registering OutputCommitCoordinator
Spark master: yarn, Application Id: application_1705355889358_0002
spark-sql> create table employee_catalog_sync_test \
  (eid int,ename string,age int);
24/01/15 22:03:59 WARN ResolveSessionCatalog: A Hive serde table will be created
as there is no table provider specified. You can set spark.sql.legacy.
createHiveTableByDefault to false so that native data source table will be
created instead.
24/01/15 22:04:01 WARN SessionState: METASTORE_FILTER_HOOK will be ignored,
since hive.security.authorization.manager is set to an instance of
HiveAuthorizerFactory.
Time taken: 8.645 seconds
spark-sql> create table dept_catalog_sync_test (deptid int,name \
  string,location string,rank int);
24/01/15 22:04:50 WARN ResolveSessionCatalog: A Hive serde table will be created
as there is no table provider specified. You can set spark.sql.legacy.
createHiveTableByDefault to false so that a native data source table will be
created instead.
Time taken: 1.126 seconds
spark-sql>
```

Navigate to the Dataplex home page by searching for the Dataplex service and clicking on it, as shown in Figure 6-7.

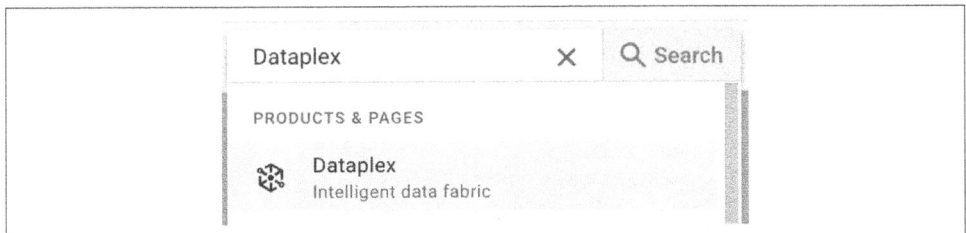

Figure 6-7. Searching for the Dataplex service in the Google Cloud console

Search with the string "*_catalog_sync_test" to view tables matching the pattern in Dataplex, as shown in Figure 6-8.

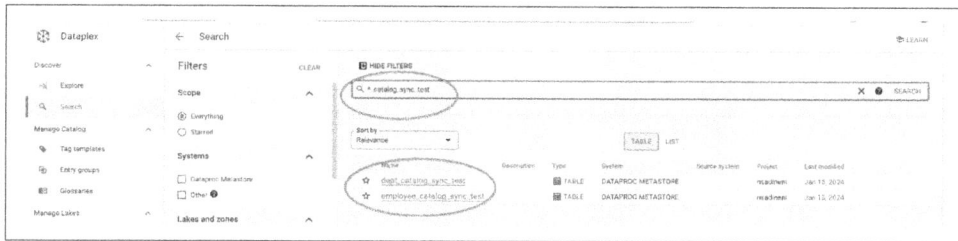

Figure 6-8. Searching Dataplex with a wildcard pattern

Click on the link to the table name to view the definition of the table. The Details tab, shown in Figure 6-9, provides information about the time the table was created, the location of the underlying data, and so on.

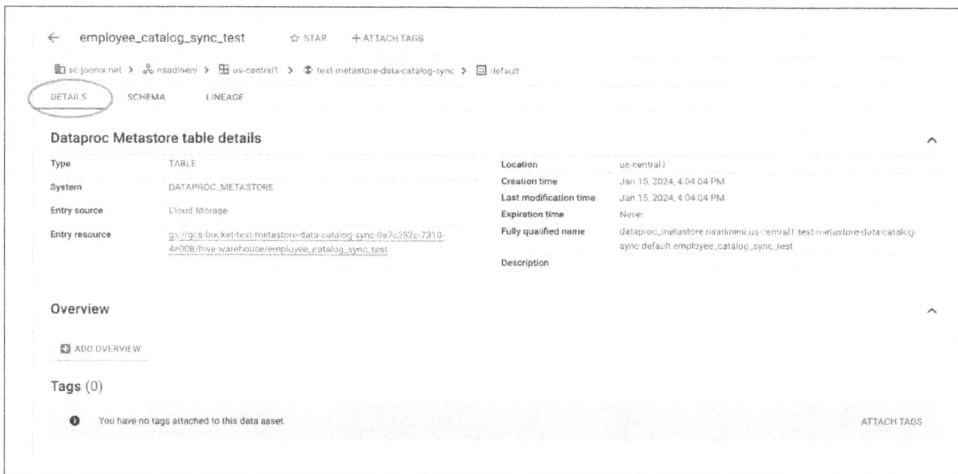

Figure 6-9. Details tab providing information about the table schema

Click the Schema tab to view the schema of the table, shown in Figure 6-10.

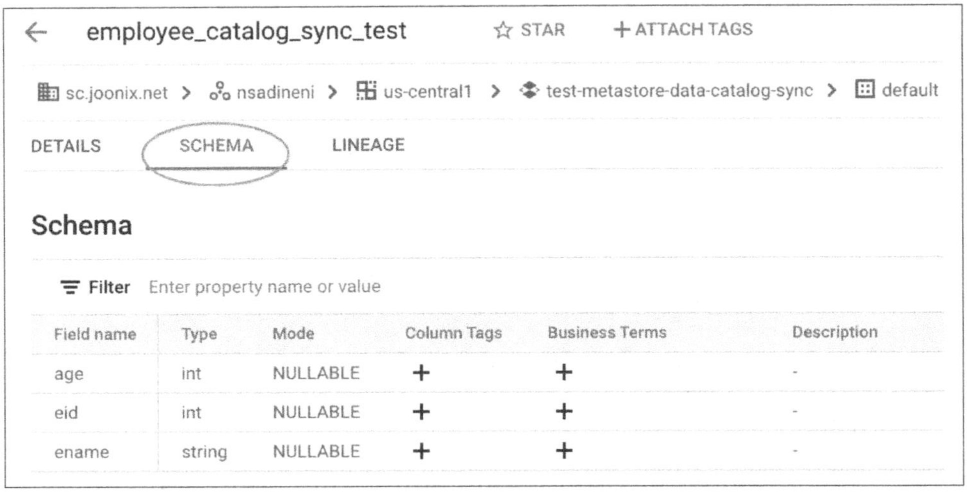

Figure 6-10. Schema tab providing column information

6.7 Automating the Backup of a DPMS Instance

Problem

You want to implement a mechanism to regularly back up your DPMS metadata for disaster-recovery purposes.

Solution

Create a custom process to achieve the following:

- A Cloud Function to trigger a DPMS export operation and store output in a GCS bucket
- A Cloud Scheduler to trigger Cloud Functions at scheduled intervals

Discussion

It is best practice to create a backup of the DPMS to protect against any unforeseen failure, application errors like deleting tables, and human errors like someone accidentally deleting a DPMS instance.

DPMS offers two types of backups:

- In-place backup that takes a backup of the instance configuration and metadata (tables definition)
- Export of the metadata to a file in GCS; however, this doesn't include the configuration specific to your DPMS instance

A metadata export dump can be in Avro format or a MySQL file. The export format depends on the replay requirements. For instance, if the requirement is to restore the metadata dump in an open source Apache metastore running on MySQL, then use the MySQL file format.

Triggering a backup or exporting the metadata are API operations that can be achieved via the web UI, gcloud, REST API, or programming (Python, SDK, and others). The backup-process architecture is shown in Figure 6-11.

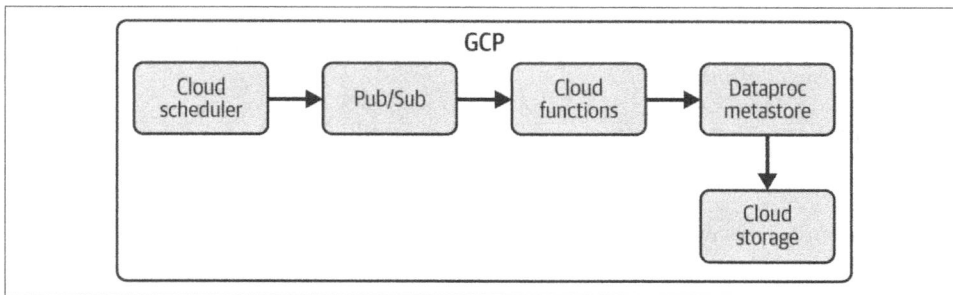

Figure 6-11. A view of the customizable, high-level architecture needed to automate the DPMS backup process

Prerequisites

Users who are creating these components require a project-level editor role or all of the following predefined IAM roles:

- `roles/cloudscheduler.admin`
- `roles/cloudfunctions.developer`
- `roles/pubsub.admin`
- `roles/metastore.editor`
- `roles/storage.objectAdmin` on the GCS bucket storing backup data

Create a service account and grant IAM privileges for exporting the DPMS backup. The following are the gcloud commands to create an IAM service account and grant the required permissions:

```
gcloud iam service-accounts create metastore-export-user

gcloud iam service-accounts keys create credential.json \
  --iam-account=metastore-export-user@dataproc-samples.iam.gserviceaccount.com

gcloud projects add-iam-policy-binding dataproc-samples \
  --member='serviceAccount:metastore-export-user@dataproc-samples.iam.
gserviceaccount.com' \
  --role='roles/metastore.metadataOperator'

gsutil iam ch serviceAccount:metastore-export-user@dataproc-samples.iam.
gserviceaccount.com:roles/storage.objectCreator gs://metastoretestb
```

Creating a Cloud Function

A Cloud Function requires two mandatory files along with any optional files needed as part of the execution:

- A *main.py* file that is the entry point for the Cloud Function execution
- A *requirements.txt* file with package dependency information

In this example, we will have three files, as shown in Figure 6-12: a *main.py* file for the Python code, a *requirements.txt* file with dependency information, and a *credentials.json* file with the service account credentials.

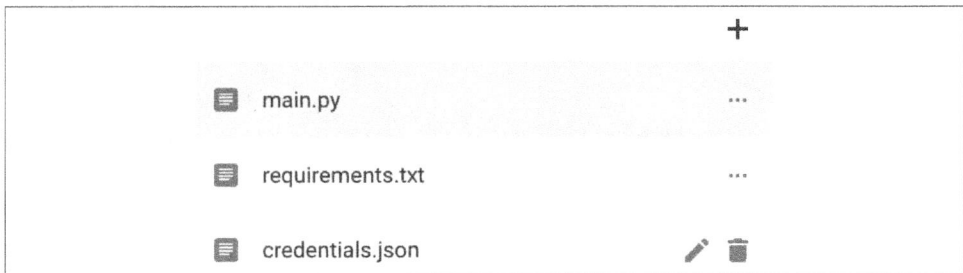

Figure 6-12. Files required for creating a Cloud Function

Here is sample code to trigger a DPMS export using the Python code (*main.py* file):

```
import requests
import google.auth
import google.auth.transport.requests
from google.oauth2 import service_account
import json
import base64
```

```python
def export_function(event, context):
    """Responds to any HTTP request.
    Args:
        request (flask.Request): HTTP request object.
    Returns:
        The response text or any set of values that can be turned into a
        Response object using
        `make_response
        <http://flask.pocoo.org/docs/1.0/api/#flask.Flask.make_response>`.
    """
    pubsub_message = base64.b64decode(event['data']).decode('utf-8')
    print("pubsub message is : ")
    print(pubsub_message)
    request_json = json.loads(pubsub_message)
    print(request_json)
    projectId = request_json["projectId"]
    locationId = request_json["locationId"]
    serviceId = request_json["serviceId"]
    print("Parameters:")
    print(projectId)
    print(locationId)
    print(serviceId)
    destinationGcsFolder = request_json["destinationGcsFolder"]
    databaseDumpType = request_json["databaseDumpType"]


    API_ENDPOINT = (
        f"https://metastore.googleapis.com/v1/projects/{projectId}"
        f"/locations/{locationId}/services/{serviceId}:exportMetadata"
    )
    print("API End point: "+API_ENDPOINT)

    #Generate authentication token
    auth_req, credentials = generate_token("credential.json")

    headersVal = {
      'Accept': 'application/json',
      'Content-Type': 'application/json',
      'Authorization': 'Bearer ' + credentials.token
    }
    jsonObject = {
      'databaseDumpType': databaseDumpType,
      'destinationGcsFolder': destinationGcsFolder
    }


    initial_req  = requests.post(
      url = API_ENDPOINT,
      headers=headersVal,
      json=jsonObject
    )
    print(initial_req.content)
```

```
print(initial_req.status_code)
print(initial_req.content)
if(initial_req.status_code == 200):
    print("export completed successfully")
else:
    initial_req.raise_for_status()


def generate_token(filePath):
    credentials = service_account.Credentials.from_service_account_file(
        filePath,
        scopes=['https://www.googleapis.com/auth/cloud-platform']
    )
    auth_req = google.auth.transport.requests
    credentials.refresh(auth_req.Request())
    return auth_req, credentials
```

Here is an example *requirements.txt* file with dependency information:

```
# Function dependencies, for example:
# package>=version
requests==2.27.1
google-auth==2.6.0
```

Next, zip all the files. The following UNIX command packages the files in ZIP format:

```
zip metastore-export.zip main.py requirements.txt credential.json
```

Then, copy your ZIP file to a GCS location:

```
gsutil cp metastore-export.zip  gs://metastoretestb/
```

Finally, deploy your Cloud Function:

```
gcloud functions deploy export-gcloud-function \
  --runtime python37 \
  --trigger-topic metastore-export-function-trigger \
  --source gs://metastoretestb/metastore-export.zip \
  --entry-point export_function
```

Creating a Pub/Sub topic

Create a Pub/Sub topic that will act as a trigger for Cloud Functions:

```
gcloud pubsub topics create metastore-export-function-trigger
```

Creating a Cloud Scheduler job

Use a gcloud command to create a Cloud Scheduler job that will trigger Cloud Functions at 12-hour intervals:

```
gcloud scheduler jobs create pubsub metastore-export-scheduled-job \
    --schedule "0 */12 * * *" \
    --topic metastore-export-function-trigger \
    --message-body '{"projectId":"dataproc-samples","locationId":"us-central1", \
    "serviceId":"test-metastore" "destinationGcsFolder":"gs://metastoretestb", \
    "databaseDumpType":"MYSQL"}' \
    --location us-central1
```

Once the job is successfully created, the scheduler will automatically run every 12 hours and export the backup to the configured GCS folder. Navigate to the Cloud Scheduler service home page to view the list of jobs, as shown in Figure 6-13.

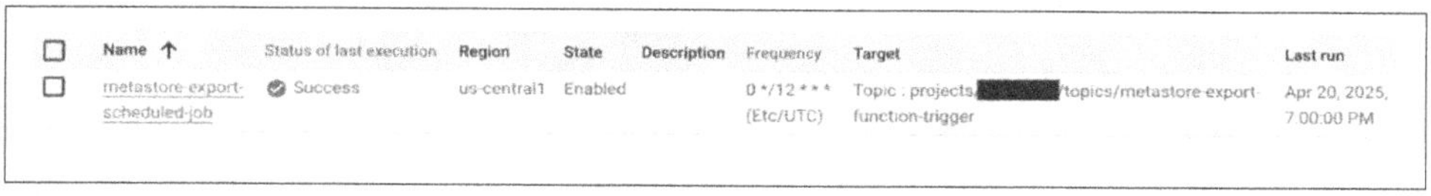

Figure 6-13. Successful creation of a Cloud Scheduler job that will run every 12 hours

## CHAPTER 7

Connecting from Dataproc to GCP Services

Dataproc provides a powerful framework for running Hadoop and Spark jobs, allowing users to connect and interact with GCP services efficiently. In this chapter, we'll explore various ways to connect Dataproc with popular GCP services like Cloud SQL, BigQuery, Bigtable and Pub/Sub Lite. We will also see how to configure Delta Lake tables on Dataproc and read from BigLake seamlessly.

In this chapter, you'll get hands-on experience and insights into the following connectors:

Spark-BigQuery connector
    A specialized connector for high-performance Dataproc-BigQuery transfers

Spark JDBC interface
    An interface to connect Dataproc to Cloud SQL and other relational databases

Pub/Sub Lite–Spark connector
    A connector to integrate Dataproc with Pub/Sub Lite's real-time messaging

Dataproc templates
    Preconfigured templates for common data tasks

Delta Lake on Dataproc
    Used to create Delta writes

BigLake integration
    Used to query Delta Lake tables using BigLake

7.1 Reading from GCS and Writing to a BigQuery Table

Problem

You need a Spark job running on Dataproc to read CSV data from GCS, process it, and write the results to the BigQuery table.

Solution

To achieve this, you can leverage the Spark-BigQuery connector (*https://oreil.ly/tdZ7L*). The Spark-BigQuery connector is preinstalled on Dataproc. No additional setup is required.

Here is the code snippet to write a DataFrame to BigQuery in append mode from Spark:

```
outputdf.write \
  .format("bigquery") \
  .option("writeMethod", "direct") \
  .mode("append") \
  .save(outputTable)
```

There are a few prerequisites to consider:

- Your Dataproc VM service account must have the necessary BigQuery access permissions.
- The BigQuery table must exist, and your DataFrame schema should align with the target table's schema.

Discussion

There are several flexible ways to connect BigQuery and Spark using the Spark-BigQuery connector. We'll show you how to write custom Spark code to achieve this.

> Write your own PySpark or Scala code for tailored interactions with BigQuery tables. This offers maximum control over data transformations and query logic.

If you are writing custom PySpark code, leverage the write method and options to specify how the Spark DataFrame is written to BigQuery. Here is the PySpark code that reads a CSV file and writes to the BigQuery table in append mode:

```
import sys
from pyspark.sql import SparkSession
from pyspark.sql.functions import *
```

```python
from pyspark.sql.types import *
from pyspark.storagelevel import StorageLevel

# Create a SparkSession (entry point to Spark functionality)

spark = SparkSession.builder.appName("wikiinsights") \
    .master("yarn") \
    .getOrCreate()

#Get the input path and output
inputPath = sys.argv[1]
outputTable = sys.argv[2]

print("Reading the CSV input file")

# Read the CSV, inferring the schema and assuming a header row

df = spark.read.option("header", True).csv(inputPath)

# Perform aggregation: Group by 'language' and get counts

outputdf=df.groupBy("language").count()

# Write the aggregated DataFrame to BigQuery and "direct" mode doesn't write
# intermediate files

outputdf.write \
  .format("bigquery") \
  .option("writeMethod", "direct") \
  .mode("append") \
  .save(outputTable)

print("Application Completed!!!")

# Closing the Spark session
spark.stop()
```

When writing to BigQuery, most people prefer to use the "direct" write method for efficiency because it bypasses intermediate files.

There are various mode options that control how data is written, including:

append
: Adds new rows to an existing table, preserving existing data.

overwrite
: Replaces the entire table.

ignore
: If the table already exists, no data is written. This helps prevent accidental overwrites.

7.2 Reading from a Cloud SQL Table

Problem

Your team is onboarding a new application in Dataproc and wants to read data from a Cloud SQL table using Spark on Dataproc.

Solution

You can use Spark's JDBC interface to read data from Cloud SQL:

```
df = (
  spark.read.format("jdbc")
  .option("url", jdbcUrl)
  .option("driver", driverClass)
  .option("dbtable", "Persons")
  .option("user", username)
  .option("password", password)
  .load()
)
```

Discussion

First, to secure the Cloud SQL instance, configure the instance with private IP (*https://oreil.ly/xFe8C*).

Next, note the IP range for Dataproc. In your VPC network settings, find the IP range allocated for the region where your Dataproc cluster resides, as shown in Figure 7-1. You'll want to copy this range or write it down because you'll need it again soon.

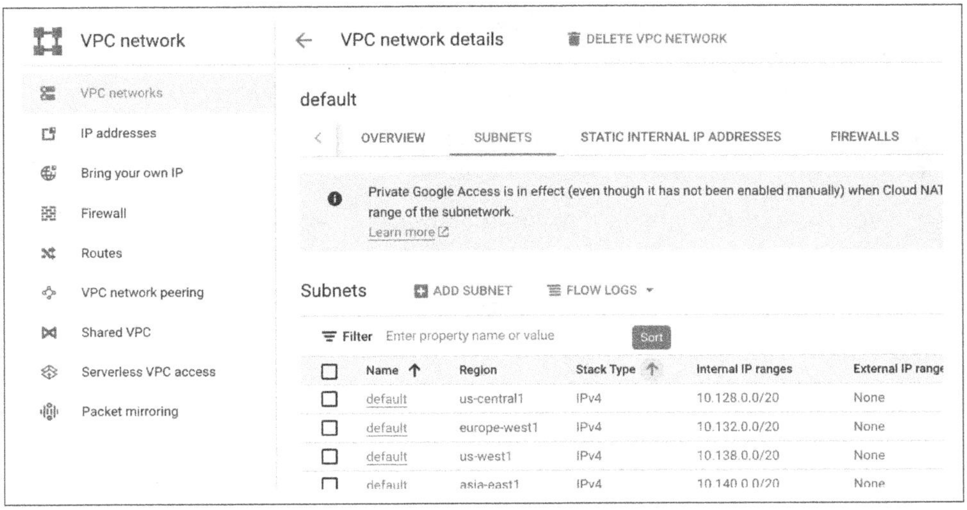

Figure 7-1. Allocated IP ranges per region in VPC networks

Then, authorize Dataproc's IP range in Cloud SQL:

- If it's a new instance, during instance creation select the Private IP option and choose your desired VPC network.
- If it's an existing instance, navigate to your Cloud SQL instance's settings, edit the Connectivity section, and enable Private IP. Select the appropriate VPC network.

To set up a private services access connection for the first time, click Setup Connection, as shown in Figure 7-2. Enter the Dataproc IP range you noted previously.

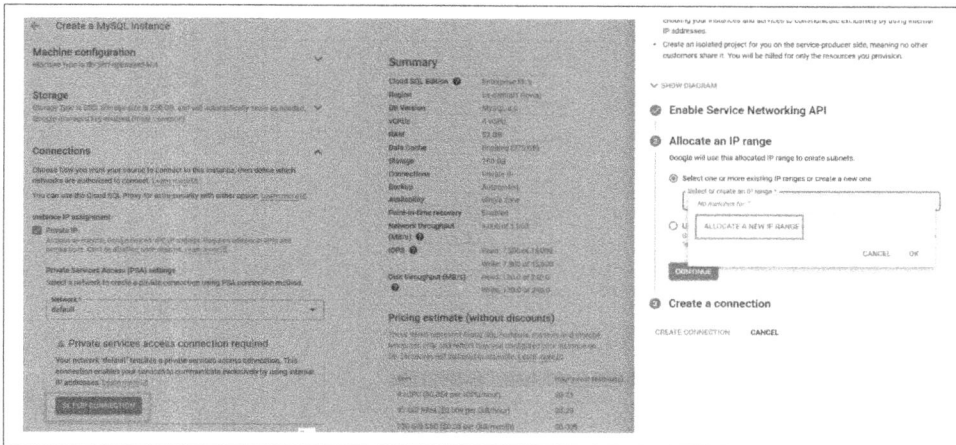

Figure 7-2. Configuring connectivity in a Cloud SQL instance

For enhanced security and authentication management, follow the instructions in the documentation (*https://oreil.ly/s7L1Y*) to set up the Cloud SQL proxy on your Dataproc cluster.

Connectivity to Cloud SQL

To connect to Cloud SQL using Spark's JDBC interface, you can choose Dataproc templates (refer to Recipe 7.6) or custom Spark code. There are four JDBCs that target Dataproc templates:

- JDBC to BigQuery
- JDBC to GCS
- JDBC to JDBC
- JDBC to Spanner

You can also leverage custom Spark code. Here is an example that connects with Cloud SQL instances via private IP and presents the output:

```python
import sys
from pyspark.sql import SparkSession
import pyspark.sql.functions as F

# Create a SparkSession
spark = (
  SparkSession.builder
  .appName("CloudSQLReader")
  .master("yarn")
  .getOrCreate()
)

# Replace with your Cloud SQL connection details
jdbcUrl = "jdbc:mysql://10.89.32.3:3306/enterprise"
username = "root"
password = ""
driverClass = "com.mysql.jdbc.Driver"

# Read data from Cloud SQL using JDBC
df = (
  spark.read
  .format("jdbc")
  .option("url", jdbcUrl)
  .option("driver", driverClass)
  .option("dbtable", "Persons")
  .option("user", username)
  .option("password", password)
  .load()
)


# Print or save the results
df.show()

spark.stop()
```

Use this example as a starting point and adapt it to your specific connection parameters and logic.

 Use secure credential management by leveraging a key management service (KMS) for password storage and retrieval.

7.3 Writing to GCS in Delta Format

Problem

You want to create a Delta table in GCS using a Dataproc Spark job.

Solution

These are the three steps you need to take when creating a Delta table in GCS from Dataproc:

1. Add the dependency io.delta:delta-core_2.12:1.2.1 (or the latest compatible version) to your Dataproc cluster. This provides the necessary functionality to work with Delta tables.

2. Replace `gcs_delta_path` with your desired GCS bucket and path (e.g., *gs://your-bucket/delta-data*). This should look something like:

   ```
   df.write.format("delta").mode("append").save("gcs_delta_path")
   ```

3. Set GCS permissions by ensuring that the Dataproc service account (or the user account if running locally) has both read and write permissions on the specified GCS bucket.

Discussion

Delta Lake is an open source storage layer that supports ACID transactions and time travel. This simplifies both batch and streaming inserts and analytics. Delta Lake tables are stored as open Apache Parquet format files in GCS.

Delta Lake generates two important files for ACID compliance and time travel:

- Transaction logs (JSON)
- Manifest (Parquet checkpoint file)

For each operation on the Delta table, JSON log files are created for time travel, as shown in Figure 7-3.

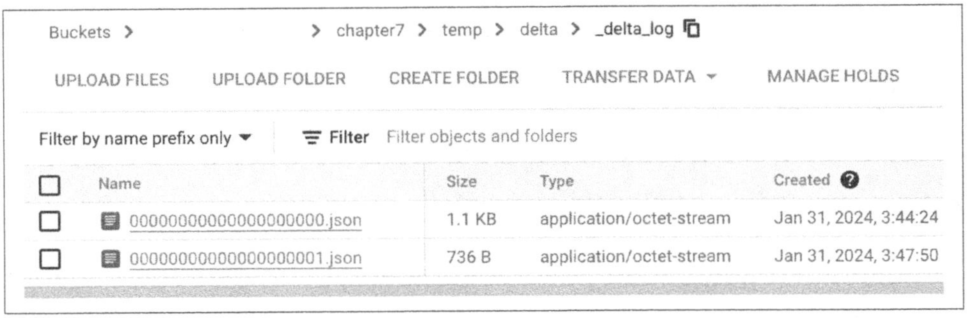

Buckets >		> chapter7 > temp > delta > _delta_log 🗐		
UPLOAD FILES	UPLOAD FOLDER	CREATE FOLDER	TRANSFER DATA ▾	MANAGE HOLDS

Filter by name prefix only ▾   ≡ Filter Filter objects and folders

	Name	Size	Type	Created ❓
☐	🗐 00000000000000000000.json	1.1 KB	application/octet-stream	Jan 31, 2024, 3:44:24
☐	🗐 00000000000000000001.json	736 B	application/octet-stream	Jan 31, 2024, 3:47:50

Figure 7-3. JSON log files in the _delta_log directory in GCS

Additionally, for every few transactions, a Parquet checkpoint file is created. It will keep the current state of the table.

Here is the sample code that reads from the BigQuery table, performs aggregation, and writes to a Delta Lake:

```
from pyspark.sql import functions as f
from pyspark import SparkConf
from pyspark.sql import SparkSession
from delta.tables import *

# Create a SparkSession
spark = SparkSession.builder.appName("CreateDeltaLakeTable").getOrCreate()

# read data from Big Query
wiki_data = (
    spark.read.format("bigquery")
    .option("table", "bigquery-samples.wikipedia_pageviews.200801h")
    .load()
)

# Aggregations
wiki_filtered = wiki_data.filter(
    "month = 1"
)

wiki_insights_report = wiki_filtered.groupBy("language").agg(
    f.sum("views").alias("TotalViews")
)

# Write to Delta Lake

wiki_insights_report.write
    .format("delta")
    .mode("append")
    .save("gs://dataproc-cookbook/chapter7/temp/delta")
```

We have taken a BigQuery table as a source in this example, although you could use Pub/Sub Lite or any other streaming or batch sources.

When submitting this PySpark job in the Dataproc job console or gcloud command, pass `io.delta:delta-core_2.12:1.2.1` (or the latest compatible version) via `spark.jars.packages`, as shown in Figure 7-4.

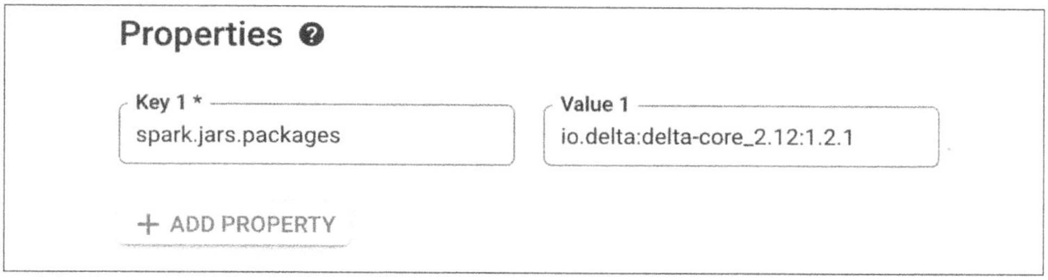

Figure 7-4. Adding the Delta core maven coordinate

7.4 Integrating a Dataproc-Managed Delta Lake with BigLake

Problem

You have implemented a lake house architecture where your Dataproc jobs process data through bronze, silver, and gold zones. The silver zone contains refined data stored in Delta Lake format within GCS. You want to leverage BigLake to access and analyze the latest state of your silver Delta Lake data in GCS.

Solution

Generate a manifest file if it is not already present. Execute the following data definition language (DDL) in the BigQuery console to create a BigLake table, replacing placeholders with your project details:

```
CREATE EXTERNAL TABLE IF NOT EXISTS `my-project.mydataset.myDeltaTable`
WITH CONNECTION `projects/anu-psodata-lab/locations/us/connections/delta-3`
OPTIONS (
  hive_partition_uri_prefix = "gs://dataproc-cookbook/chapter7/temp/delta/",
  uris = [
   'gs://dataproc-cookbook/chapter7/temp/delta/_symlink_format_manifest/manifest'
  ],
  file_set_spec_type = 'NEW_LINE_DELIMITED_MANIFEST',
  format="PARQUET"
);
```

Ensure that the BigLake service account has either storage object creator or storage object viewer permissions for your Delta Lake data. This allows BigLake to read the manifest file and Parquet data files.

Discussion

Multiple steps are needed to integrate a Dataproc-managed Delta Lake with BigLake. A Delta Lake table needs to be in GCS, and you should have the necessary IAM permissions to create BigQuery resources and write objects in GCS.

Syncing a Delta Table on Dataproc

If you don't have a manifest file, here is the PySpark code needed to generate a manifest for a Delta table:

```
from delta.tables import *

#Replace "gs://your_gcs_bucket/delta_folder" with the actual GCS path
#to your Delta table.

deltaTable = DeltaTable.forPath(spark, "gs://your_gcs_bucket/delta_folder")

#Creates necessary manifest file
deltaTable.generate("symlink_format_manifest")
```

When submitting your PySpark job (via Dataproc or gcloud), ensure that you include the Delta Lake dependency (`spark.jars.packages: io.delta:delta-core_2.12:1.2.1`). Adjust 1.2.1 to the latest compatible Delta Lake version.

Whenever your Dataproc jobs modify the Delta Lake table, run the preceding manifest-generating code to regenerate the manifest file and ensure that BigLake reflects the latest changes.

Creating a BigLake connection

Navigate to the BigQuery console. In the Explorer panel, click +Add and select "Connections to external data sources," as highlighted in Figure 7-5.

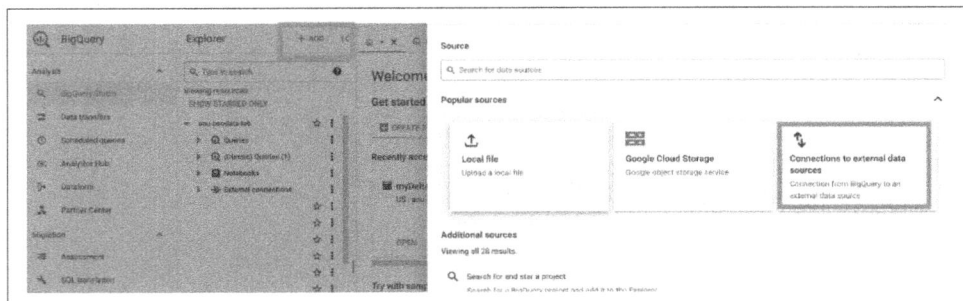

Figure 7-5. Creating a BigLake connection

Then, select "Vertex AI remote models, remote functions and BigLake (Cloud Resource)," as shown in Figure 7-6. Provide a descriptive name for your connection (e.g., "delta_lake_connection"). Click "Create connection."

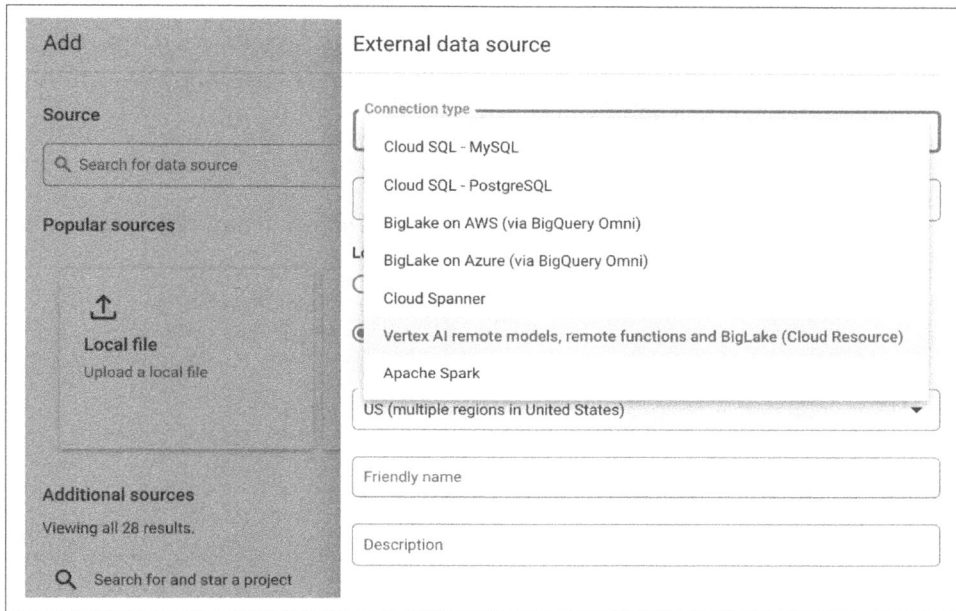

Figure 7-6. Choosing a BigLake connection type

Locating connection information

From the BigQuery console, find the newly created connection under "External connections." Note the connection ID and service account ID, as shown in Figure 7-7. These details will be needed in the next step.

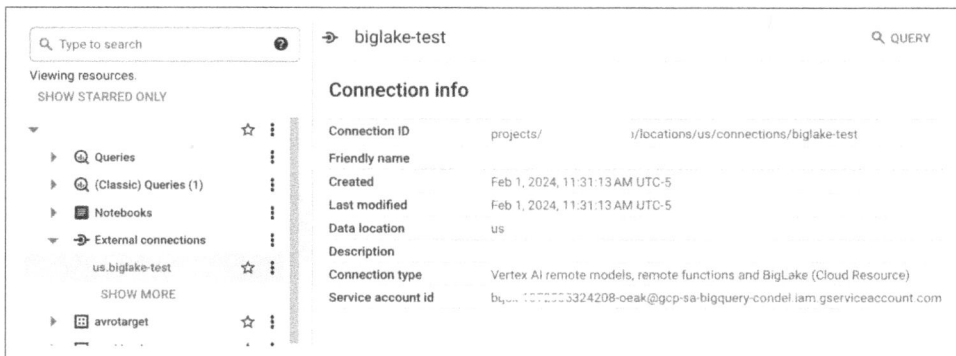

Figure 7-7. Connection ID and service account ID under connection info

Granting GCS permissions for BigLake access

To query your Delta Lake data using BigLake, you need to ensure that the BigLake connection has the necessary permissions to read objects within your GCS bucket. Navigate to the IAM console. Grant either the storage object creator or storage object viewer role to the service account that you created in the previous step. The viewer role provides read-only access. Once permission is granted, users will have access to Delta objects like manifest and Parquet files.

Creating the BigLake table

In the BigQuery query console, use the following DDL as a template to create your BigLake table:

```
CREATE EXTERNAL TABLE IF NOT EXISTS `my-project.mydataset.myDeltaTable`
WITH CONNECTION `projects/project_id/locations/us/connections/delta-3`
-- Replace with your connection ID
OPTIONS (
  hive_partition_uri_prefix = "gs://gcs_bucket_name/delta/",
  uris = ['gs://gcs_bucket_name/delta/_symlink_format_manifest/manifest'],
  file_set_spec_type = 'NEW_LINE_DELIMITED_MANIFEST',
  format="PARQUET"
);
```

Replace placeholders in the SQL query with your actual project ID, dataset name, GCS bucket name, and manifest. In this template, WITH CONNECTION specifies the BigLake connection you created earlier, hive_partition_uri_prefix is the base directory in GCS where your Delta Lake table is stored, and uris is the path to the _symlink_format_manifest/manifest file.

7.5 Connecting to GCP Services Using Dataproc Templates

Problem

The organization has users with minimal Spark development experience who want to leverage Dataproc or Spark for a data-migration project between platforms.

Solution

The Dataproc templates solution created by the Google team offers easy-to-use Spark code for frequent operations (*https://oreil.ly/xMEV4*).

Discussion

The Google Cloud team has developed a set of templates focused on moving data from one place to another. These templates provide a great starting point for those who are new to GCP but want to quickly run Dataproc jobs for data migration. The Dataproc templates (*https://oreil.ly/xMEV4*) are Spark based and are available in both Python and Java versions.

Some popular templates can:

- Read data from GCS and loading it to BigQuery
- Convert data from one format to another, such as CSV to Parquet
- Read from transactional databases with a JDBC connection and load to BigQuery

These templates can be run programmatically or via the web UI. The templates can also be extended for adding any custom logic. Simply clone the template project from GitHub and add your custom logic.

Let's look at the steps to convert data from CSV to a Parquet file using Dataproc templates via the web UI. First, navigate to the Dataproc home page in the Google Cloud console and select Batches. Then, select the "Create from template" option, as shown in Figure 7-8.

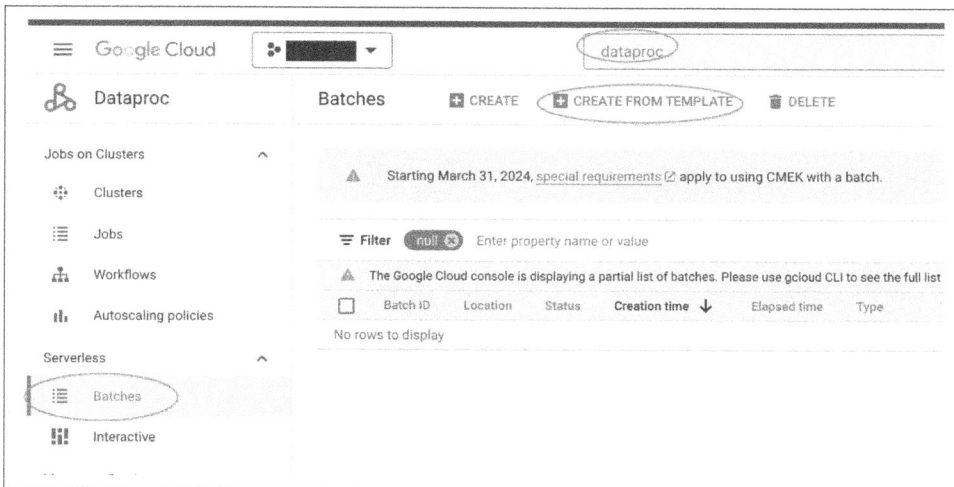

Figure 7-8. Creating a new job using Dataproc templates

Every job, or batch, has to have a unique batch ID assigned. Select the type of template you want to run, as shown in Figure 7-9.

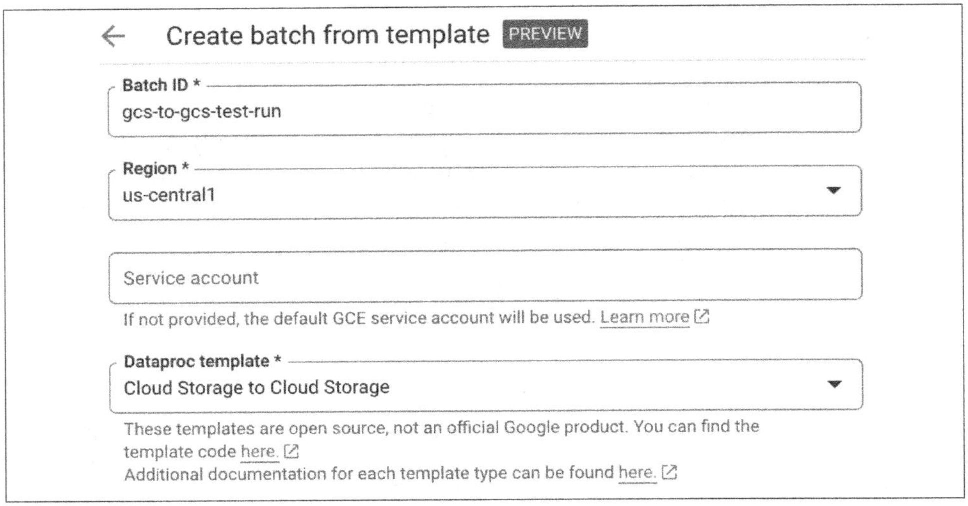

Figure 7-9. Populating a batch ID and selecting the template type

Choosing "Cloud Storage to Cloud Storage" requires the parameters of input location, input data format, output location, and output data format, as shown in Figure 7-10.

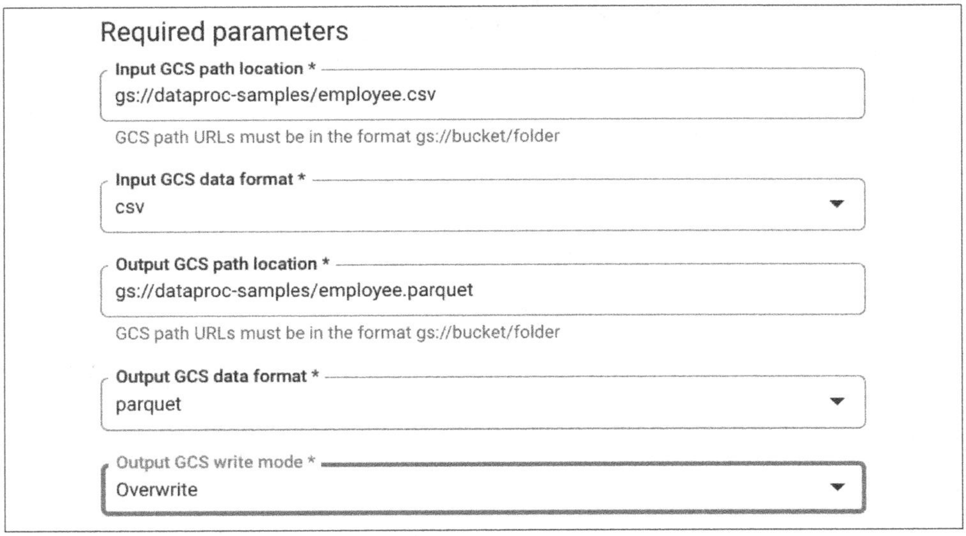

Figure 7-10. Adding required parameters to the template

Next, choose the network where the Dataproc Serverless job will execute, as shown in Figure 7-11.

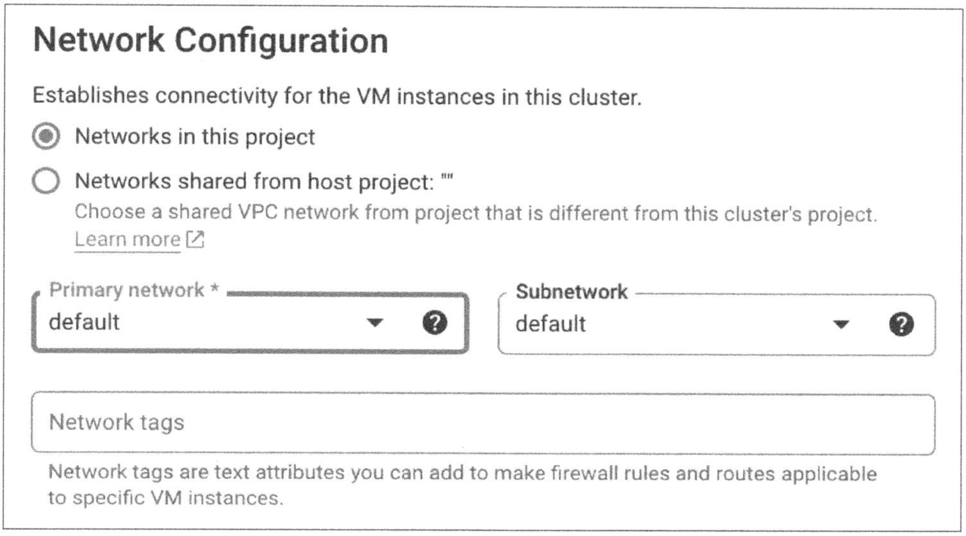

Network Configuration

Establishes connectivity for the VM instances in this cluster.

⦿ Networks in this project

◯ Networks shared from host project: ""
Choose a shared VPC network from project that is different from this cluster's project.
Learn more ↗

Primary network *
default ▾ ❓

Subnetwork
default ▾ ❓

Network tags

Network tags are text attributes you can add to make firewall rules and routes applicable to specific VM instances.

Figure 7-11. Adding the primary network and subnetwork

Optionally select the history server for preserving the job logs. Click the Submit button to launch a serverless batch job, as shown in Figure 7-12.

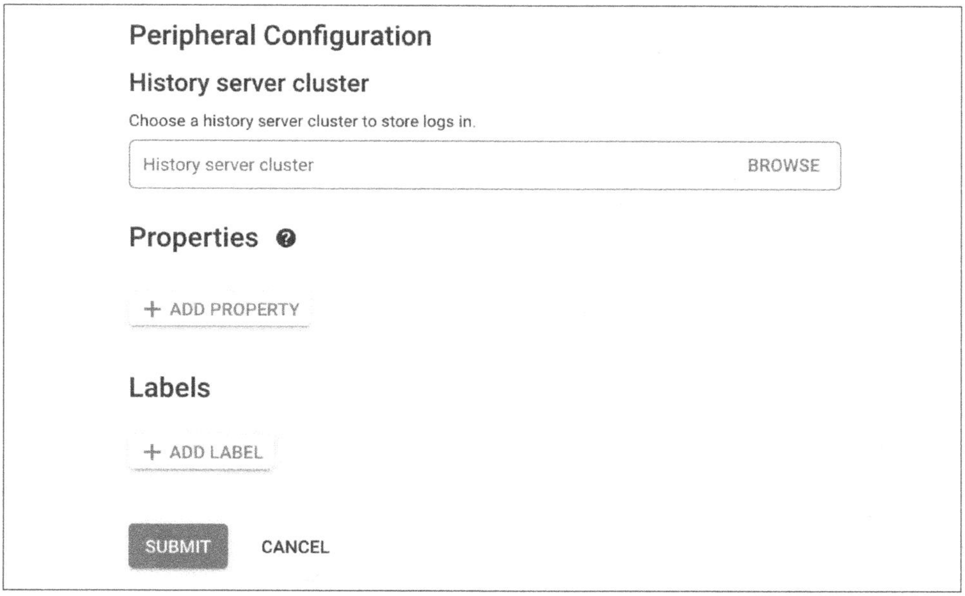

Peripheral Configuration

History server cluster

Choose a history server cluster to store logs in.

History server cluster   BROWSE

Properties ❓

＋ ADD PROPERTY

Labels

＋ ADD LABEL

SUBMIT   CANCEL

Figure 7-12. Configuring the history server in the template

Dataproc templates can be run as a serverless job or on an existing Dataproc cluster. When using the web UI route, only the serverless batch option is supported. For running templates programmatically, refer to the GCP GitHub project for Dataproc templates (*https://oreil.ly/46fso*).

> Running Dataproc templates on the premium tier is not yet supported.

For any queries or to request new features, reach out to *dataproc-templates-support-external@googlegroups.com*.

7.6 Spark Job Running on Dataproc Reading from GCS and Writing to Bigtable

Problem

You have CSV files stored in GCS and want to load their data into Cloud Bigtable. You prefer to use a Spark job running on Google Cloud Dataproc.

Solution

There are two ways to write to Bigtable from Dataproc:

- Use the Bigtable-Spark connector (*https://oreil.ly/hl1Pc*) to write from Spark to Bigtable. Here is the gcloud command to write to an existing table in Bigtable:

```
gcloud dataproc batches submit --project <PROJECT_ID_HERE> --region \
  <REGION> pyspark --batch <BATCH_JOB_NAME> <GCS_PYSPARK_CODE> \
  --version 1.1 \
  --jars gs://spark-lib/bigtable/spark-bigtable_2.12-0.2.1.jar \
  --subnet default --service-account <COMPUTE_ENGINE_SERVICE_ACCOUNT> \
  --properties spark.jars.packages=org.slf4j:slf4j-reload4j:1.7.36,
spark.dataproc.appContext.enabled=true
```

- Use Dataproc templates, specifically the GCS-to-Bigtable template (*https://oreil.ly/Mcbf4*).

Discussion

There are various ways to write to Bigtable using Spark, with different programming language choices available. One option is to use the Bigtable-Spark connector (*https://oreil.ly/LHtzE*) to write from Spark to Bigtable. Another option is using Dataproc templates.

For the Bigtable-Spark connector, here is sample PySpark code that writes to an existing table in Bigtable:

```python
from pyspark.sql import SparkSession

spark = SparkSession.builder \
    .appName("BigtableExample") \
    .getOrCreate()

bigtable_project_id = "dataproc-cookbook-425300"
bigtable_instance_id = "test-instance"
bigtable_table_name = "wordinsights"

catalog = ''.join(("""{
        "table":{"name":" """ + bigtable_table_name + """
        ", "tableCoder":"PrimitiveType"},
        "rowkey":"wordCol",
        "columns":{
          "word":{"cf":"rowkey", "col":"wordCol", "type":"string"},
          "count":{"cf":"example_family", "col":"countCol", "type":"long"}
        }
        }""").split())

csv_file_path = "gs://bucket_name/sample.csv"
input_data = spark.read.csv(csv_file_path, header=True, inferSchema=True)

input_data.write \
    .format('bigtable') \
    .options(catalog=catalog) \
    .option('spark.bigtable.project.id', bigtable_project_id) \
    .option('spark.bigtable.instance.id', bigtable_instance_id) \
    .option('spark.bigtable.create.new.table', "false") \
    .save()
```

You can create a batch job in Dataproc and pass the PySpark code path. Choose the supported runtime version and Bigtable-Spark connector JAR, as shown in Figure 7-13.

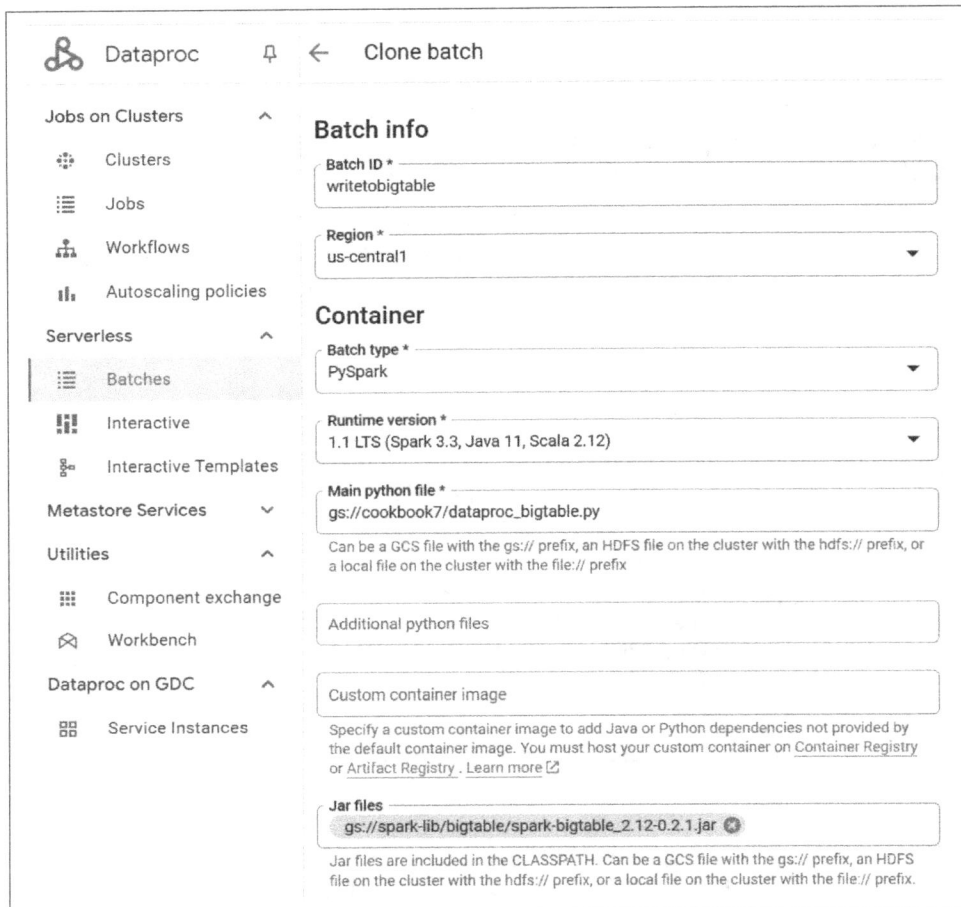

Figure 7-13. Adding a PySpark filepath and Bigtable-Spark connector to a Dataproc batch

Add the slf4j package to `spark.jars.packages`, as shown in Figure 7-14, and submit the job. This package allows for different logging frameworks to coexist.

If you prefer a template-based approach, leverage Dataproc templates. Here are some reasons why you may choose this approach:

- It's easier to use. Templates provide preconfigured code and setup to simplify complex tasks like installing appropriate JARs to connect from Spark to Bigtable.

- It's flexible. You can choose between Java and Python templates and customize them if necessary or even use a code-agnostic method.

- It has active support. Dataproc templates are regularly maintained, ensuring compatibility and bug fixes.

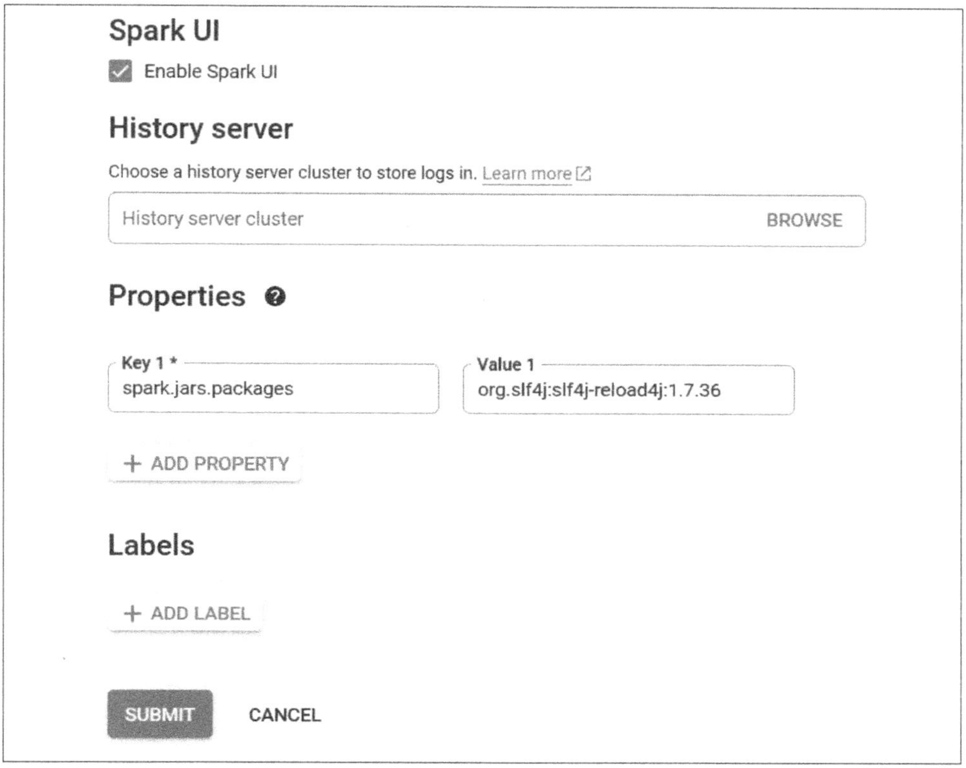

Spark UI

☑ Enable Spark UI

History server

Choose a history server cluster to store logs in. Learn more ☑

| History server cluster | BROWSE |

Properties ❓

Key 1 *
spark.jars.packages

Value 1
org.slf4j:slf4j-reload4j:1.7.36

+ ADD PROPERTY

Labels

+ ADD LABEL

SUBMIT   CANCEL

Figure 7-14. Enabling Spark UI

There are two Dataproc templates for writing to Bigtable:

- Pub/Sub to Bigtable
- GCS to Bigtable

Here are the steps for using Dataproc templates to easily read CSV files from GCS and write them to Bigtable:

1. Clone the Dataproc templates repository (shown in Figure 7-15) using Google Cloud shell or a compute instance with the Google SDK installed:

   ```
   git clone https://github.com/GoogleCloudPlatform/dataproc-templates.git
   ```

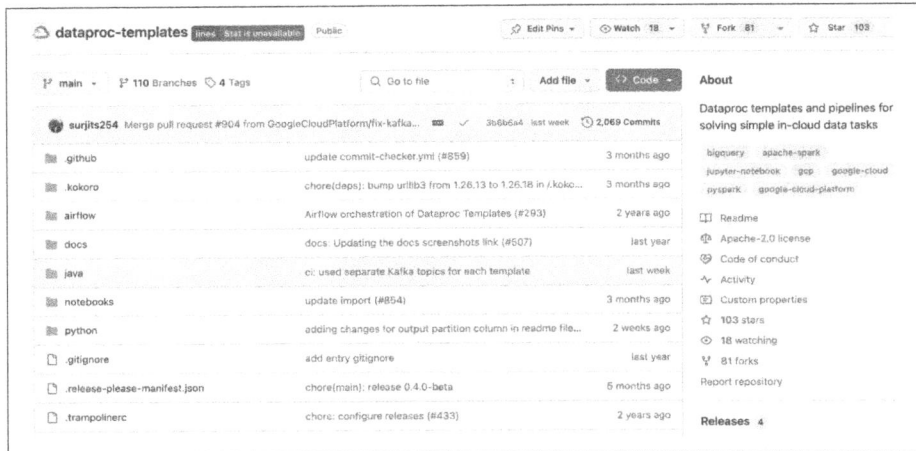

Figure 7-15. Dataproc templates public repository

2. Navigate to the *Java* directory. This is where you'll work with the GCS-to-Bigtable template. If you need to make any customizations to the template's logic or additional data processing logic, modify the source code here:

```
anuyogam@big-table-new-test-m:~/dataproc-templates/java$ ls
JAVA_LICENSE_HEADER  dependency-reduced-pom.xml  target
READ.md              pom.xml
bin                  src
```

3. Before running the template, set the required environment variables:

```
export GCP_PROJECT={project_name_here}
export REGION={region_here}
export GCS_STAGING_LOCATION={gcs_staging_path}
export JOB_TYPE=CLUSTER
export CLUSTER=dataproc-large-1
```

Choose between Dataproc Serverless (the default) or an existing job-scoped Dataproc cluster. For Serverless, leave JOB_TYPE unset. For an ephemeral Cluster, set export to JOB_TYPE=CLUSTER. See Chapter 4 for help deciding which mode is best for you.

4. Execute the following command, replacing placeholders with your specific values:

```
bin/start.sh -- \
--template GCSTOBIGTABLE \
--templateProperty project.id={project_name_here} \
--templateProperty gcs.bigtable.input.location={gcs_file_path}/file.csv \
--templateProperty gcs.bigtable.input.format=csv \
--templateProperty gcs.bigtable.output.instance.id={bigtable_instance_id} \
--templateProperty gcs.bigtable.output.project.id={project_name_here} \
--templateProperty gcs.bigtable.table.name={bigtable_name} \
--templateProperty gcs.bigtable.column.family={bigtable_column_family}
```

5. When it's successful, you will see the status in the terminal:

```
status:
  state: DONE
  stateStartTime: '2024-01-31T15:22:20.459604Z'
  statusHistory:
    - state: PENDING
      stateStartTime: '2024-01-31T15:21:47.113369Z'
    - state: SETUP_DONE
      stateStartTime: '2024-01-31T15:21:47.141097Z'
    - details: Agent reported job success
      state: RUNNING
      stateStartTime: '2024-01-31T15:21:47.359996Z'
  yarnApplications:
    - name: GCS to Bigtable load
      progress: 1.0
      state: FINISHED
```

6. Use the cbt command-line tool to check if the data loaded into Bigtable:

```
cbt -project=GCP_PROJECT -instance={bigtable_instance_id} \
  read {bigtable_name}
```

See Also

- Bigtable documentation on using the Bigtable-Spark connector (*https://oreil.ly/ L5ca1*)
- Blog on how to use the GCS-to-Bigtable template (*https://oreil.ly/VJTbI*)
- Video on how to use the GCS-to-Bigtable template (*https://oreil.ly/MuELV*)

## CHAPTER 8

Configuring Logging in Dataproc

In the world of distributed data processing, logging is an essential tool that empowers you to monitor the health of your clusters, pinpoint bottlenecks, and rapidly diagnose issues. However, the sheer volume and variety of logs generated across the Dataproc ecosystem can be overwhelming. This chapter equips you with the knowledge and strategies you need to navigate the Dataproc logging landscape effectively.

Before we dive in, let's set the stage a bit. First, let's briefly explore why logging matters. Logging is more than just a stream of text. It's your window into the inner workings of Dataproc. Logging provides:

Visibility

See what's happening at each stage of your cluster's lifecycle, from creation to job execution:

Performance Optimization

Identify resource-intensive operations and fine-tune your configurations for maximum efficiency.

Debugging

Quickly isolate the root causes of errors and failures, saving you valuable time and effort.

Security

Monitor for suspicious activity or unauthorized access attempts.

There are challenges with Dataproc logging, though. For example, Dataproc generates logs from multiple sources, including:

Cluster logs
> Capture events related to cluster creation, configuration, and operation

Initialization scripts
> Record the output of scripts that customize your cluster environment

Service logs
> Provide insights into the behavior of core Dataproc services (master, workers, etc.)

Application logs
> Contain messages from your Spark, Hadoop, or other processing jobs

To be successful, it's important to strike the right balance. Too much logging can lead to excessive storage costs and difficulty extracting meaningful information. Too little logging leaves you in the dark when problems arise.

This chapter will guide you through understanding and managing Dataproc logs. We'll cover everything from the basics of different log types to advanced analysis techniques. You'll learn how to set up logging, customize it to your needs, and get the most out of your log data. By the end, you'll have the skills to troubleshoot issues and make your Dataproc workloads run smoothly.

8.1 Understanding Different Types of Logs in Dataproc

Problem

You want to know the different types of logs that are generated within a Google Cloud Dataproc cluster and how to use them effectively for monitoring and troubleshooting.

Solution

The Dataproc service generates logs at different levels. Here are the high-level categories:

Dataproc service logs
> Logs about cluster creation, cluster health monitoring, autoscaling, and so on

Service component logs
> Logs from components like HDFS NameNode, YARN RM, Spark, and so on

Application logs
> Logs generated by applications and jobs running in the cluster

Discussion

The Dataproc service generates several logs from its activities. These logs are valuable for troubleshooting issues, monitoring cluster health, and tracking usage. Some of the service logs include:

Cluster creation logs
> Provide information about the cluster's configuration, the VMs created, installed Dataproc components, and any custom bootstrap script output

Cluster monitoring logs
> Generated while the cluster is running and contain details about the cluster's health and components

Autoscaling logs
> Created by clusters with an autoscaling policy and include information about resource demand and scale-up or scale-down decisions

Cloud audit logs
> Capture administrative activities like resource access, cluster creation, deletion, and the like

Within the Dataproc cluster, multiple components generate their own sets of logs. Most of these services run across all nodes in the cluster in distributed mode, which generates two types of logs: master daemon logs and worker daemon logs.

Table 8-1 lists the component logs that are classified as master or worker.

Table 8-1. Dataproc logs by component type

Component	Master daemon logs	Worker daemon logs
HDFS	NameNode logs Secondary NameNode logs	DataNode logs
YARN	RM logs MapReduce history server logs YARN Application Timeline Server logs	NodeManager logs
Spark	Spark history server logs	N/A
Hive	Hive server logs Hive Metastore server logs	N/A
ZooKeeper	ZooKeeper service log	N/A

> The jobs we run in the cluster also generate logs at various levels. For example, Spark jobs that read from a file and store output in BigQuery can create logs about the number of records they have processed.

Now that we have a better understanding of what these logs are, where do they get stored? Are they all stored in a single place or file? Here are the various places where your logs may be stored (illustrated in Figure 8-1):

Google Cloud Logging
A fully managed logging service designed to ingest, store, and analyze large volumes of log data.

GCS
Provides a scalable, cost-effective way to store large volumes of log files.

Log files on Compute Engine
Can be generated or stored inside the Compute Engine where the services or applications are running

Custom locations
Logs can also be integrated with Elasticsearch and similar services for storage and analysis.

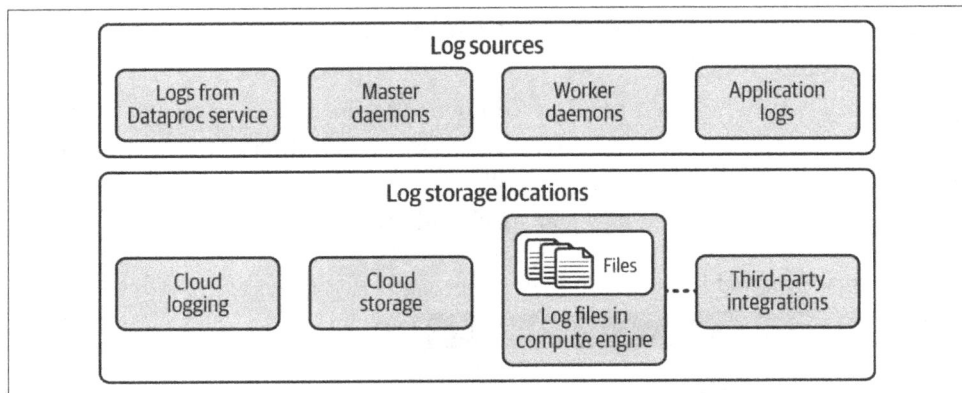

Figure 8-1. Different types of log source and log storage locations

In the following sections, we will learn more about when to use which log storage location and how to effectively manage this.

8.2 Understanding Cloud Logging

Problem

What is Google Cloud Logging, and how does it work?

Solution

Google Cloud Logging is a solution for storing and processing logs in a serverless manner. Google Cloud abstracts the underlying complexities and lets you manage storing, routing, and accessing logs using the Logging API. The logs are stored in a logging bucket, as shown in Figure 8-2. By default, every GCP project gets two logging buckets named _Default and _Required.

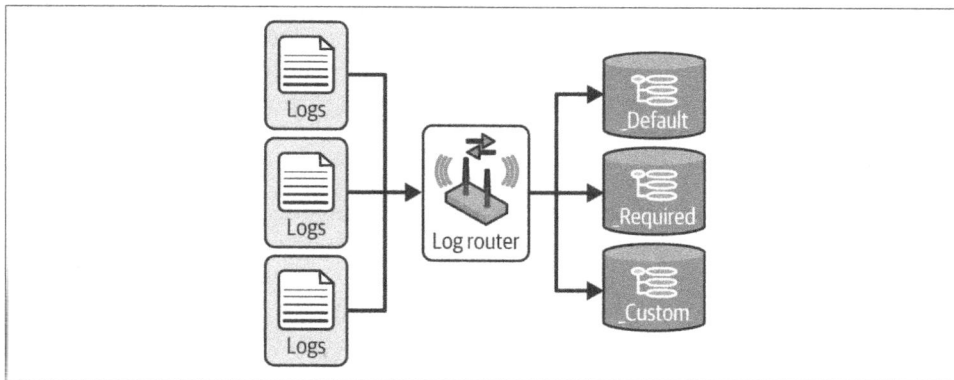

Figure 8-2. Log Router routing logs from various sources to logging buckets

Discussion

Let's dive into the key concepts of Cloud Logging.

Log source

A *log source* in Cloud Logging represents the origin of log messages. It's the entity or application responsible for generating the log data. These can encompass a wide range of sources, including Google Cloud services like Compute Engine instances, Cloud Functions, and App Engine applications as well as custom applications running within or outside Google Cloud. Specifically within Dataproc, the logs from master and worker daemons, along with the logs generated by Spark or Hadoop applications running on the cluster, serve as the origins for the log source.

Log Router

The Cloud Logging Log Router acts as a traffic director for your log data. It examines incoming log entries and applies user-defined rules, called *sinks*, to determine their destination. These sinks can direct logs to various locations like Cloud Logging buckets, BigQuery, GCS, or Pub/Sub. The Log Router allows you to selectively route logs using filters, which optimizes storage costs and enables targeted analysis across different GCP services. Essentially, it ensures that your logs reach the right place for storage, analysis, or real-time processing.

Log sink

A *log sink* defines the destination for your routed log entries. It's the endpoint where logs are stored or processed, enabling you to manage and analyze your data. Sinks direct logs to various locations, such as durable Cloud Logging buckets for storage and analysis, BigQuery for powerful querying, GCS for archival, or Pub/Sub for real-time streaming to other systems. Essentially, a sink is the final destination, chosen based on your logging and analysis needs.

Logging buckets

Logging buckets are storage locations within Cloud Logging where log entries are held. They provide a structured environment for managing and querying log data. When you create a new GCP project, it automatically creates two logging buckets (_Default and _Required) for that project, and you can define additional buckets:

_Default bucket
> This is a system-managed bucket that receives logs that do not go to the _Required bucket or to user-defined buckets. The retention period of this bucket is customizable. The default retention window is 30 days. You can customize the log sink and choose what messages should go to _Default bucket.

_Required bucket
> This is a system-managed bucket that stores critical audit logs, including admin activity and system event audit logs. This bucket has a fixed log-retention window that is currently set to four hundred days and is not configurable. You cannot modify the sink configuration to change the logs routed to the _Required bucket.

User-defined buckets
> These are user-created custom logging buckets for organizing log data. The retention period is customizable per user requirements.

> ## What Is the Rationale Behind Having Three Distinct Types of Logging Buckets?
>
> Cloud Logging uses three logging bucket types to address varied logging needs and compliance requirements. The _Required bucket is specifically designed for immutable audit logs, ensuring that critical security and administrative activity records are retained for regulatory purposes with a fixed retention period. The _Default bucket serves as a general-purpose repository for operational logs, offering customizable retention and a catchall for logs not explicitly routed elsewhere, thus separating operational logs from strict compliance logs. Finally, user-defined buckets provide the flexibility to tailor log storage based on specific application, environment, or regional needs, allowing for customized retention policies and cost optimization. This separation of concerns provides robust security, compliance adherence, and efficient log management.

8.3 Viewing Logs in Cloud Logging

Problem

You want to access logs that are stored in Cloud Logging.

Solution

There are three ways to access Cloud Logging logs:

- View and interact with logs directly in the Google Cloud console.
- Use the gcloud logging commands to retrieve and manage logs from the command line.
- Programmatically access and manage logs using the Cloud Logging API.

Discussion

Google Cloud allows you to access the Cloud Logging service in multiple ways: using the web UI, gcloud, and the REST API. Let's walk through each option.

Accessing logs from the web UI

To view logs from the web UI, search for the Logging API and click "Logs explorer," as shown in Figure 8-3.

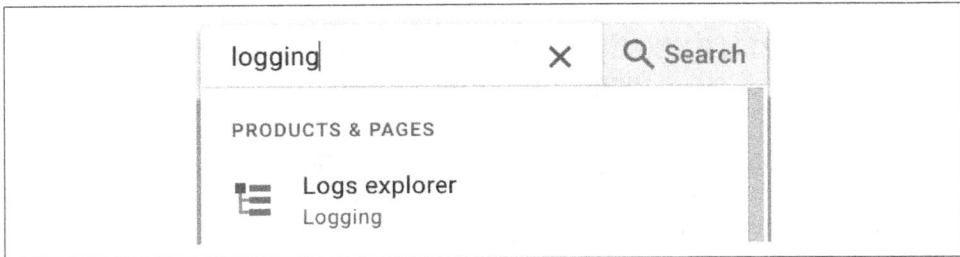

Figure 8-3. Searching for the logging service

Figure 8-4 shows the home screen of Logs Explorer.

Figure 8-4. Home screen of Logs Explorer

> Searching for text without any filters can be an expensive operation in terms of performance. The more search filters you add, the better the performance will be. You will notice the difference especially when the time-range window and log volume are larger.

The key options available on the Logs Explorer home screen are:

Time range

This option lets you choose the time range for when the log search should execute. You can specify the search to take place during a range of time, such as *x* minutes to hours, or you can specify exact start- and end-time ranges.

Search for specific fields

You can look for specific logs by filtering logs using the log fields. For instance, you can search all logs belonging to the Dataproc service, or you can search all logs from a Compute Engine. Do this either by typing your field choices in Search or by using the drop-down menu provided.

Query text

You can also enter query searches directly. For example, the following formats can be used to execute queries:

- Search for exact text: "search_text_here"

- Search for text with a "like" condition: Log_Field ~= "search_text_here"

- Search for text that does not match a given condition: Log_Field != "search_text_here"

Run query

Once you're done entering your queries and selecting filters (time range, fields, etc.), click "Run query" to view the results. If there are any matching results for your search query, they will be displayed in the results section.

Figure 8-5 highlights the number of results your search criteria matched. You can drill down further on log fields through the left menu pane.

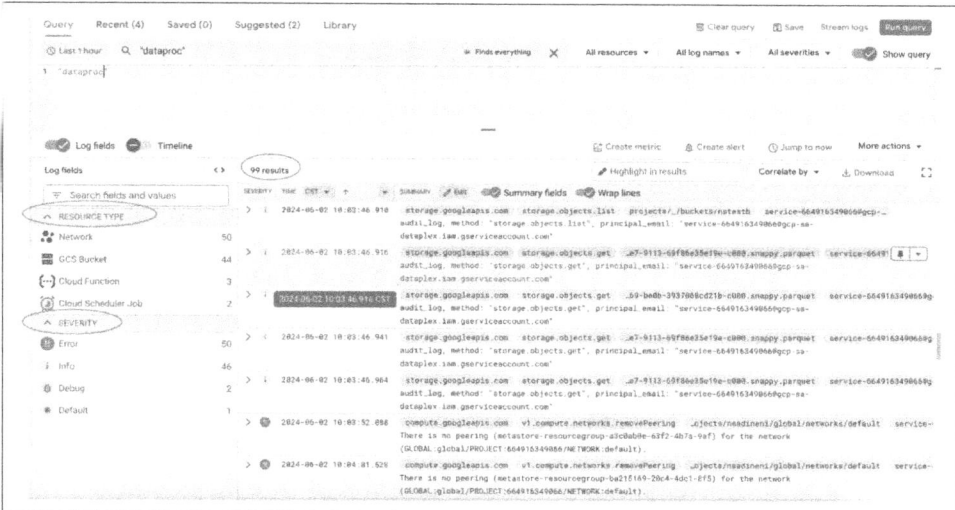

Figure 8-5. Log search results

Another option to narrow your search results is to expand the log message, as shown in Figure 8-6. To do this, right-click on the field you want to search on and click "Show matching entries," as shown in Figure 8-7.

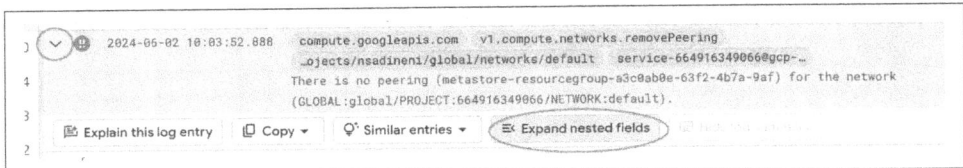

Figure 8-6. Expanding a log entry to expand all nested fields

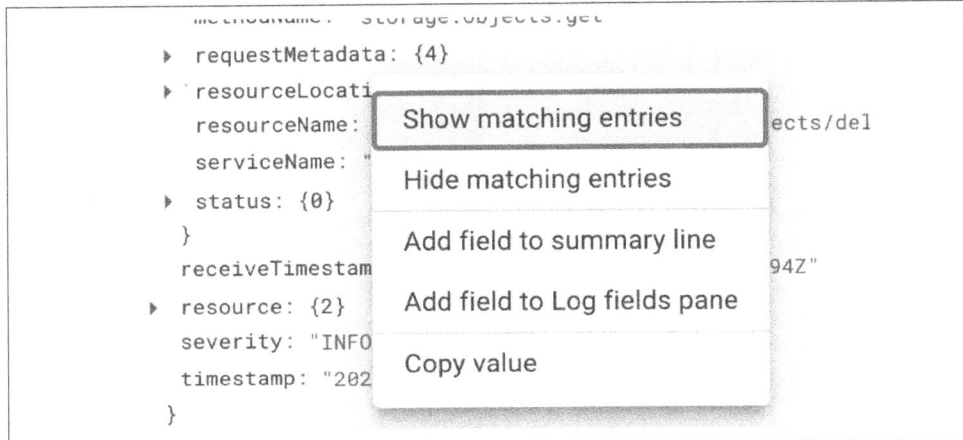

Figure 8-7. Adding a condition to results by clicking "Show matching entries"

To summarize, you can apply filters before your search in the form of log fields or drop-down options as well as within the query itself. Alternatively, you can apply filters on top of the results you've already extracted. If you're unsure about the search pattern, start with a broader search and then refine it based on the results you get.

Viewing logs using gcloud command

If you have the gcloud utility installed and authenticated, then you can use it to view your logs. For example, if you want to view log messages from the Dataproc service that occurred over the past hour, use the following gcloud command:

```
gcloud logging read 'resource.type="cloud_dataproc_cluster" severity=INFO \
resource.labels.cluster_name="dataproc-samples" --limit 10 --freshness 1h
```

It's worth noting that you can customize the variables at the end of the prompt. For instance, limit refers to the number of results. In this case, the number of results is limited to 10. Additionally, freshness configures the search window to a specific duration (one hour for this example).

Searching for logs using the REST API

Let's consider an example of how to search for logs using the REST API. Here is a sample curl command that will search for similar search queries, as in the previous example. Be sure to replace *<PROJECT_ID>* with your own unique ID:

```
curl -X POST \
  -H "Authorization: Bearer $(gcloud auth print-access-token)" \
  -H "Content-Type: application/json" \
  -d '{
        "resourceNames": [
          "projects/<PROJECT_ID>"
        ],
        "filter": "resource.type=\"cloud_dataproc_cluster\" \
        AND severity=INFO AND resource.labels.cluster_name=\"dataproc-samples\" \
        AND \"startup\"",
        "orderBy": "timestamp desc",
        "pageSize": 10
      }' \
  https://logging.googleapis.com/v2/entries:list
```

In an enterprise environment, it is a best practice to centralize logs in a common project. Google Cloud allows you to search logs in Project B while you are working in Project A. You can define the scope of logs to be searched using the "Refine scope" option shown in Figure 8-8.

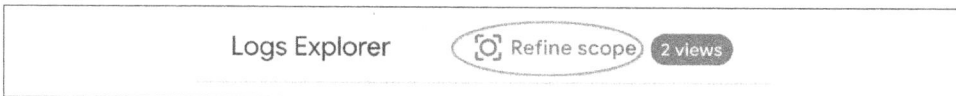

Figure 8-8. Refining the log scope in Log Explorer

8.4 Routing Dataproc Logs to Cloud Logging

Problem

You want to selectively manage which specific Dataproc logs are forwarded to Cloud Logging.

Solution

Logging properties can be configured at the time of cluster creation by passing parameters or by editing Fluentd configuration with `init` actions.

Discussion

Google Cloud uses the open source Fluentd service to unify logging across various sources. A Fluentd agent is installed on the Dataproc master and worker nodes to control log synchronization with the Cloud Logging service. Fluentd's configuration files determine which logs are sent to Cloud Logging. These configuration files can be found at the following folder: */etc/google-fluentd/config.d*.

By default, the following logs are configured to be sent to Cloud Logging from the Dataproc service:

- All daemon logs from the master and worker nodes, including:
  - Resource manager, NameNode, Hive server, and so on from the master node
  - Node manager, DataNode, and so on from the worker node
- Dataproc agent logs
- Dataproc cluster initialization logs
- YARN application logs (logs of applications that run on top of YARN, such as Spark, Sqoop, and Hive)
- Logs from any services added at the time of cluster creation, such as Ranger, Jupyter Notebook, and so on

Since these logs are enabled by default when a cluster is created, they will be sent to Cloud Logging automatically. If you would like to disable any of these logs to keep them from going to Cloud Logging, you have to edit the Fluentd configuration for the service.

The Fluentd configuration has three main components:

Source
> This is where Fluentd will look for log data.

Filter
> Fluentd filters modify, enrich, or filter log data between collection (source) and storage/forwarding (sink), helping you manage and refine logs in your pipeline.

Output
> This is where Fluentd is going to send logs it has received, typically to the Cloud Logging service.

Here is a simple code snippet for Fluentd that will read logs from a local file, apply filters, and then send the filtered logs to Cloud Logging:

```
# Input: Read logs from a local file
<source>
  @type tail
  path /var/log/myapp.log
  pos_file /var/log/myapp.log.pos
  tag app.logs
  format json
</source>
```

```
# Filter: Match only messages with "error" (case-insensitive)
<filter app.logs>
  @type grep
  <regexp>
    key message
    pattern /error/i    # Case-insensitive match for "error"
  </regexp>
</filter>


# Output: Send filtered logs to Google Cloud Logging
<match app.logs>
  @type google_cloud
  # (Optional) Project ID if not running on GCP
  # project <YOUR_PROJECT_ID>
  # Use the default log name if unspecified
  log_name myapp_error_logs
  # Structured logging format
  use_structured_logging true
  # (Optional) Resource type if not auto-detected
  resource
    type custom_resource_type
    labels
      instance_id custom_instance_id
      zone custom_zone
</match>
```

As you can see, the source configuration is set to monitor the file */var/log/myapp.log* for new messages. A filter condition has been applied to accept only messages with the term *error* in them. The <match> block defines the output destination (or "sink"). This configuration specifies Google Cloud Logging as the sink and enriches the log entries with custom labels.

If you want to disable all logs from Dataproc from going to Cloud Logging, then set the property `dataproc.logging.stackdriver.enable` as false. Here is the sample gcloud command that you can use to create a cluster and disable all logs from being sent to Cloud Logging:

```
gcloud dataproc clusters create log-test2 --enable-component-gateway \
  --region us-central1 --no-address --master-machine-type n2-standard-4 \
  --master-boot-disk-type pd-balanced --master-boot-disk-size 500 \
  --num-workers 2 --worker-machine-type n2-standard-4 --worker-boot-disk-type \
  pd-balanced --worker-boot-disk-size 500 --image-version 2.2-debian12 \
  --properties dataproc:dataproc.logging.stackdriver.enable=false \
  --project <replace_project_here>
```

Disabling logs from going to Cloud Logging will disable all logs from the Dataproc cluster except audit logs (data access and activity).

8.5 Attaching Custom Labels to Logging

Problem

You want to write a log message from a Python application with a custom label attached to the log message.

Solution

When you create your log entry, populate the labels option with the required values.

Discussion

You can filter log messages using various attributes, as covered in Recipe 8.3. Filtering log messages by labels will improve search performance.

Here is some sample Python code that will write log messages to Cloud Logging with a custom label attached to them:

```python
import google.cloud.logging
import logging

def write_to_cloud_logging(
    message,
    log_name='my_python_log',
    severity='INFO',
    **labels
):
    """Writes a log message to Google Cloud Logging with custom labels.

    Args:
        message: The log message text.
        log_name (optional): The name of the Cloud Logging log to write to.
                             Defaults to 'my_python_log'.
        severity (optional): The severity level of the log message. Defaults to
                             'INFO'. Possible values: 'DEBUG', 'INFO', 'WARNING',
                             'ERROR', 'CRITICAL'.
        **labels: Additional keyword arguments representing custom labels
                  (key-value pairs).
    """

    # Initialize the Cloud Logging client
    logging_client = google.cloud.logging.Client(project=projectID)
    logger = logging_client.logger(log_name)

    # Create a LogEntry object
    log_entry = google.cloud.logging.entries.LogEntry(
        payload=message,
        severity=severity,
        logger=logger,
```

```
        labels=labels
    )

    # Write the log entry to Cloud Logging
    logger.log_struct(log_entry.to_api_repr())

# Example usage
write_to_cloud_logging(
    "Hello from Cloud Logging!",
    severity='INFO',
    environment='production',
    user='admin'
)

write_to_cloud_logging(
    "This is a warning message.",
    severity='WARNING',
    component='database',
    error_code='404'
)
```

> You will need to install the google-cloud-logging Python package
> to use this code.

The write_to_cloud_logging function creates a log entry with a message, severity, and optional labels, then sends it to the specified Cloud Logging log. It initializes a Cloud Logging client and formats the log entry for submission. The following two example calls demonstrate its usage with varying severity and metadata:

Informational message

write_to_cloud_logging("Hello from Cloud Logging!", severity='INFO', environment='production', user='admin') sends an info-level message, tagging it with environment and user details.

Warning message

write_to_cloud_logging("This is a warning message.", severity='WAR NING', component='database', error_code='404') sends a warning-level message, providing context about a database component and a specific error code.

To run this Python code, you'll need to save it to a *.py* file (e.g., *logging_example.py*) and then execute it from your terminal using the Python interpreter. Here's the general command:

```
python logging_example.py
```

After successful execution, you can search for these log entries in the Google Cloud Logging UI. You can search by the log message itself ("Hello from Cloud Logging!" or "This is a warning message.") or, more specifically, by the custom labels you've added. For example, you can filter logs by `environment:production`, `user:admin` or `error_code:404` for more precise log analysis and troubleshooting.

8.6 Optimizing Cloud Logging Costs

Problem

You would like to optimize Cloud Logging for performance and cost.

Solution

Cloud Logging is vital for monitoring and troubleshooting, but unmanaged logging can lead to surprising costs. Effectively managing these costs requires a combination of up-front configuration, ongoing monitoring, and regular optimization.

Discussion

There are two types of pricing with logging:

Log streaming
    This one-time cost involves sending logs to Cloud Logging for storing, indexing, and making them ready for query.

Log retention
    This is the cost of retaining logs until their maximum life, and it occurs once ingestion is done.

In Cloud Logging, log streaming incurs higher costs than basic log retention because of the real-time processing and indexing required for immediate log availability and search.

> Be aware that any prices mentioned within this recipe are provided for illustrative purposes only. The actual pricing of GCP services can vary depending on the specific region you select for your resources. Furthermore, GCP service pricing is subject to change over time. For the most up-to-date and accurate pricing information, refer directly to the official Google Cloud pricing documentation (*https://cloud.google.com/pricing*). It is always recommended to consult the official documentation before making any decisions based on cost.

For instance, if you want to store 100 GB of logs in Cloud Logging, here is what the pricing may look like:

- Log streaming cost: $0.5 per GB (with the first 30 days of storage cost included)
- Log retention cost: $0.01 per GB per month

An example of costs over a three-month period using these variables is given in Table 8-2.

Table 8-2. Monthly log-retention costs in Cloud Logging

Month 1	100 (GB) × $0.5 (per GB cost) = $50
Month 2	100 (GB) × 0.01 (per GB cost) = $1
Month 3	100 (GB) × 0.01 (per GB cost) = $1
Total cost for three months	**$50 + $1 + $1 = $52**

At the time of writing, there is no cost involved for querying logs.

On the other hand, if you are storing logs in a GCS bucket instead of Cloud Logging, here is what pricing may look like:

- Standard storage pricing: $0.02 per GB per month

To figure out the cost of storing 100 GB of logs in GCS, multiply 100 by $0.02, which equals $2 per month. A month-by-month breakdown is shown in Table 8-3.

Table 8-3. Monthly log-retention costs in GCS

Month 1	100 (GB) × $0.02 (per GB cost) = $2
Month 2	100 (GB) × 0.02 (per GB cost) = $2
Month 3	100 (GB) × 0.02 (per GB cost) = $2
Total cost for three months	**$2 + $2 + $2 = $6**

GCS involves other types of costs for creating, listing, and deleting files. These are categorized as Class A and Class B types of operations. However, Cloud Logging currently allows you to query stored data at no additional cost.

Some of the common reasons why logging costs increase are:

- Leaving the default settings when running a large number of applications
- Making underlying changes to the Dataproc image or frameworks (Spark, YARN, HDFS, etc.)
- Enabling debug and trace-level logging
- The nature of the application logic (for instance, job handling with a large number of small files can generate millions of logs that are sent to Cloud Logging)

While GCS offers a more cost-effective solution for log retention, Cloud Logging provides valuable features like optimized string- and tag-based search, which can significantly expedite troubleshooting compared to the potentially slower and more costly retrieval from Cloud Storage files. Therefore, your choice depends on your specific use case. If frequent searches for specific strings or tags are crucial for rapid resolution of issues, Cloud Logging is ideal. However, if your primary need is infrequent viewing of entire log files for archival purposes, Cloud Storage offers a more economical option. For example, a cluster executing Spark jobs can utilize a persistent history server to store all application logs in GCS rather than storing them in Cloud Logging.

8.7 Sinking Logs to BigQuery

Problem

You want to perform analysis on log metrics generated by a Dataproc application.

Solution

BigQuery is a potent data warehouse used for storing and analyzing large volumes of data, including log data. It can handle the high velocity and volume of log data while ensuring fast, scalable querying. Log sinks are used to specify the destinations for log messages, and filters can be applied to determine which logs are sent to the sink. BigQuery offers versatile query operators for analyzing log data, enabling you to identify trends, monitor performance, and troubleshoot. Its capabilities empower organizations to extract valuable insights and enhance the performance of their applications. The relationship between these components is shown in Figure 8-9.

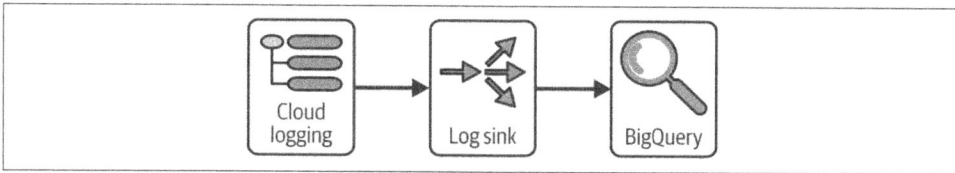

Figure 8-9. Log routing from Cloud Logging to BigQuery via a log sink

Discussion

Log sinks enable you to export log messages beyond their default storage. These sinks act as channels, channeling your log data to various destinations like BigQuery, GCS, or Pub/Sub. This centralization opens up a world of possibilities, enabling you to perform in-depth analysis, set up comprehensive monitoring systems, and gain deeper insights from your logs.

Before you can establish a log sink, you need to lay the groundwork by creating a suitable destination. This destination could be a Cloud Storage bucket for long-term archiving, a BigQuery table for powerful querying and analysis, a Pub/Sub topic for real-time processing, or even a logging bucket in another project for centralized management. Each destination caters to different needs, so choose wisely based on your requirements.

With your destination ready, you can proceed to create the log sink itself. During this process, you'll have the option to define a filter that acts as a gatekeeper, ensuring that only relevant logs are exported. This filtering capability adds a layer of precision to your logging strategy. You can also set a retention period for the logs at the destination, striking a balance between data availability and storage costs.

Some of the benefits of using log sinks include:

- Centralizing your logs in a single location
- Performing analysis and monitoring on your logs
- Complying with regulatory requirements
- Improving the security of your logs

Here are some examples of how you can use log sinks:

- Export logs to BigQuery to perform analysis on them. For example, you can use BigQuery to identify trends in your logs or to monitor the performance of your applications.
- Export logs to GCS to archive them. This can be useful for compliance purposes or for long-term storage of logs.

- Export logs to Pub/Sub to trigger events. For example, you can use Pub/Sub to trigger a notification when a critical error is logged.

To configure log routing to BigQuery, first create a BigQuery dataset from the gcloud command (skip this step if you already have the BigQuery dataset you want to use):

```
bq mk --dataset <PROJECT_ID_HERE>:<Name_OF_Log_SINK_Dataset_Here>
```

Create a log sink to route all error messages from the Dataproc service to the BigQuery dataset:

```
gcloud logging sinks create errors-to-bigquery-table \
  bigquery.googleapis.com/projects/<project_id>/datasets/log_sink_ds/tables/
error_logs \
  --log-filter='severity=ERROR'
```

Behind the scenes, the log sink uses a service account to write logs to a BigQuery table. Therefore, this service account must be granted the necessary write permissions to the BigQuery dataset you created. You can identify the default service account that the log sink uses with the following command:

```
gcloud logging sinks describe <name_of_log_sink_here>
```

From the command output, look for the value of the field `writerIdentity`. Alternatively, you can configure the log sink to use a custom service account.

Now, let's grant the service account the BigQuery editor role. While this grants editor permissions at the project level, best practice dictates granting the editor role only to the specific BigQuery dataset where logs will be written:

```
gcloud projects add-iam-policy-binding <project_id> \
  --member=serviceAccount:service-644926244067@gcp-sa-logging.iam.
gserviceaccount.com \
  --role=roles/bigquery.dataEditor
```

After successfully creating the log sink and granting the service account write permissions to the BigQuery table, log messages matching the configured filter (`severity=ERROR`) will flow to the BigQuery table as soon as they are generated. Refer to Recipe 8.5 for generating new log messages to Cloud Logging.

## CHAPTER 9

Setting Up Monitoring and Dashboards

In this chapter, we will explore Dataproc's built-in metrics, which offer crucial information about cluster status, resource consumption, and job performance. We will share techniques for using Metrics Explorer to create custom visualizations and dashboards that enable you to analyze and interpret these metrics effectively. We will also cover the predefined metrics charts available within Dataproc, providing a quick but comprehensive overview of cluster performance.

We will outline the process of setting up alerts based on key metrics thresholds to ensure proactive problem solving, which will help you detect and address issues before they affect operations. For those who are working across multiple projects, this chapter includes steps for migrating dashboards from one project to another, facilitating efficient collaboration and resource sharing.

We will also discuss creating custom log-based metrics, which enables you to develop specific metrics tailored to your unique workloads and monitoring needs. This approach offers deeper insights into system behavior and logs, providing greater control over performance optimization and troubleshooting.

By the end of this chapter, you will have developed a thorough understanding of Dataproc's monitoring capabilities and will know how to optimize performance, maintain cluster health, and address operational challenges with confidence.

9.1 Monitoring Cluster Status

Problem

You want to monitor the health of a Dataproc cluster. What key metrics should you track to ensure optimal performance, availability, and early issue detection?

Solution

Think of the cluster health metric as a quick checkup for your Dataproc cluster. It confirms that your cluster is up and running and that Dataproc hasn't found any major problems. However, the definition of *cluster health* can be subjective and depends on your application's specific needs. Therefore, a more comprehensive approach to monitoring is often required that includes the following:

Comprehensive service monitoring
This involves monitoring the underlying services of your Dataproc cluster, such as HDFS, YARN, and Hive Metastore.

Application-centric monitoring
This approach uses your application or job running on the cluster as the primary indicator of cluster health. By monitoring the performance and behavior of your application, you can indirectly infer the health of the underlying cluster.

Discussion

The cluster status reported on the cluster's home page provides a useful high-level view of your Dataproc cluster's status but shouldn't be relied upon as the sole indicator of overall health. While it can signal successful cluster creation and basic functionality, it might mask underlying issues affecting individual components.

For instance, a cluster may appear "healthy" despite experiencing problems with a critical service like the HDFS NameNode, which could significantly affect your application's performance. Hitting resource quotas might also prevent cluster scaling even if the overall health status looks fine.

Effective monitoring involves a deeper, more granular approach. This requires understanding your application's dependencies and actively monitoring the health of the services those components rely on. For example, if your application relies on Spark (e.g., Spark jobs run on top of the YARN framework), you should track metrics related to the YARN service.

You can build your monitoring using two approaches: comprehensive service monitoring and application-centric monitoring.

Comprehensive service monitoring

This method involves systematically tracking the health and performance of all individual services within your Dataproc cluster. For example, you'd monitor:

- Core Hadoop services HDFS (NameNode, DataNode) and YARN (Resource-Manager, NodeManagers)
- Optional components such as Hive Metastore and ZooKeeper
- Dataproc-specific services such as cluster creation and autoscaling

The pros of this method are:

- It provides in-depth visibility: a detailed picture of every service's behavior.
- You can identify problems in individual components before they significantly affect the entire cluster or applications.
- It helps pinpoint the sources of issues for targeted troubleshooting.

Some cons include:

- It requires setting up and maintaining monitoring for numerous services, which can be a complex process.
- It can generate many alerts, requiring careful filtering and prioritization.

Application-centric monitoring

This approach focuses on observing the behavior of your applications running within the cluster. For instance, you'd monitor:

Job status
    Success and failure rates, duration, and resource utilization

Long-running jobs
    Detection of jobs exceeding expected runtimes

Application-specific metrics
    Custom metrics related to your application's performance or business logic

Some examples of this type of monitoring include:

- Tracking Spark job completion times, stage durations, and executor resource usage
- Monitoring query execution times, success rates, and resource consumption

The pros of this method are:

- It focuses on the most critical aspect: your application's health.
- It signals problems directly affecting your workload, providing actionable insights.
- It is easier to implement than comprehensive service monitoring.

Some cons include:

- It may not detect issues in underlying services that haven't yet affected your application.
- You may need to add custom metrics or logging to your applications.

Both approaches have their merits. Ideally, a combination of both provides the most comprehensive monitoring strategy. The following are some recommendations for building your monitoring strategy:

- Start with application-centric monitoring to ensure that your workloads are running smoothly.
- Supplement with key service monitoring to catch underlying issues before they escalate.
- Leverage monitoring tools and alerts to proactively address problems and maintain a healthy Dataproc cluster.

9.2 Exploring Predefined Metrics Charts

Problem

You have just created a new Dataproc cluster and want to monitor its performance and health. What predefined metrics charts are readily available in the Dataproc service to help you with this task?

Solution

Once you create the cluster, the following metrics charts are available predefined:

- YARN memory
- YARN pending memory
- YARN NodeManagers
- HDFS capacity
- CPU utilization
- Network bytes
- Network packets
- Disk bytes
- Disk operations

There may be changes to the predefined metrics charts in future releases.

Discussion

When you create a cluster while it is active and running, you can access some important metrics using predefined charts. To view these charts, navigate to the Dataproc service screen and click on the specific cluster you want to monitor. The cluster home screen will have multiple tabs such as Monitoring, Jobs, VM Instances, Configuration, and Web Interfaces. Click the Monitoring tab to view the predefined charts, as shown in Figure 9-1.

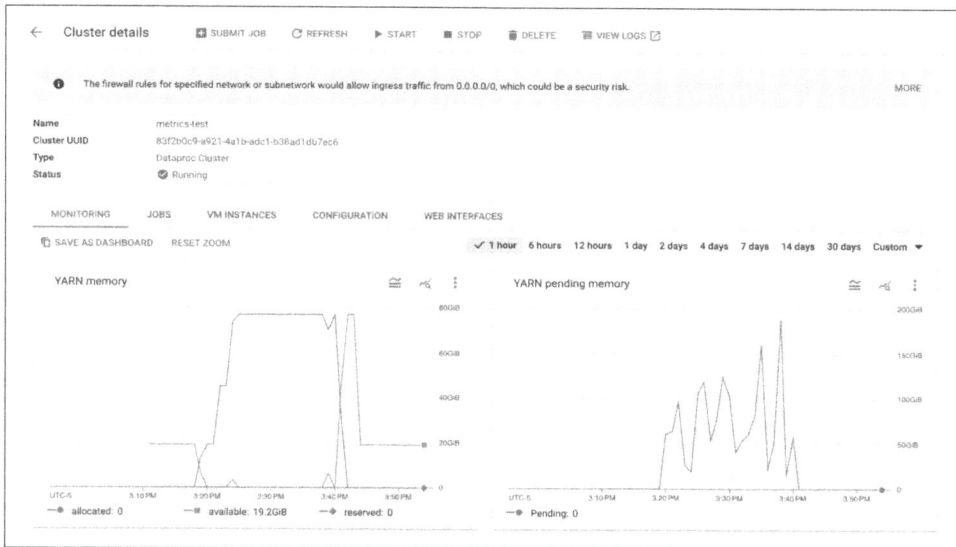

Figure 9-1. Viewing metrics from the cluster's home page

This image displays the metric's value over the last hour. To customize the time range, you have two options:

Select a custom time range
    Click Custom and specify the desired start and end times, as shown in Figure 9-2.

Zoom in on a specific period
    Click and drag within the chart to select the desired time window, as shown in Figures 9-3 and 9-4.

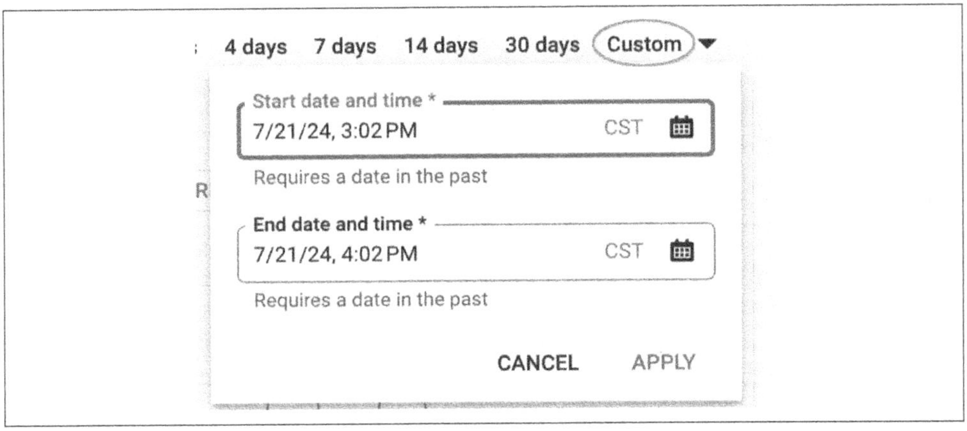

Figure 9-2. Choosing a specific start and end time from the Custom option

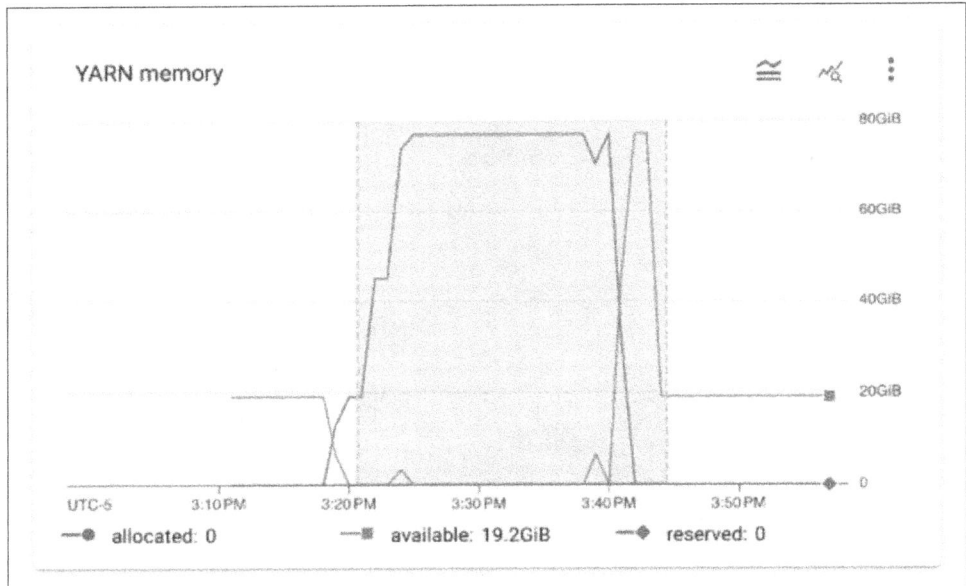

Figure 9-3. Selecting a specific area in a chart to zoom in on

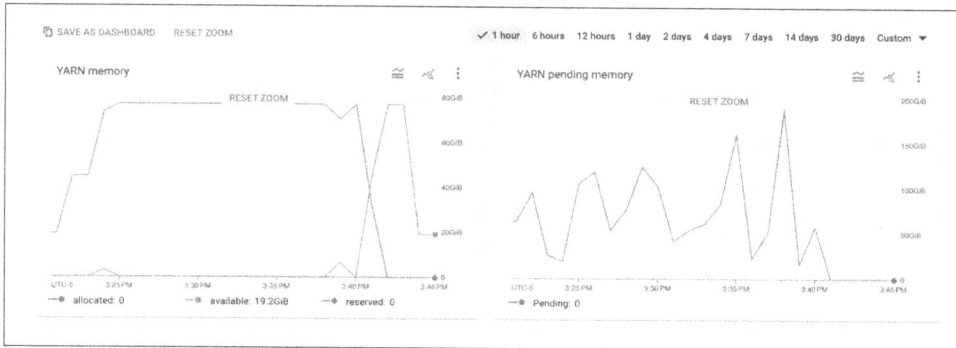

Figure 9-4. All charts in the view zoomed to the selected time-window range

These charts show the values for a selected time range. For instance, you can choose to see the metric's values for the last hour. Let's understand the significance of each chart metric:

YARN memory

This chart displays the YARN memory currently allocated, available, and pending in the Dataproc cluster. This memory information is aggregated across the entire cluster. Several scenarios can affect these values during a cluster's lifetime, including adding or removing nodes (through manual updates or autoscaling) and nodes becoming unhealthy.

YARN pending memory

This metric shows how much additional memory is needed by currently running applications in the cluster. If autoscaling is not configured, applications will receive memory allocations only after currently running tasks complete and release their resources.

YARN NodeManagers

This chart displays the worker nodes in your cluster categorized by their status, such as active, unhealthy, or shutdown. YARN NodeManagers are responsible for executing the individual tasks that make up your applications. They manage resources, monitor task progress, and ensure that your applications run smoothly across the cluster.

HDFS capacity

This metric tracks the overall storage capacity of your HDFS. It shows how much disk space is available across your cluster for storing data used by your applications. Monitoring this metric is essential to ensure that you have sufficient storage for your processing needs and to prevent your cluster from running out of space.

CPU utilization

This metric displays the aggregate CPU utilization across all nodes in your Dataproc cluster. It provides a valuable insight into the overall processing load on your cluster. High CPU utilization suggests that your running applications or jobs are CPU intensive and might benefit from optimization or additional resources. Conversely, low CPU utilization could indicate underutilization of your cluster's processing capabilities. By monitoring this metric, you can ensure that your cluster is adequately sized for your workload and identify potential performance bottlenecks.

Network bytes

This metric represents the total amount of data transferred over the network across all the nodes in your cluster. It includes both incoming and outgoing network traffic. By monitoring network bytes, you can gain insights into the network activity of your applications and identify potential bottlenecks or unexpected spikes in data transfer. This information can be crucial for optimizing network performance and troubleshooting network-related issues in your Dataproc cluster. For example, a sudden increase in network bytes might indicate a change in application behavior or a potential network issue that needs investigation.

Network packets

This metric tracks the number of packets transmitted to and from your Dataproc cluster across all its nodes. Network packets are the fundamental units of data transmitted over a network. Monitoring this metric can help you understand the volume of network communication within your cluster and identify potential issues, such as network congestion or packet loss. For example, a sudden increase in network packets with a corresponding drop in network bytes might suggest an issue where data is being fragmented into an excessive number of small packets, leading to inefficiency.

Disk bytes

This metric measures the aggregate amount of data read from and written to the disks of all machines in your Dataproc cluster. It's a key indicator of disk I/O activity within your cluster. Higher values for disk bytes indicate that your applications are performing a significant number of disk read and write operations. While some disk I/O is expected, excessively high values might point to potential performance bottlenecks. In such cases, optimizing data-access patterns, using more efficient storage formats, or increasing disk capacity could improve overall performance.

Disk operation

This metric monitors the number of disk operations performed across all nodes in your Dataproc cluster. It encompasses various operations, including reading data, writing data, deleting files, creating directories, and listing directory contents. A high number of disk operations can indicate that your applications are heavily interacting with the disk subsystem, which could lead to performance bottlenecks if the disk I/O capacity is exceeded. Analyzing this metric in conjunction with disk bytes can provide a more complete picture of disk usage and help identify potential areas for optimization. For example, if you observe a high number of disk operations but relatively low disk bytes, that might suggest that your applications are performing many small reads and writes, which can be less efficient than fewer, larger operations.

Although many additional metrics are available to capture and monitor, the ones we have listed here provide a solid foundation for monitoring your Dataproc cluster.

9.3 Creating Charts Using Metrics Explorer

Problem

You want to create custom charts to visualize specific Dataproc cluster metrics.

Solution

Metrics Explorer provides a user-friendly, visual interface for creating custom charts. You can select the desired metrics, apply filters to refine the data, and choose from various aggregation methods like average, sum, or maximum. This allows you to tailor the charts to your specific monitoring needs and gain a clearer understanding of your cluster's performance.

Discussion

Metrics Explorer in GCP is a powerful tool that enables you to visualize and analyze the performance of your cloud resources. It provides a graphical interface where you can select metrics from various GCP services, apply filters, and create charts and dashboards to gain insights into your applications and infrastructure. You can use Metrics Explorer to identify trends, troubleshoot issues, and optimize the performance and cost of your cloud deployments.

The following predefined IAM roles grant access to building charts in the Google Cloud Monitoring service:

Monitoring viewer (`roles/monitoring.viewer`)
This role grants read-only access to monitoring data and dashboards, which includes the ability to view existing charts.

Monitoring editor (`roles/monitoring.editor`)
This role provides broader access, allowing you to create, modify, and delete dashboards and charts as well as read monitoring data.

To create a metric chart, search for Metrics Explorer and select the service, as shown in Figure 9-5.

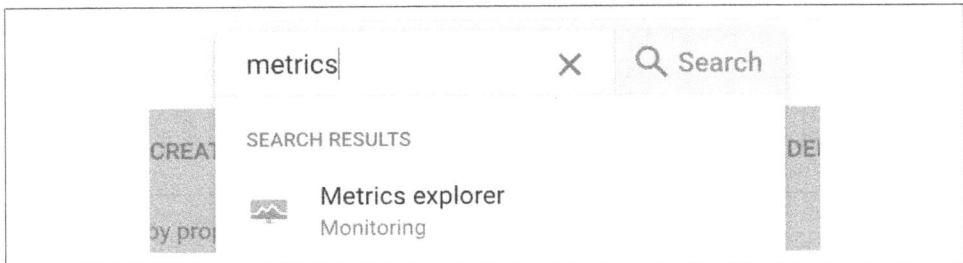

Figure 9-5. Searching for Metrics Explorer in the Monitoring service

This will take you to the home screen of Metrics Explorer with the option to add queries and build charts, as shown in Figure 9-6.

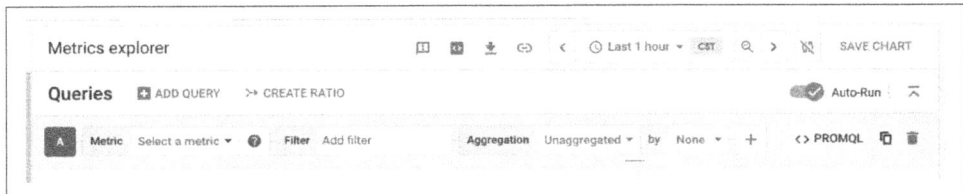

Figure 9-6. Home screen of Metrics Explorer

Metrics Explorer uses specific terminology. Here are some key terms to understand:

Query
A query specifies the metric you want to visualize. It can include filters to refine the data and aggregation logic to group and summarize it over a specific time frame.

Metric
This represents a specific data point related to your cloud services, measured over time. Examples include CPU utilization, network traffic, and disk I/O.

Filter
Filters allow you to refine the metric's data by applying conditions. For example, you could filter by resource type, location, or specific labels.

Aggregation
Aggregation methods group a metric's values to provide a summarized view. Common aggregations include average, sum, minimum, and maximum.

CGP captures metrics from various services by default, which are grouped by the service name. When you select a specific metric, it is easy to navigate to the group first and then choose the specific metric you want to visualize.

In Figure 9-7, you can see that a primary group of metrics for a Cloud Dataproc cluster has two subgroups: cluster and logs-based metrics.

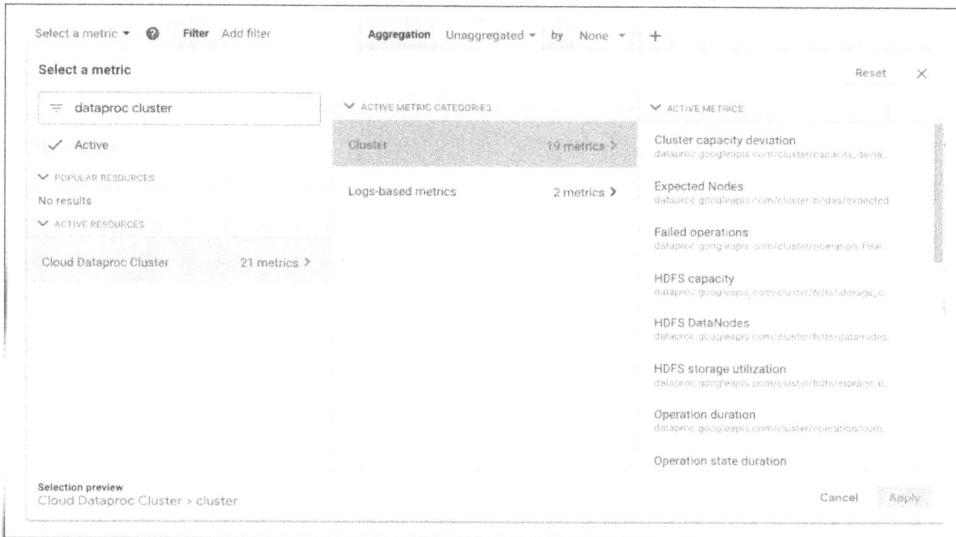

Figure 9-7. Viewing active metrics under Cluster category

Cluster metrics are predefined by Google Cloud Monitoring and provide insights into the performance and resource utilization of your Dataproc clusters. Some key examples include:

CPU utilization
Monitors the percentage of CPU resources used by the cluster

Memory utilization
Tracks how much memory the cluster is consuming

HDFS operations
Monitors filesystem-related activities like bytes read and written

YARN metrics
Includes metrics related to job scheduling, such as the number of running containers, memory allocated to jobs, and job failure counts

Logs-based metrics are derived from log data generated by Dataproc services and applications running on the cluster (e.g., Hadoop, Spark). Out of the box, the standard logs-based metrics are:

Log bytes
> The total number of bytes written to logs for the cluster

Log entries
> The number of log entries recorded by the system

Select any of the metrics as shown in Figure 9-8 to view their data in the form of a chart.

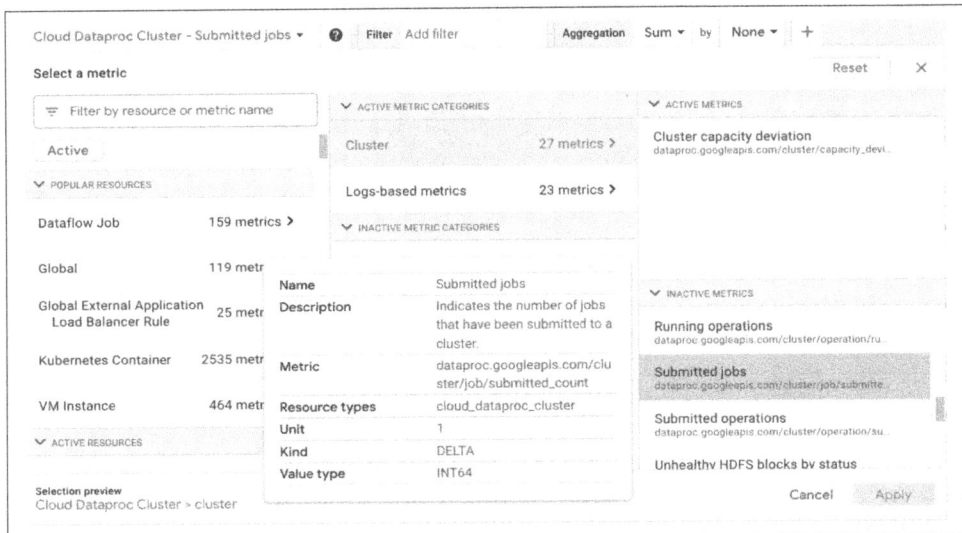

Figure 9-8. Selecting a metric to view its data in chart form

Add a filter condition to filter on, as shown in Figure 9-9.

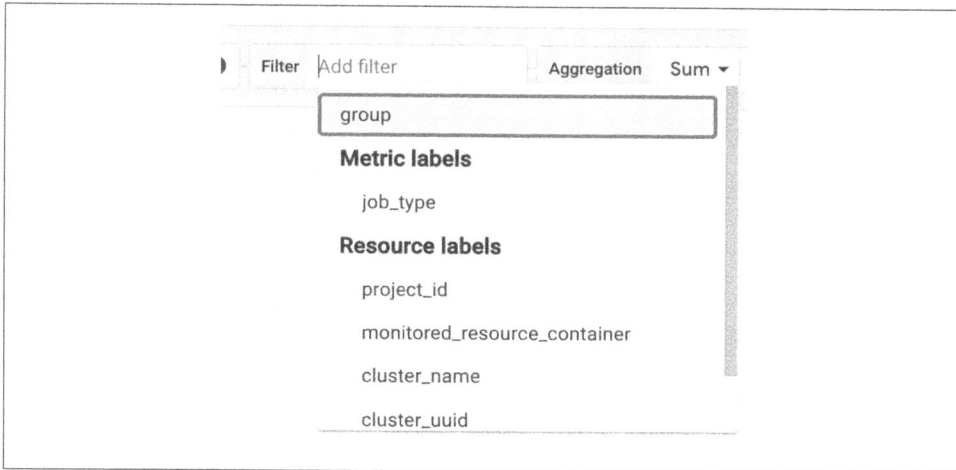

Figure 9-9. Applying a filter to a metric's chart

Your Dataproc cluster is constantly generating data about its performance, such as CPU usage, memory consumption, and disk activity. This data is collected as individual data points at very frequent intervals. When you look at a chart in Metrics Explorer, you're usually not interested in seeing every single data point. Instead, you want to see trends and patterns over a longer period, such as an hour, a day, or a week.

That's where aggregation comes in (see Figure 9-10). Aggregation takes all those individual data points within your chosen time range and combines them into a single representative value for each time interval on the chart. For example, if your chart shows hourly data, aggregation might calculate the average of all the data points collected within each hour. Or it could show the minimum or maximum value within that hour. The type of aggregation you choose depends on what you want to see in your chart. If you're interested in overall trends, an average might be useful. If you want to see peak usage, you might choose maximum.

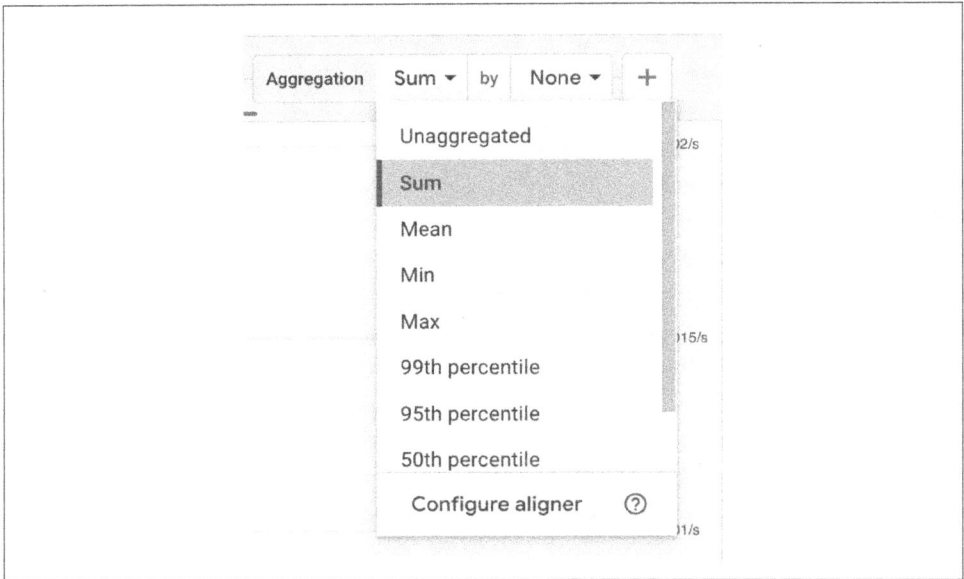

Figure 9-10. Applying aggregation to the metric's data

Figure 9-11 shows a custom metric chart created in Dataproc's Metrics Explorer. This chart specifically tracks the number of running Hadoop jobs within your cluster.

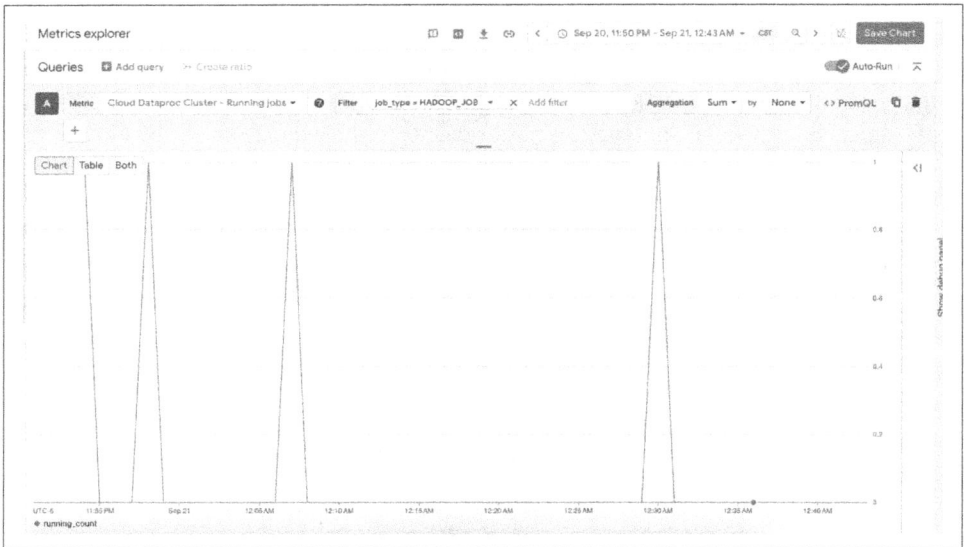

Figure 9-11. Metrics Explorer chart showing a running jobs count

To achieve this, the following configurations were applied:

Metric

The "Running jobs" metric was selected, which likely provides a count of active jobs.

Filter

A filter was applied to include only jobs of the type HADOOP_JOB, ensuring that the chart focuses specifically on Hadoop jobs and excludes other job types (e.g., Spark, Hive).

Aggregation

The Sum aggregation method was used. This means that for each time interval displayed on the chart (e.g., every hour), the chart shows the total number of Hadoop jobs that were running during that period.

9.4 Creating Dashboards Using Metrics Explorer

Problem

You find it challenging to keep track of the health and performance of each cluster by individually examining various metrics and logs. You need a centralized view that provides a comprehensive overview of your clusters' key performance indicators (KPIs) and allows you to quickly identify potential issues. How can you create dashboards in Dataproc to effectively monitor your clusters, gain insights into their performance, and proactively address potential problems?

Solution

Google Cloud Monitoring dashboards provide a customizable visual interface to display metrics, logs, and other KPIs of your cloud resources, allowing you to gain insights into their health and performance at a glance.

Discussion

Google Cloud Monitoring's dashboards offer a centralized, customizable view of your cloud resources, allowing you to efficiently filter and analyze data across multiple charts simultaneously. This streamlines monitoring, troubleshooting, and decision making, empowering you to proactively manage the health and performance of your cloud environment.

The following are some benefits of working with Monitoring dashboards:

Custom view

You can group charts based on specific criteria, such as data source, metric type, or project; create multiple dashboards, each with a focused set of charts for a specific purpose; and easily navigate between dashboards to quickly access the information you need.

Collaboration

You can share dashboards with other users, such as team members or stakeholders; specify permissions for each user, such as view only or edit access; and collaborate on dashboards in real time, making it easy to discuss insights and make decisions.

Real-time view

Dashboards are automatically refreshed with the latest metrics information. You can see the most up-to-date data at a glance, without having to manually refresh the page, and quickly identify and respond to changes in your environment.

Centralization

You can monitor multiple services from a single location, get a comprehensive overview of your entire infrastructure, and easily compare data across services to identify trends and patterns.

Easy management

You can apply filters and time-series conditions at the dashboard level. These conditions are automatically populated to all underlying charts. You can also easily manage and maintain dashboards without having to manually configure each chart.

Easy import and export

Dashboards can be exported as JSON format files. Having dashboards in JSON format makes it easy to back them up and restore them when needed. It also helps with promoting dashboards from low-level environments to high-level environments.

To create a new dashboard, navigate to Monitoring in the search window, as shown in Figure 9-12.

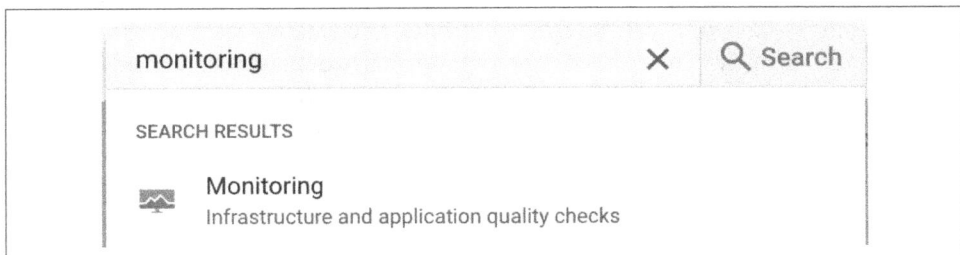

Figure 9-12. Searching for the Monitoring service

Click Dashboards, as shown in Figure 9-13.

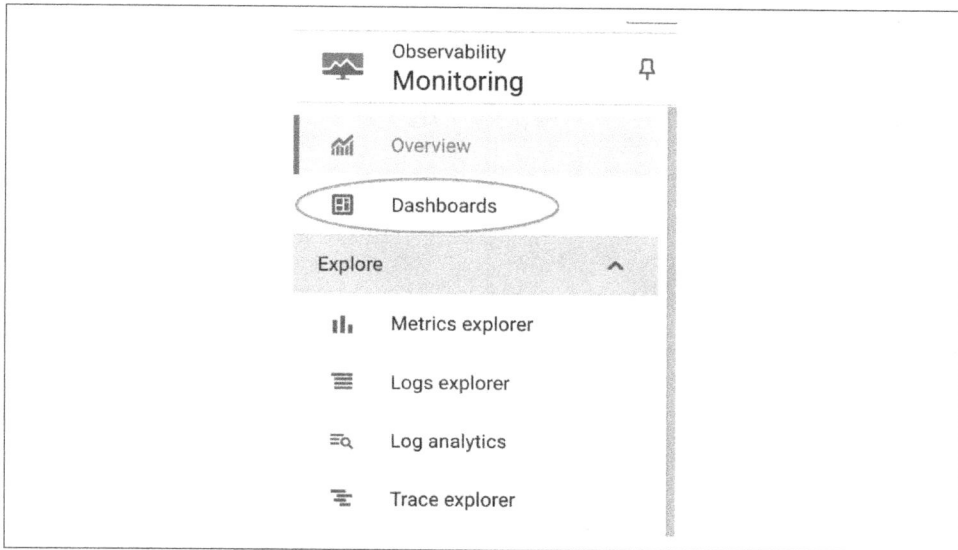

Figure 9-13. Selecting Dashboards from the Monitoring home page

Click +Create Dashboard, as shown in Figure 9-14.

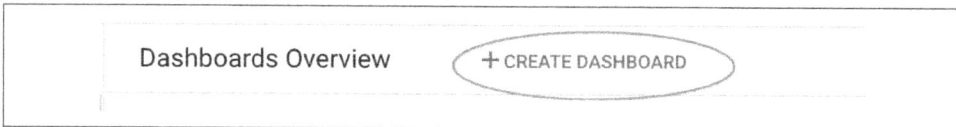

Figure 9-14. Selecting the option to create a dashboard

Once the dashboard is created, click its name to edit it as needed, as shown in Figure 9-15.

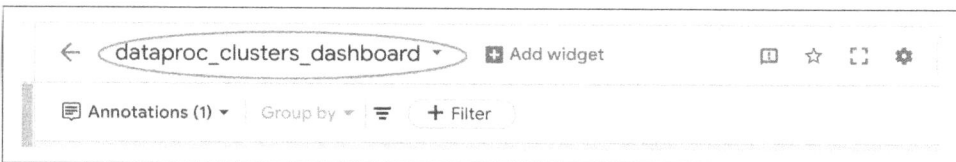

Figure 9-15. Selecting the name of a dashboard to edit it

Click +Add Widget to add new widgets (charts, tables, etc.) to the dashboard, as shown in Figure 9-16.

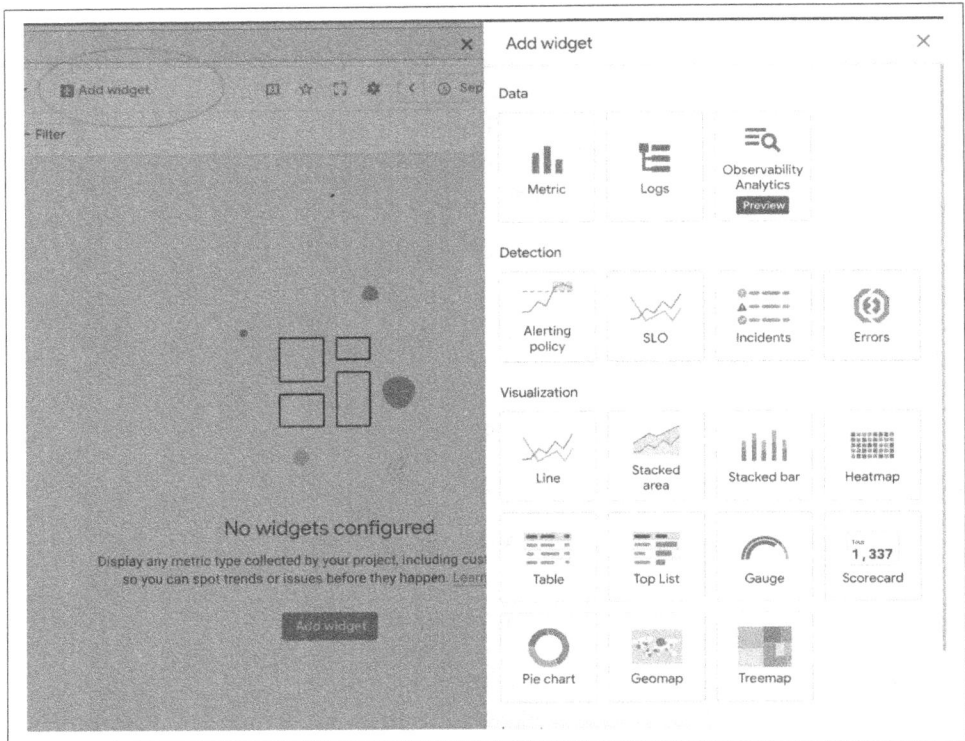

Figure 9-16. Adding a new widget to the dashboard

Select a chart widget type and follow the instructions in Recipe 9.3 to create the chart and add it to the dashboard.

Another nice feature with dashboards is you can share them with a group of users with read/edit permissions. Click the Share button shown in Figure 9-17 to share the dashboard.

Figure 9-17. The option to share a dashboard with other user groups

Add the required people's email addresses, as shown in Figure 9-18, to grant them access and notify them by email.

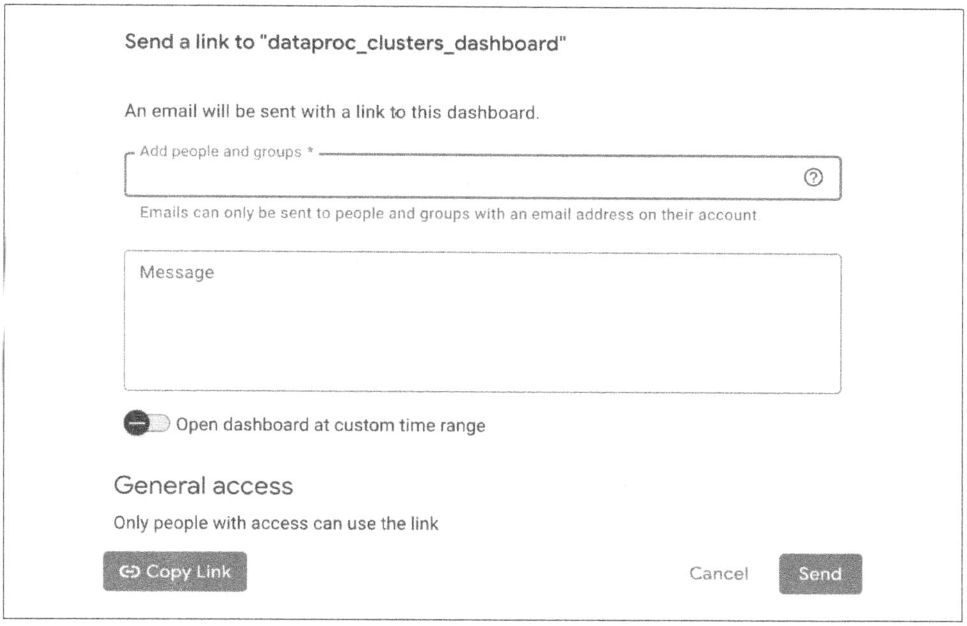

Send a link to "dataproc_clusters_dashboard"

An email will be sent with a link to this dashboard.

Add people and groups *

Emails can only be sent to people and groups with an email address on their account.

Message

Open dashboard at custom time range

General access

Only people with access can use the link

Copy Link   Cancel   Send

Figure 9-18. Populating users and groups and sending messages when sharing a dashboard

9.5 Setting Up Alerts

Problem

You want to receive alerts when a Dataproc cluster has pending YARN memory requests for more than 15 minutes. Having pending memory for a prolonged period is a sign of resource exhaustion.

Solution

To get alerted when Dataproc jobs wait for resources for more than 15 minutes, use the pending memory metric in Google Cloud Monitoring. Here are the steps:

1. Create a chart visualizing this metric and convert it into an alert with a threshold that triggers when pending memory remains above zero for 15 minutes or more.
2. Configure notifications through notification channels (email, SMS, or webhooks) to be informed proactively and ensure efficient job execution.

Discussion

To create an alert in GCP, you need to understand the following components:

Alerting policy
> This defines the conditions that trigger an alert and the notification channels used to send the alert.

Alert conditions
> These specify the criteria for triggering an alert. You can select a metric and set thresholds for it. For example, you could create an alert condition that triggers if CPU utilization exceeds 80%.

Notification channels
> These determine how you receive alerts. Common notification channels include:

> *Email*
>> Sends alerts to your email address

> *Google Chat*
>> Sends alerts to a Google Chat room

> *Pub/Sub topic*
>> Publishes alerts to a Pub/Sub topic, allowing for integration with other services

> *Third-party integrations*
>> Integrates with services like Slack or PagerDuty to receive alerts in those platforms

> *Webhook*
>> Sends an HTTP POST request to a specified URL when an alert is triggered. This allows you to integrate with custom applications or services. It is typically used for integrating with enterprise incident management systems like ServiceNow.

In this example, we are going to send an alert to email whenever there is pending memory for a duration of 15 minutes in the cluster. First, navigate to Alerting, as shown in Figure 9-19.

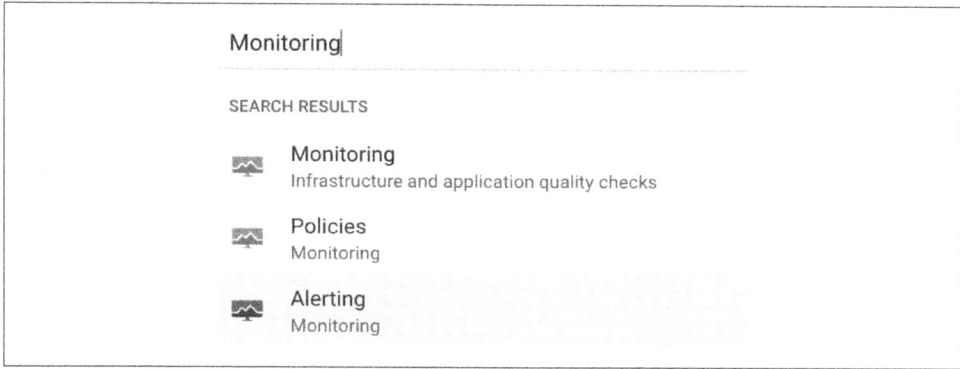

Figure 9-19. Navigating to Alerting

Click +Create Policy, as shown in Figure 9-20.

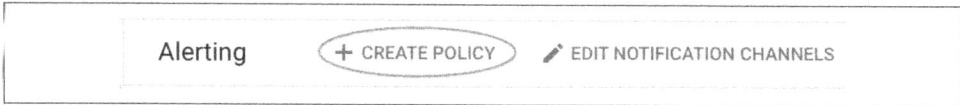

Figure 9-20. Creating a new alert policy

Select the metric, as shown in Figure 9-21.

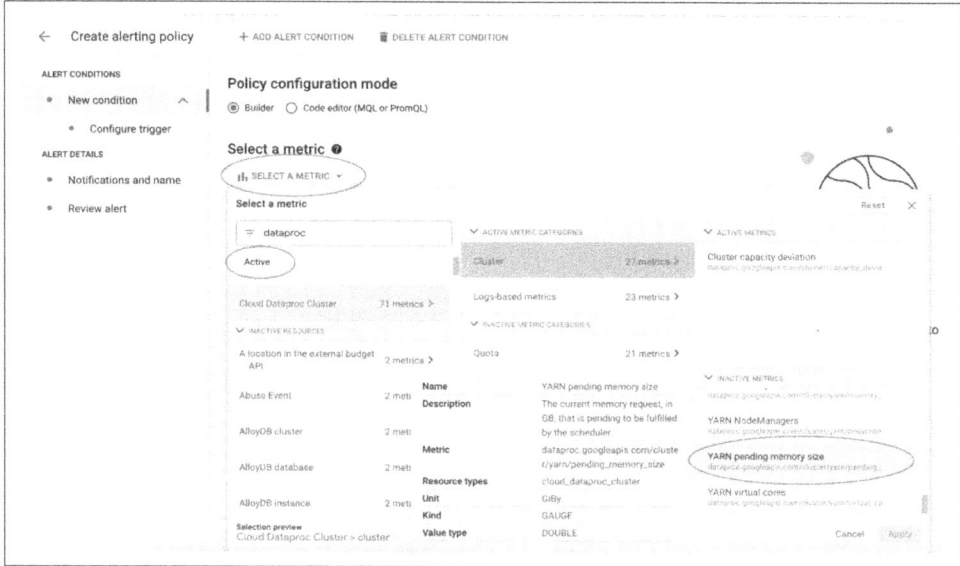

Figure 9-21. Choosing the YARN pending memory size as a metric

Alternatively, if you have an existing chart, you can convert it to an alert policy by clicking the bell icon at the top of the chart, as shown in Figure 9-22.

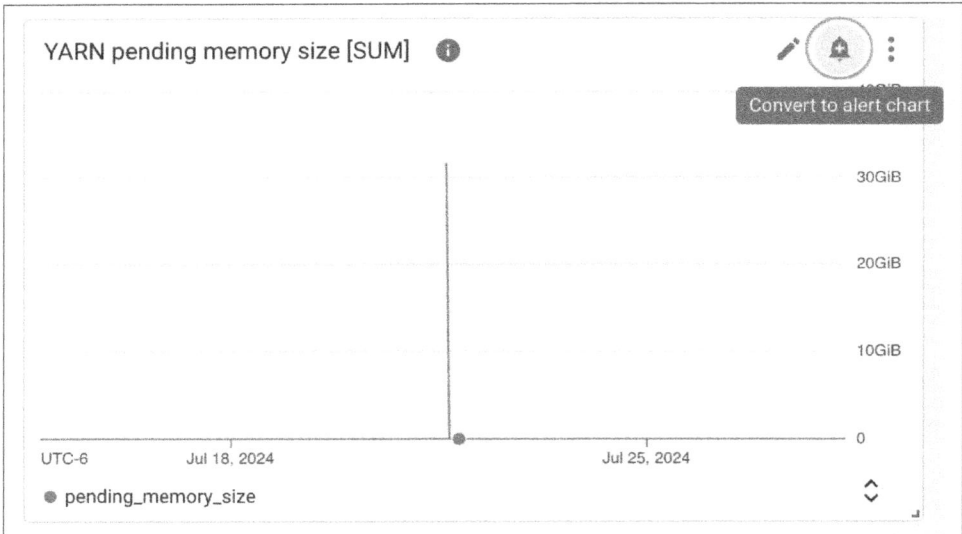

Figure 9-22. Converting an existing chart into an alert chart

Clicking "Convert to alert chart" will open the home screen of the alert chart, as shown in Figure 9-23.

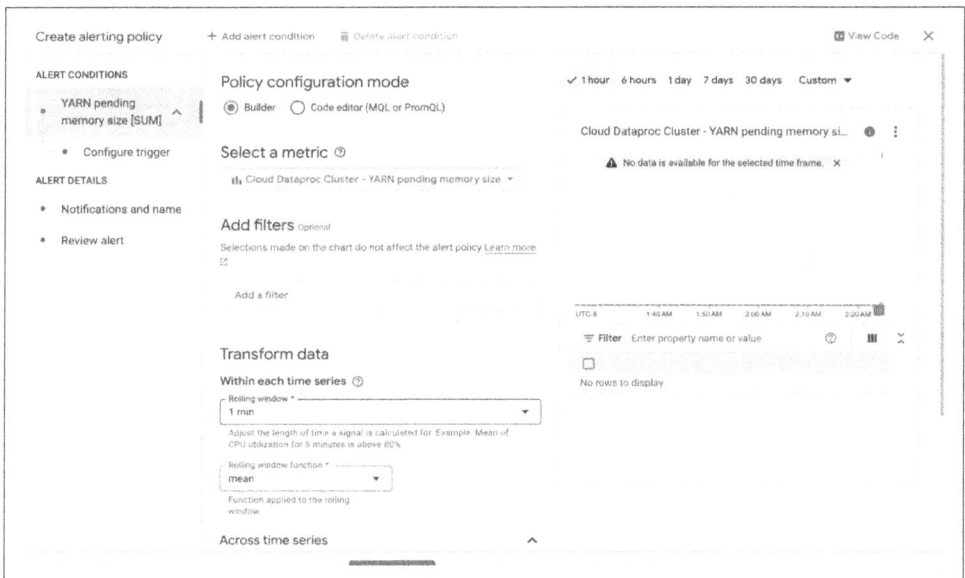

Figure 9-23. The screen for configuring an alert chart

Configure the data-transformation rules, as shown in Figure 9-24. Since we want to be notified if pending memory is present for more than 15 minutes continuously, we have selected the rolling window to be "15 min."

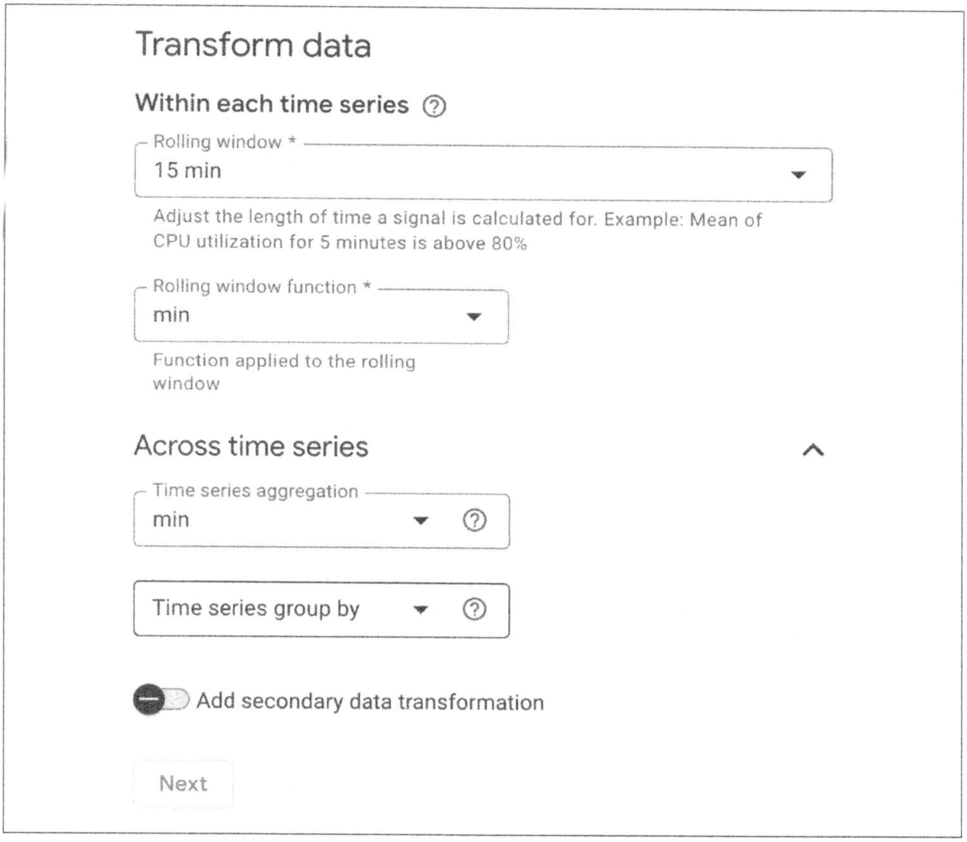

Figure 9-24. Configuring the rolling window for metric data aggregation

Configure the alert trigger to trigger the alert at any point if the threshold value is beyond 0 bytes, as shown in Figure 9-25.

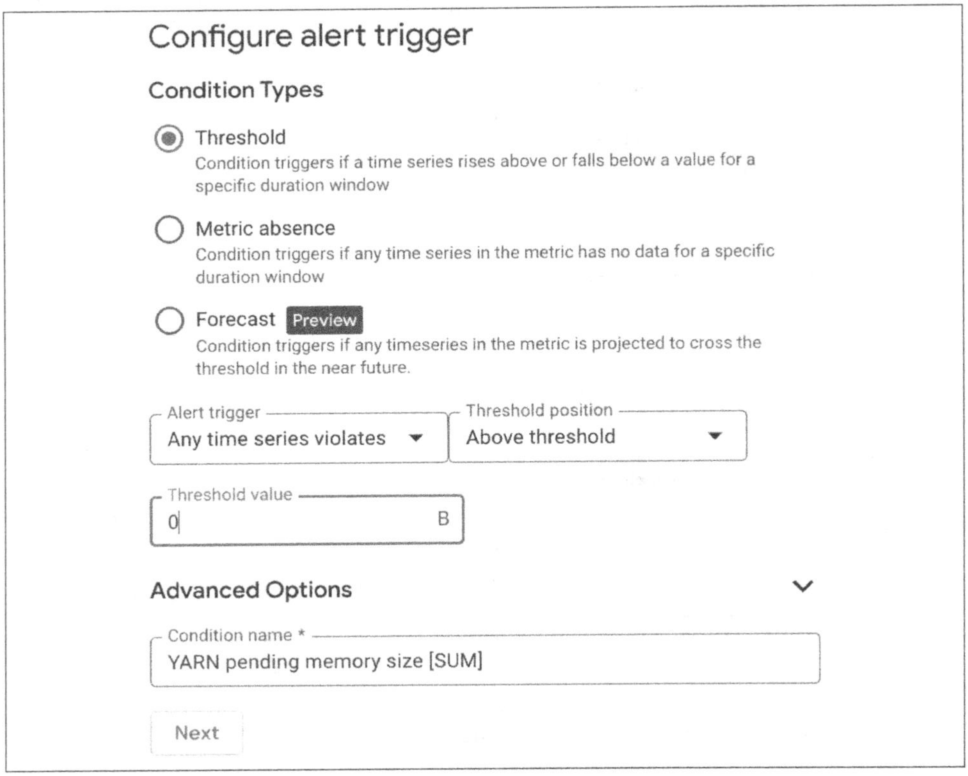

Configure alert trigger

Condition Types

- ● **Threshold**
  Condition triggers if a time series rises above or falls below a value for a specific duration window

- ○ **Metric absence**
  Condition triggers if any time series in the metric has no data for a specific duration window

- ○ **Forecast** `Preview`
  Condition triggers if any timeseries in the metric is projected to cross the threshold in the near future.

Alert trigger
Any time series violates ▼

Threshold position
Above threshold ▼

Threshold value
0   B

Advanced Options ⌄

Condition name *
YARN pending memory size [SUM]

Next

Figure 9-25. Configuring the triggering threshold values for an alert

Choose the notification channels for how to get notified when the alert is triggered, as shown in Figure 9-26.

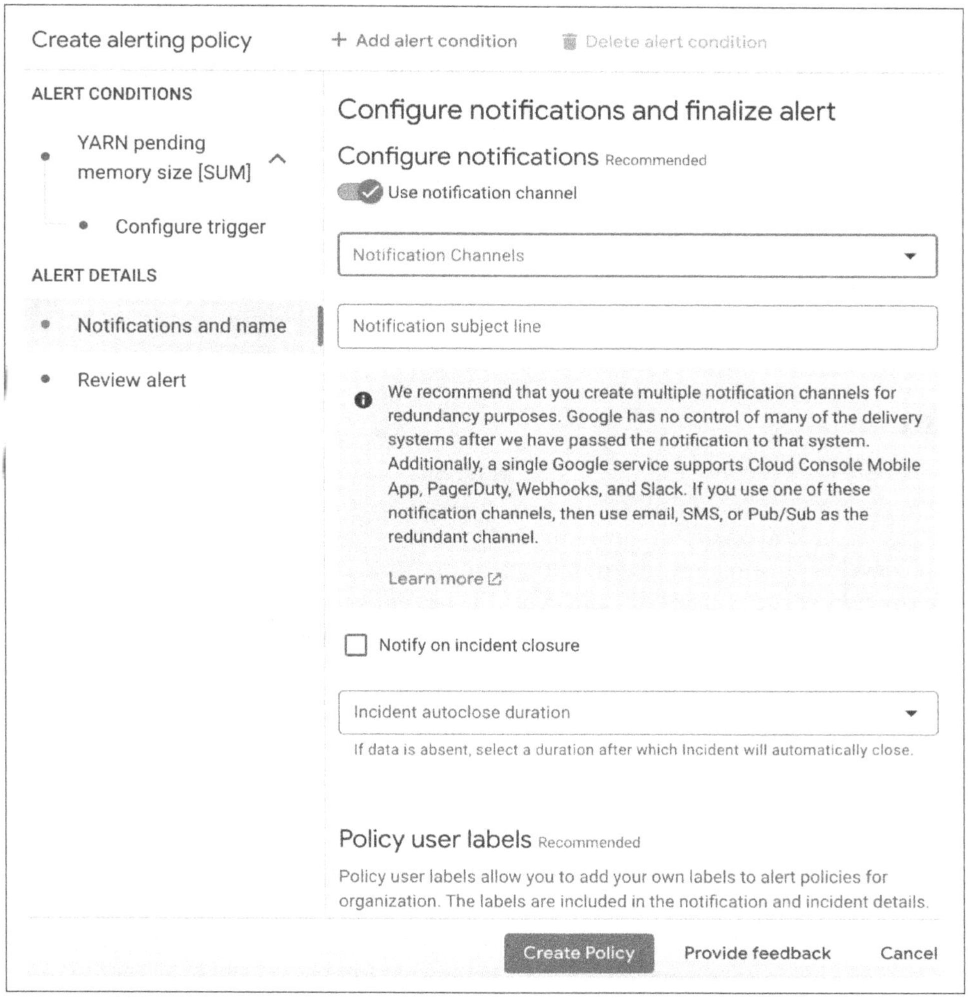

Create alerting policy   + Add alert condition   🗑 Delete alert condition

ALERT CONDITIONS

- YARN pending memory size [SUM] ∧
  - Configure trigger

ALERT DETAILS

- Notifications and name
- Review alert

Configure notifications and finalize alert

Configure notifications Recommended

✅ Use notification channel

| Notification Channels ▼ |

| Notification subject line |

ℹ️ We recommend that you create multiple notification channels for redundancy purposes. Google has no control of many of the delivery systems after we have passed the notification to that system. Additionally, a single Google service supports Cloud Console Mobile App, PagerDuty, Webhooks, and Slack. If you use one of these notification channels, then use email, SMS, or Pub/Sub as the redundant channel.

Learn more ↗

☐ Notify on incident closure

| Incident autoclose duration ▼ |

If data is absent, select a duration after which Incident will automatically close.

Policy user labels Recommended

Policy user labels allow you to add your own labels to alert policies for organization. The labels are included in the notification and incident details.

Create Policy   Provide feedback   Cancel

Figure 9-26. Selecting the notification channels and adding a subject line

Click Create Policy. Now you have an alert policy set on top of YARN pending memory to get alerted if pending memory is ever present in the cluster for more than 15 minutes.

9.6 Migrating Dashboards from One Project to Another

Problem

You have created a dashboard with all the required widgets. Now you want to promote it to higher-level environments.

Solution

Google Cloud Monitoring allows you to easily migrate dashboards between projects by exporting and importing their configurations following these steps:

1. Export the dashboards as a JSON file from the source project.
2. Use the JSON file to create a new dashboard in the target project.

Discussion

Dashboards often consist of numerous widgets, complex configurations, and various types of charts and alerts, all categorized meticulously. Re-creating these dashboards in each environment results in redundant work. Since the metrics you monitor in lower-level environments, like development, are usually the same as in higher-level environments, such as production, you can simplify the process by exporting the dashboard JSON file from one project to another. This allows you to easily create a new dashboard in another project using the exported JSON file, saving time and effort while ensuring consistency across environments.

Here are the steps to download an existing dashboard as a JSON file. First, navigate to the Dashboard home page and click the Settings button. Choose JSON and Download JSON, as shown in Figure 9-27.

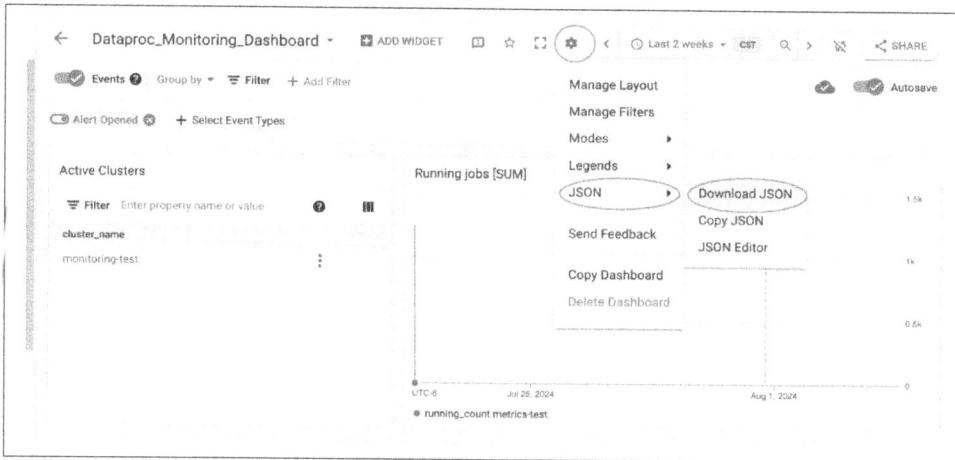

Figure 9-27. Downloading a JSON configuration from the settings option

This will download a JSON file to your machine. The exported JSON file includes all the details of your dashboard, such as the charts, filters, layout, and any customizations you've made.

Now, to create a new dashboard using the JSON file, navigate to the new project where you want to replicate the dashboard and click Create New Dashboard. Click the settings button, then the JSON and JSON Editor options, as shown in Figure 9-28.

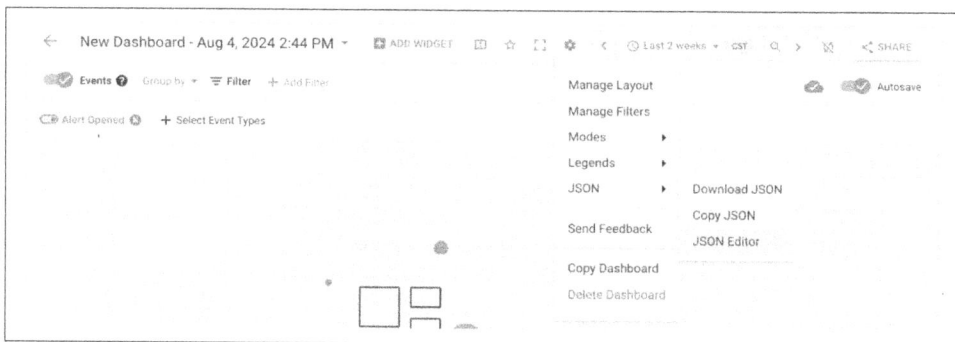

Figure 9-28. Cloning a dashboard by editing its JSON file

Click the Upload option, select the downloaded file, and click Apply Changes, as shown in Figure 9-29.

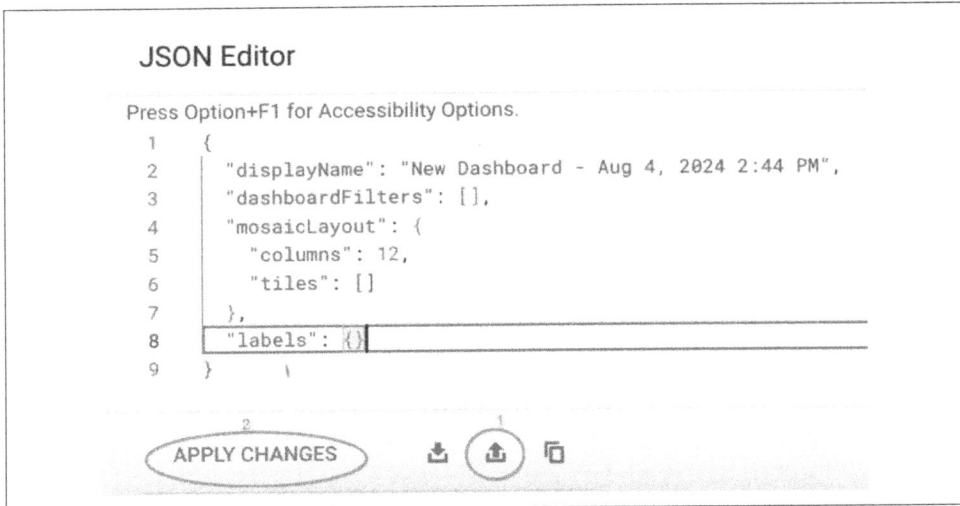

```
JSON Editor

Press Option+F1 for Accessibility Options.
1    {
2        "displayName": "New Dashboard - Aug 4, 2024 2:44 PM",
3        "dashboardFilters": [ ],
4        "mosaicLayout": {
5            "columns": 12,
6            "tiles": []
7        },
8        "labels": {}
9    }
```

Figure 9-29. Saving changes after modifying JSON content

9.7 Creating Custom Log-Based Metrics

Problem

Predefined metrics are not sufficient for your application monitoring. You also want to be alerted if a specific log message occurs in logging.

Solution

Log-based metrics can be effectively tracked by continuously monitoring your logs and capturing predefined matching patterns. This information becomes available in Metrics Explorer, where you can view the time-series data and use it to create various alert-type charts. Metrics from logs can be captured in the following ways:

Counter
Tracks the number of occurrences of a specific pattern

Distribution
Captures the counts of occurrences across multiple matching pattern groups

Discussion

Google Cloud captures predefined metrics for most of its services during usage. However, these predefined metrics might not always meet your custom monitoring needs. In such cases, log-based metrics provide a solution, allowing you to create and monitor metrics based on log messages.

The process for creating log-based metrics involves two steps:

1. Create a log message and write it to the Logging API.
2. Create a log-based metric by defining a pattern to match the log message written to the Logging API.

Let's consider a scenario where you need to capture metrics for autoscaling failures in Dataproc clusters. There are currently no predefined metrics that specifically track how many times autoscaling has failed in a given cluster. You want to be alerted whenever there is a failure in cluster autoscaling.

Autoscaling in Dataproc clusters can fail for various reasons:

- The required machine type might not be available.
- You might reach a quota limit for a specific resource (e.g., CPUs, disk space).
- There could be a failure during the execution of your custom bootstrap or initialization actions on the newly autoscaled nodes.

Typically in such cases, the cluster will keep showing that it is in a healthy state and will retry the scaling operation in the next iteration after the cooldown duration is met. By setting up log-based metrics, you can track such failures and ensure that you're notified when they occur so that you can respond quickly to these issues.

You can capture this metric by examining the log messages. Whenever a scaling operation occurs, Dataproc logs a message about the operation's status, indicating whether it was successful. The following is a sample message from a failed scaling operation due to the requested resources being beyond quota limit:

```
Insufficient 'CPUS' quota. Requested 64.0, available 48.0. Your resource request
exceeds your available quota.
```

Creating a Log-based metric

Navigate to "Log-based metrics," as shown in Figure 9-30.

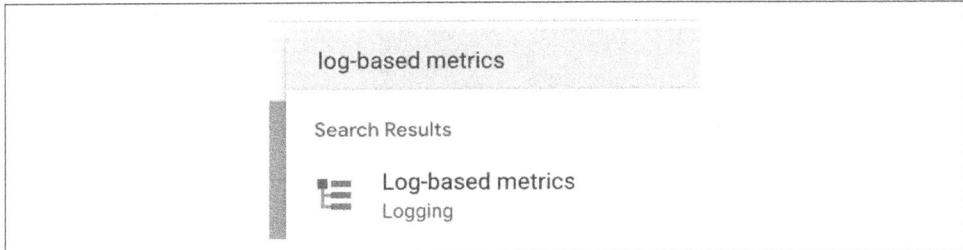

Figure 9-30. Searching for log-based metrics

Click "Create metric," as highlighted in Figure 9-31.

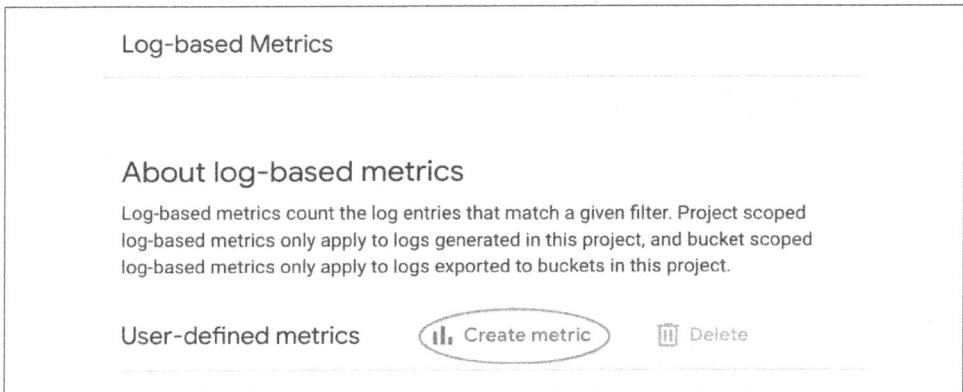

Figure 9-31. Creating a new log-based metric

Configure the log-based metric by choosing Counter as the type and giving it a name and description, as shown in Figure 9-32.

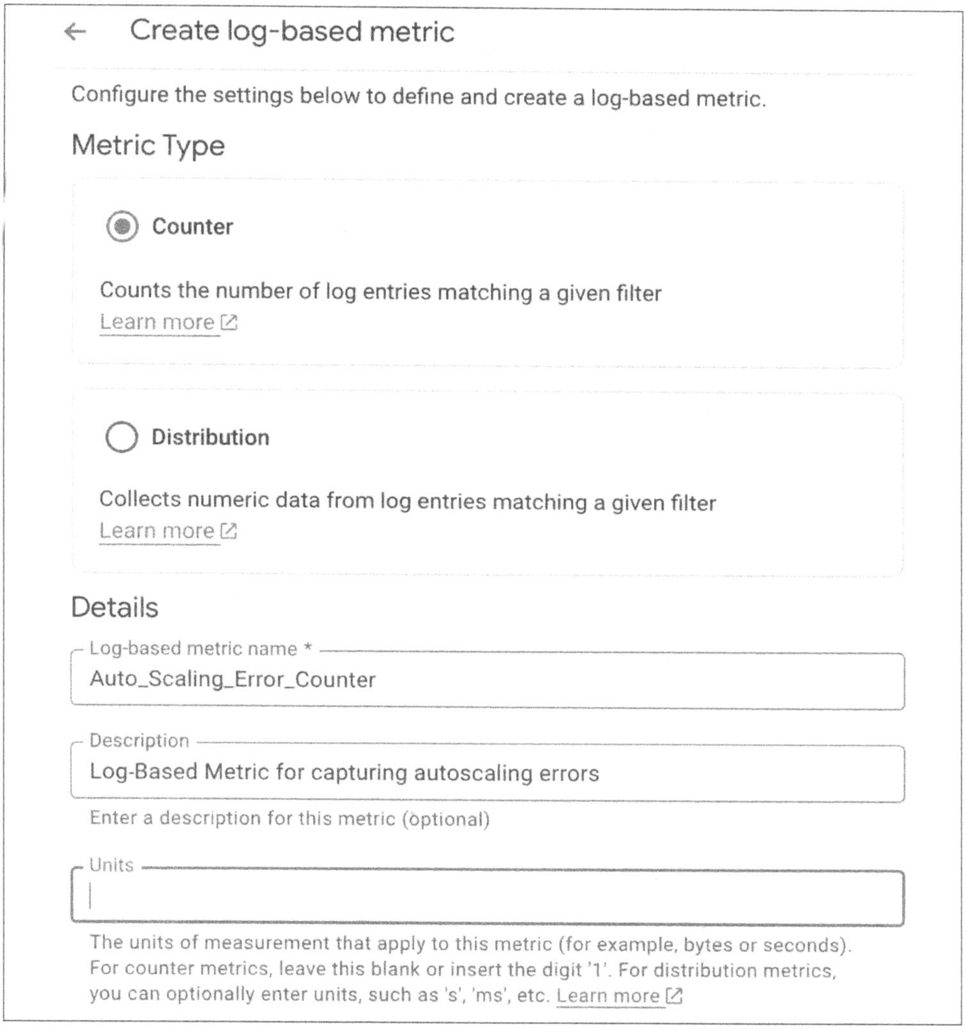

Figure 9-32. Selecting the Counter type for a log-based metric

Now configure the filter pattern for identifying log messages as follows:

```
resource.type="cloud_dataproc_cluster" AND severity="ERROR" AND \
    protoPayload.status.message=~("Insufficient" AND "'CPUS'" AND "quota")
```

This filter targets specific log entries within Google Cloud, as shown in Figure 9-33. The filter fields are:

```
resource.type="cloud_dataproc_cluster"
```
Focuses on logs generated by Cloud Dataproc clusters

```
severity="ERROR"
```
Filters for log entries classified as errors, indicating potential problems

```
protoPayload.status.message=~("Insufficient" AND "'CPUS'" AND "quota")
```
Examines the error message within the log entry, specifically looking for messages related to insufficient CPU quota. The =~ operator means "contains."

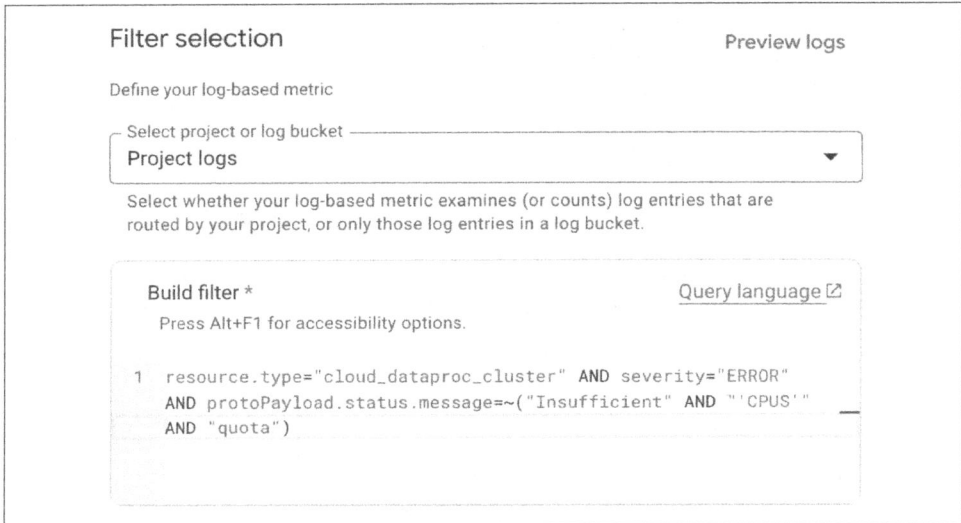

Figure 9-33. A filter condition to match log messages by pattern

Finally, click "Create metric."

Viewing metrics data from Metrics Explorer

From the log-based metrics screen, click on the metric and select the option to view the metrics data in Metrics Explorer, as shown in Figure 9-34.

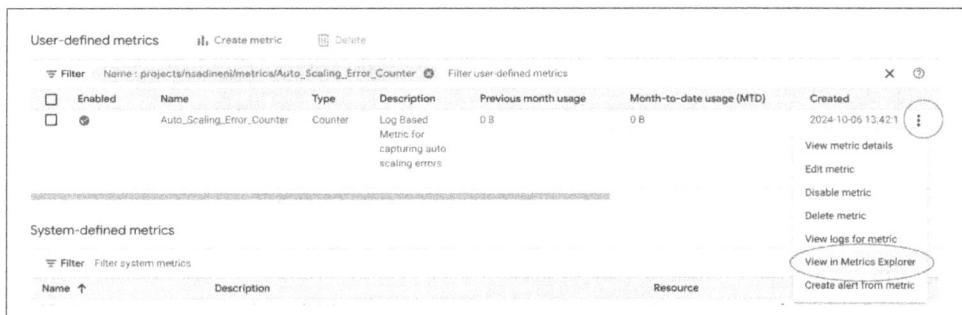

Figure 9-34. Viewing log-based metric data in Metrics Explorer

Metrics Explorer view gives you the details of the log-message occurrence. From Figure 9-35, you can see that a log message about insufficient quota has occurred at around 1:45 P.M.

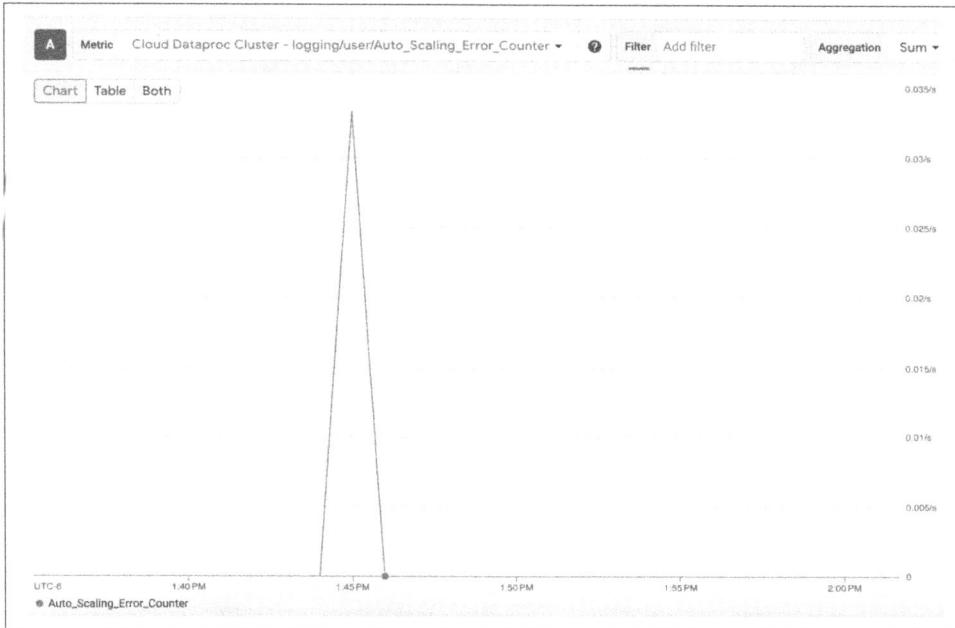

Figure 9-35. Visualizing log-based metrics data in a Metrics Explorer chart

Follow the instructions in Recipe 9.5 to convert this Metrics Explorer chart into an alert chart.

## CHAPTER 10

Dataproc Security

Security is typically implemented at multiple levels, using a variety of techniques to ensure comprehensive protection, as shown in Figure 10-1. When securing a Google Cloud Dataproc environment, the first consideration is perimeter security, which controls who can even attempt to access the resources. This can be achieved through firewall rules, network access control lists (ACLs), or more advanced solutions like VPC Service Controls (VPC SCs), which create a security perimeter around sensitive data and services.

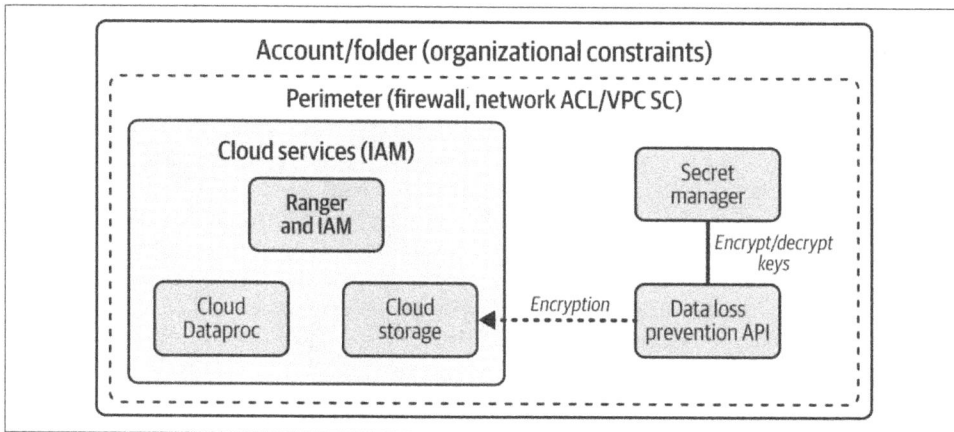

Figure 10-1. Tools and techniques to implement security at multiple levels

Once a user gains access to the perimeter, the focus shifts to service-level security. At this stage, two critical tasks come into play: authentication and authorization. Authentication ensures that the user has the correct credentials to prove their identity, which can be implemented using Kerberos within Dataproc clusters or through

Google Cloud's built-in authentication mechanisms (IAM) when accessing other services.

After authenticating, authorization verifies that the authenticated user is permitted access to specific resources. This can be managed using Google Cloud's IAM for many services, while Apache Ranger can provide fine-grained access control within the Dataproc cluster itself. To enforce policies across multiple projects or folders, organization constraints in Google Cloud offer another layer of control, ensuring that security policies are consistent throughout the environment.

In this chapter, we will delve into the critical aspects of safeguarding your Dataproc clusters and the data processed within them. We will explore various security mechanisms, best practices, and tools that you can implement to protect your Dataproc environment from unauthorized access, data breaches, and other security threats.

By the end of this chapter, you will be able to:

- Understand the different ways to manage identities and access within Dataproc clusters
- Enforce restrictions using organization rules
- Configure authentication methods using Kerberos and IAM
- Securely store and manage credentials using Secret Manager
- Implement data tokenization using data loss prevention (DLP) to protect sensitive information
- Configure authorization rules using Ranger and IAM

> This chapter provides an overview of key Dataproc security topics. It is not intended to be an exhaustive list of all available security features and configurations. The specific security requirements for your Dataproc environment may vary depending on your application needs, data sensitivity, and compliance obligations. It is essential to conduct a thorough security assessment and customize your Dataproc security measures accordingly.

10.1 Managing Identities in Dataproc Clusters

Problem

Your team is new to using Dataproc and wants to explore different strategies for managing identities within Dataproc clusters. You are considering the implications of using clusters for single users versus multiple users and how to securely access other services in the Google Cloud ecosystem. What are the best approaches for your use cases?

Solution

Managing identities in Dataproc clusters involves three key approaches:

Custom service accounts
> Create custom service accounts with appropriate permissions and assign them to Dataproc clusters. Use these service accounts for authentication in Spark or other applications.

Personal cluster authentication
> Enable personal cluster authentication to associate individual user accounts with specific Dataproc clusters. This provides a convenient way for users to have their own dedicated clusters.

Secure multitenancy
> Implement secure multitenancy by mapping multiple users or groups of users to specific service accounts.

Discussion

To better understand how to manage identities, we first have to be clear on what an identity is. *Identity* refers to a unique user or service account within a system. Each identity must have the necessary permissions to perform actions in a cloud environment. Examples of actions controlled by identity include:

- A user or service account accessing a GCS bucket
- A user or service account viewing the details of a Dataproc cluster
- A user or service account running a job in a Dataproc cluster
- A user or service account loading data into BigQuery

When you understand how to control access to your cluster and its resources, you can significantly enhance security, prevent unauthorized use, and ensure compliance with industry regulations. By controlling who can access and use Dataproc resources, you can prevent unauthorized access, protect sensitive data, comply with regulations, track user activity, allocate resources efficiently, and facilitate collaboration. This ultimately helps maintain a secure and well-governed Dataproc environment.

There are three approaches to handling identity mapping within the Dataproc cluster while it accesses the underlying resources, which are illustrated in Figure 10-2:

- Service account approach
- Personal cluster authentication
- Secure multitenancy

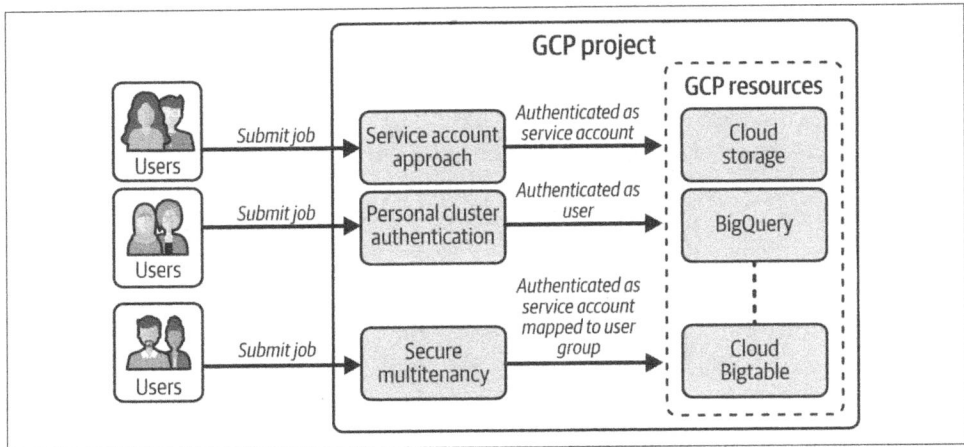

Figure 10-2. Identity mapping from a Dataproc cluster while accessing underlying GCP resources

Let's consider how each approach works.

Service account approach

In the service account approach, a designated service account is employed by the Dataproc cluster to execute actions on behalf of the users. For instance, if User A submits a job to read data from GCS and load it into BigQuery, this job utilizes the identity of the service account rather than the identity of User A who initiated the job.

> By default, Dataproc clusters use a default service account to interact with other Google Cloud services when running jobs. Recipe 1.2 discusses the two types of service accounts: VM service account and service agent service account.

Here is the gcloud command for creating a cluster with a custom service account:

```
gcloud dataproc clusters create <cluster_name_here> \
  --region <region-name>
  --service-account <custom-vm-service-account-here>
```

Benefits of this approach include the following:

- Service accounts can be managed centrally, making it easier to control permissions and audit access.

- Applications and services can easily authenticate using the service account's credentials, simplifying the authentication process.

- Using the VM's service account simplifies identity management because you don't need to explicitly manage or pass credentials for each job. The VM's service account is automatically used for authentication and authorization.

- The VM's service account integrates seamlessly with other Google Cloud services, allowing jobs to access resources like GCS, BigQuery, and others without additional configuration.

Here are some challenges of this approach:

- Managing service accounts can add some overhead, especially if you have many different clusters or jobs requiring unique permissions.

- Using the VM's service account means that all jobs running on the VM share the same set of permissions. This can be problematic if different jobs require different levels of access or if there is a need for fine-grained control over permissions.

- Using service accounts can make it harder to track individual user activity within a cluster since all actions are performed under the service account's identity.

Using the VM's service account for Dataproc is a popular approach that offers centralized management, simplified authentication, and seamless access to Google Cloud services. However, it's crucial to consider potential drawbacks, such as overhead, broad permissions, and limited user visibility.

Personal cluster authentication

Personal cluster authentication in Dataproc allows individual users to have their own dedicated clusters with their own identity for authentication. This means that when a user interacts with the cluster and its resources, including other Google Cloud services, they do so using their own user account credentials instead of a service account.

To enable personal cluster authentication, use the gcloud command with the following options:

- `--enable-personal-auth`   - `--personal-auth-user`

Here is the gcloud command for creating a cluster with personal authentication:

```
gcloud dataproc clusters create <cluster_name_here> \
    --region <region-name>
    --enable-personal-auth \
    --personal-auth-user=<user_email>
```

This approach is typically used in proof-of-concept development, for single-user data exploration, or when you have a security requirement to isolate all actions at a single-user level.

Benefits of this approach include the following:

- Users don't need to manage service accounts or worry about service account permissions.

- Actions performed on the cluster are directly linked to the user's identity, improving accountability and auditability.

- By using their own accounts, users are limited to the permissions granted to their accounts, preventing unauthorized access.

The challenge of this approach is that each cluster is tied to a single user account, making it unsuitable for shared environments. While personal cluster authentication offers simplified authentication, improved accountability, and enhanced security, it is best suited for individual use cases due to its single-user limitation.

Secure multitenancy approach

Secure multitenancy in Dataproc allows users to securely share a single Dataproc cluster among multiple users or teams while ensuring that each group's data and operations remain isolated. Secure multitenancy works by mapping different groups of users to distinct service accounts, allowing for granular access control and resource management.

Let's consider an example. The gcloud command for creating a cluster where User 1 and User 2 are mapped to Service Account 1 and User 3 and User 4 are mapped to Service Account 2 is as follows:

```
gcloud dataproc clusters create <cluster_name_here> \
  --region=<region_here> \
  --secure-multi-tenancy-user-mapping="user1@example.com:service-account-1
@project.iam.gserviceaccount.com,user2@example.com:service-account-1@project.iam.
gserviceaccount.com,user3@example.com:service-account-2@project.iam.gservice
account.com,user4@example.com:service-account-2@project.iam.gserviceaccount.com"
```

Once the cluster is created, it is accessible to four users. Any jobs submitted by User 1 or User 2 will use the identity of Service Account 1. Jobs submitted by User 3 or User 4 will use Service Account 2.

Benefits of this approach include the following:

- Mapping users to separate service accounts enables you to isolate their activities, reducing the risk of unauthorized access and ensuring that security policies are enforced for each group.

- Sharing a cluster with multiple users can help with some of the quota limits that are at the per-user/account level.

- Multitenancy allows for better resource utilization since multiple tenants can share the same cluster resources, avoiding idle capacity and maximizing efficiency.

- You can easily adjust access controls and resource allocation for different tenants as their needs change, providing flexibility and scalability.

Some challenges of this approach include:

- Managing multiple service accounts and their associated permissions adds complexity to cluster administration. It requires careful planning and maintenance to ensure that permissions are correctly configured and updated.

- There is the potential for "noisy neighbor" issues. One tenant's workload might consume excessive resources, affecting the performance of other tenants sharing the same cluster. Resource quotas and monitoring can help mitigate this.

> In a secure multitenancy setup in Dataproc, while you use different service accounts to manage user-specific access to resources, you still utilize the default VM service account for the Compute Engine VM instances. This VM service account handles permissions for the overall cluster operations, while job-specific or user-specific service accounts provide granular access control for individual tasks and resources.

10.2 Securing Your Perimeter Using VPC Service Controls

Problem

In a project, the Dataproc service and GCS buckets need to be secured so that they can be accessed only from within a service perimeter. However, one user requires access from outside the perimeter. How can this be achieved?

Solution

VPC SCs create a logical perimeter boundary, allowing you to place required services in a restricted environment. Once the perimeter is established, internal users can access resources, but anyone outside the perimeter will need ingress or egress rules to be granted access; otherwise, they will be denied access by the perimeter.

Discussion

VPC SCs act like a security perimeter around your Google Cloud resources, helping you prevent data exfiltration and control access to sensitive services. Imagine it as a virtual fence around your valuable data within Google Cloud. Here are the steps for implementing VPC SCs (illustrated in Figure 10-3):

1. Define the perimeter.
2. Include the projects that will be part of your perimeter.
3. Specify the services to be placed in restricted access mode.
4. Optionally add ingress/egress rules to allow outside users to access specific resources.

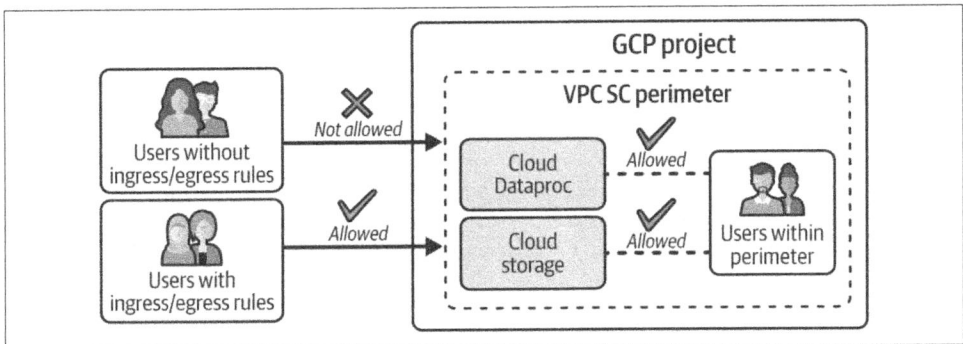

Figure 10-3. Architecture of a VPC SC perimeter and users' access

Creating a new VPC service perimeter

Users creating a VPC service perimeter need to have the access context manager admin role. Switch to organization context to be able to view VPC service controls. Search for VPC SC and select the VPC Service Controls option, as shown in Figure 10-4.

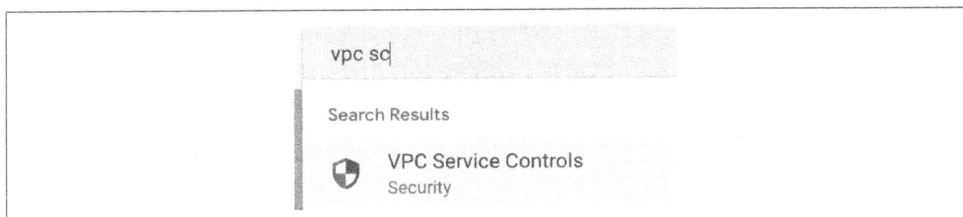

Figure 10-4. Searching for the VPC Service Controls service

Click +New Perimeter, highlighted in Figure 10-5.

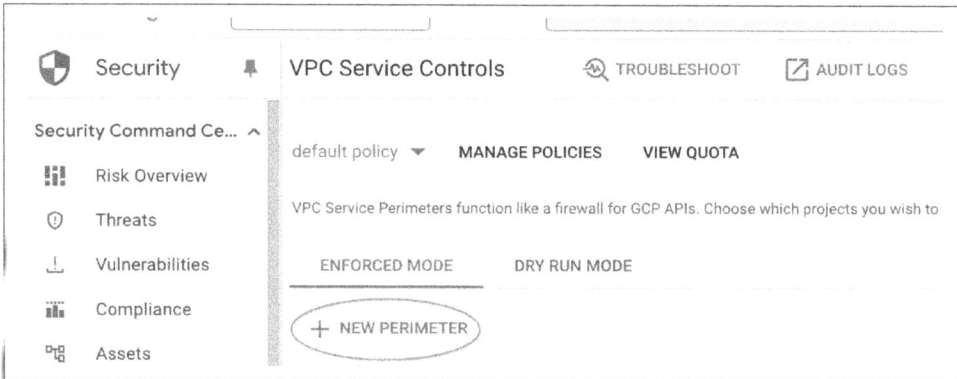

Figure 10-5. Adding a new VPC SC perimeter

Choose the name of the perimeter, as shown in Figure 10-6.

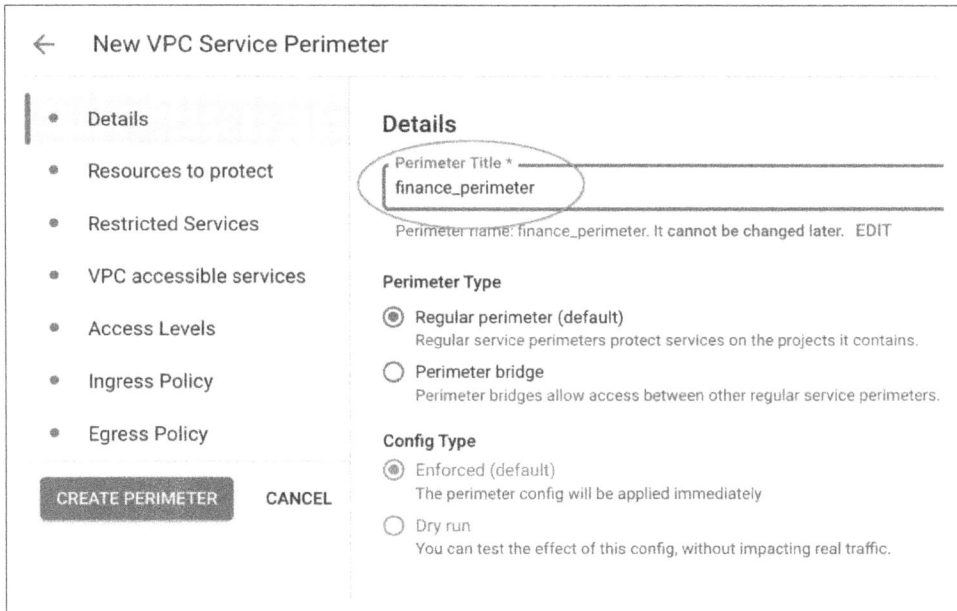

Figure 10-6. Naming the VPC SC perimeter

Select the projects to be added to the perimeter and then click Add Resources to choose a project, as shown in Figure 10-7.

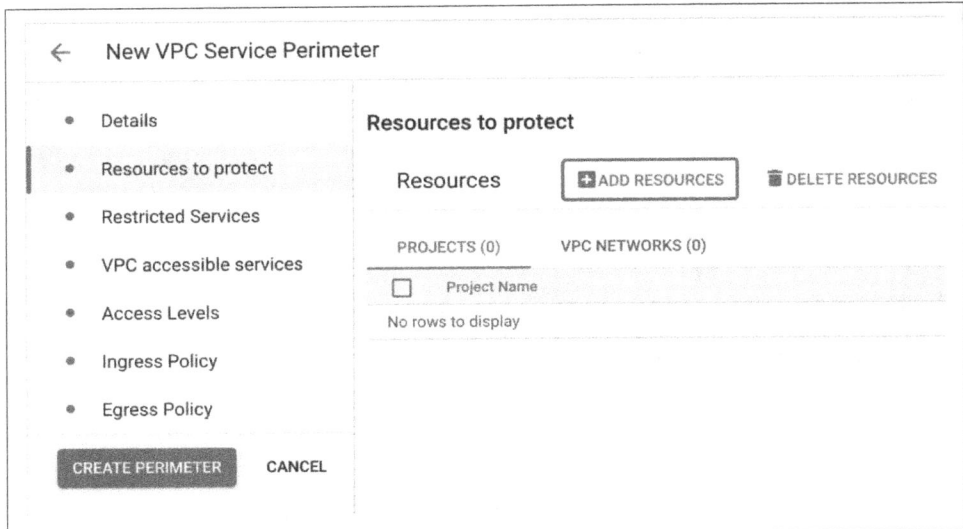

Figure 10-7. Adding resources or projects to a perimeter

Select the services to be added to the perimeter. Then, click Add Services, as shown in
Figure 10-8, and select Dataproc & Google Cloud Storage APIs.

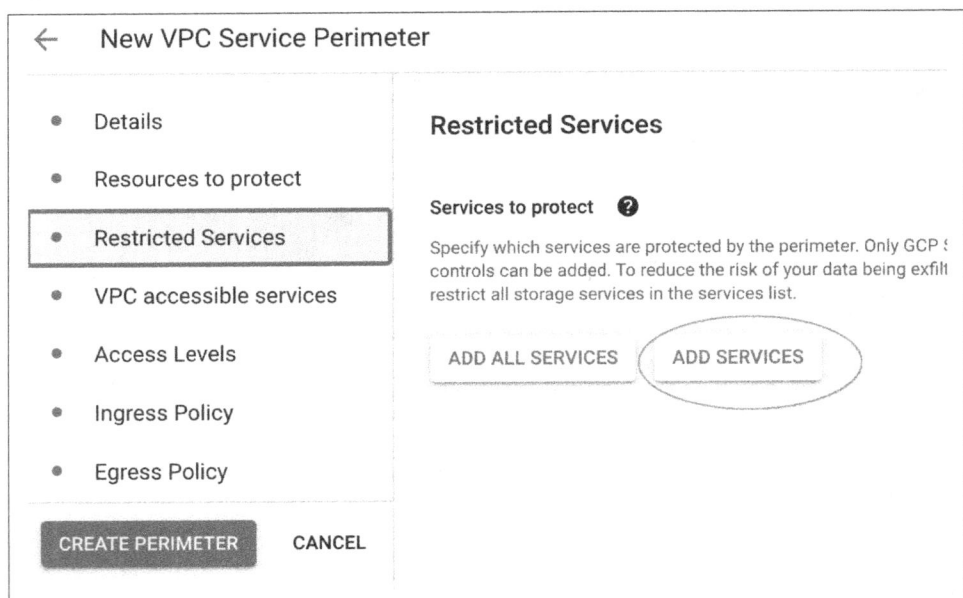

Figure 10-8. Adding services to a perimeter

Click Create Perimeter, highlighted in Figure 10-9.

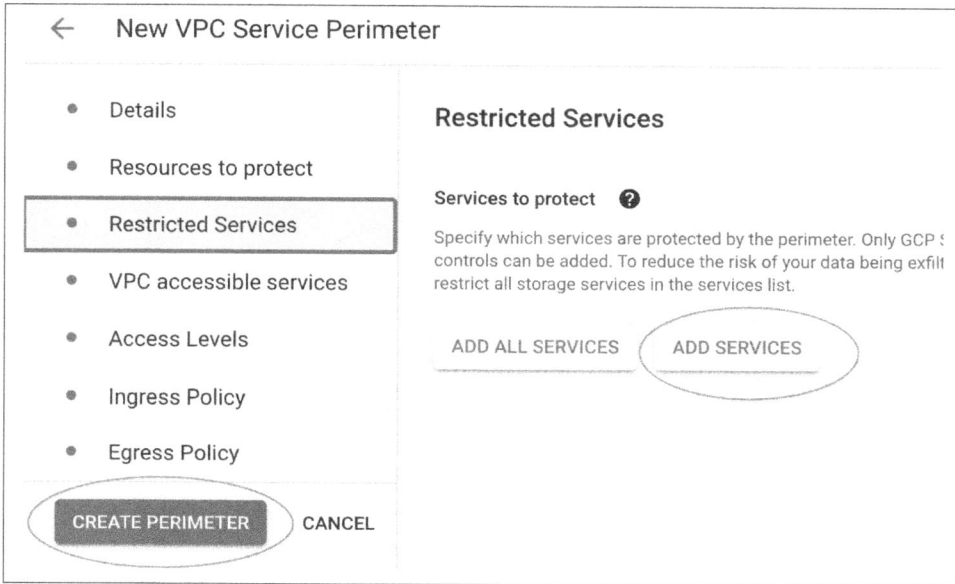

Figure 10-9. Finishing the VPC SC perimeter creation

We've established a secure perimeter and included the Dataproc service and all GCS buckets associated with this project within its boundaries. This means that any user attempting to submit a job to Dataproc or access GCS buckets must do so from within the perimeter. For example, if you try to access a GCS bucket within the project from your local machine, you will encounter an error message:

```
AccessDeniedException: 403 Request is prohibited by organization's policy.
vpcServiceControlsUniqueIdentifier: uAjbXwrSWsV-oRgiX9m8LdMK7O470Zd8mk1MDIriIPFTa
ZNpvY74sxeo9mNO49I5uRpittiRZ6Bi1QR8
```

Allowing outside users access to resources in a secured perimeter

Creating an ingress and egress rule will allow you to run a job or access a GCS bucket and get the result back. First, select the perimeter, and then click Edit Perimeter, as shown in Figure 10-10.

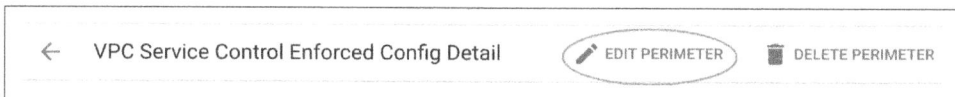

Figure 10-10. Editing the VPC perimeter configuration

Navigate to Ingress Policy and click Add Rule, as shown in Figure 10-11.

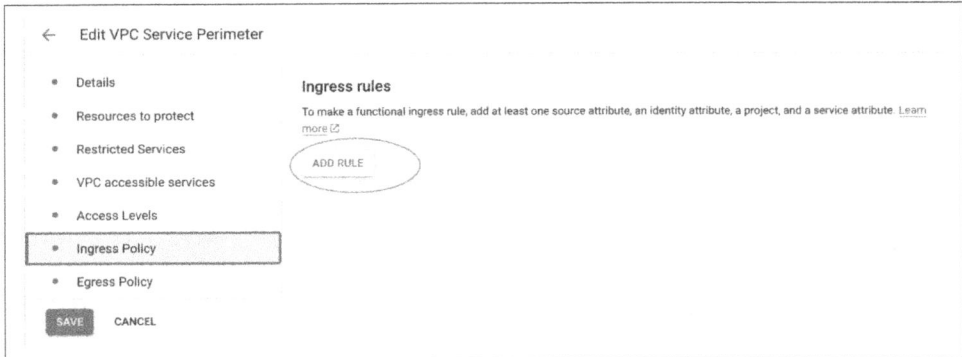

Figure 10-11. Adding an ingress rule to allow users to access the VPC SC perimeter

Configure the ingress policy for a selected user to access all services in the project from anywhere, as shown in Figure 10-12.

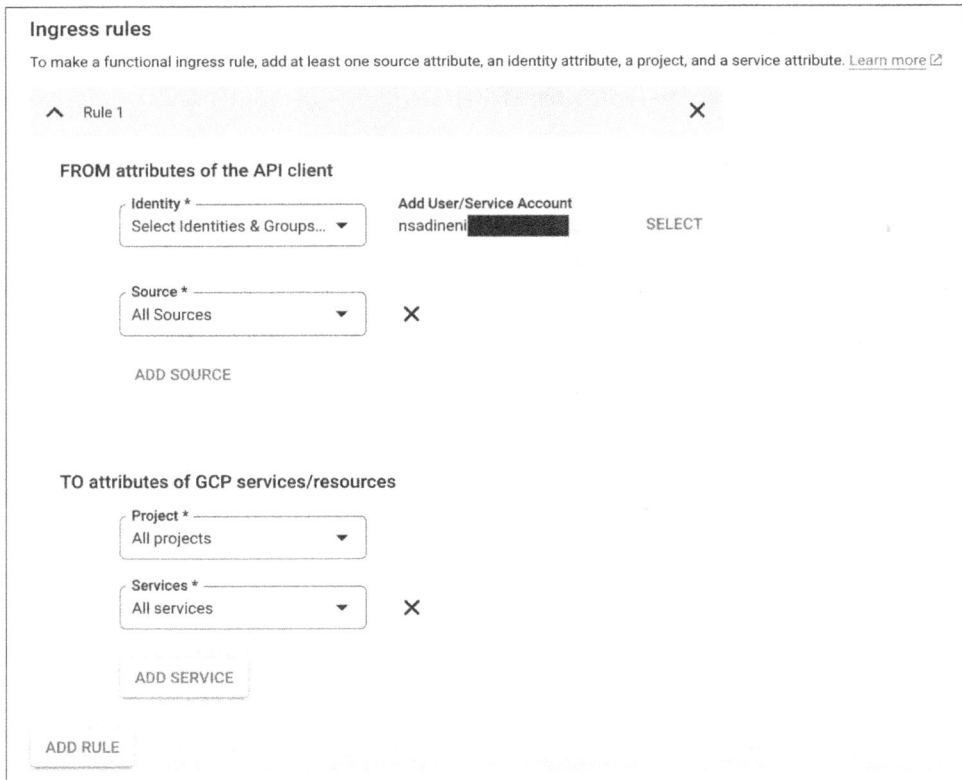

Figure 10-12. Configuring the ingress rules for VPC SC

Configure the egress policy for the selected user to get egress data from all the services in perimeter, as shown in Figure 10-13.

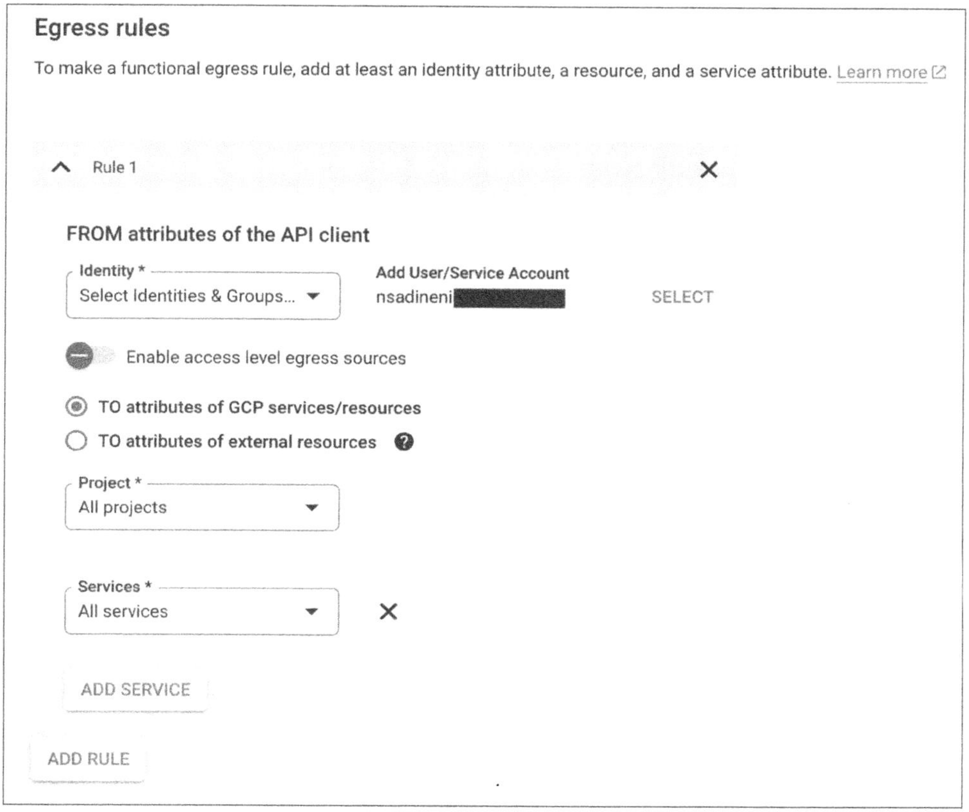

Egress rules

To make a functional egress rule, add at least an identity attribute, a resource, and a service attribute. Learn more ☑

∧   Rule 1   ✕

FROM attributes of the API client

Identity *
Select Identities & Groups... ▼   Add User/Service Account
nsadineni███████   SELECT

⊝ Enable access level egress sources

◉ **TO attributes of GCP services/resources**
○ **TO attributes of external resources** ❷

Project *
All projects ▼

Services *
All services ▼   ✕

ADD SERVICE

ADD RULE

Figure 10-13. Configuring egress rules for VPC SC

Save the updated perimeter configuration. The configured user can now access the perimeter resources from any machine.

10.3 Authenticating Using Kerberos

Problem

In a shared VPC with multiple projects, ensuring secure authentication in a Dataproc cluster is critical. Without it, users could impersonate others, tamper with data, or access sensitive information. How can you prevent impersonation and unauthorized access in this shared environment?

Solution

Kerberos is a secure network authentication protocol that enforces verified access to resources. In a Dataproc cluster, Kerberos ensures that only authenticated users and services can access Hadoop components, preventing impersonation, data tampering, and unauthorized access in a shared VPC environment (illustrated in Figure 10-14). This strengthens security across multitenant projects by safeguarding sensitive data and enforcing strict verification of user identities.

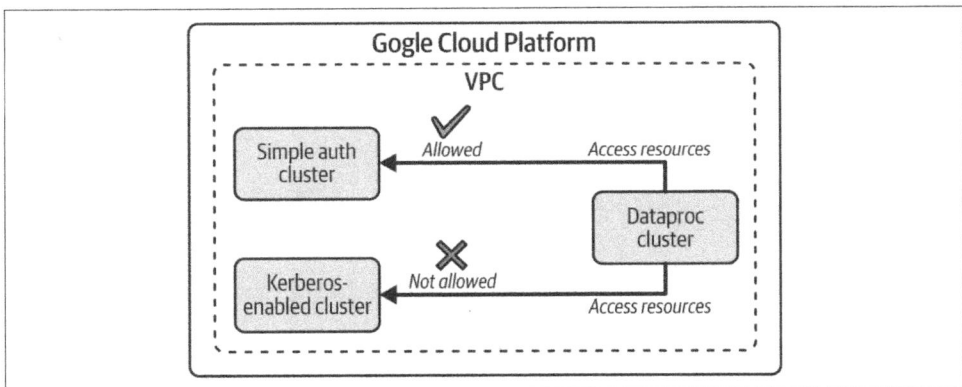

Figure 10-14. Securing cluster resources with Kerberos-based authentication

Discussion

When a Dataproc cluster is created, the services within it, such as HDFS, often rely on a basic form of authentication. For instance, HDFS might identify a user based on an environment variable like HADOOP_USER_NAME or information embedded within a job request. This simplified approach can become a security concern when multiple clusters exist within the same VPC. In such a scenario, users on one cluster could potentially access and even manipulate data or resources residing on other clusters. This lack of strict access control exposes sensitive information to unauthorized users and highlights the need for a more robust authentication mechanism like Kerberos, especially in environments with multiple clusters and varying security requirements.

Let's consider an example that demonstrates how clusters with simple authentication can expose data and/or resources to other cluster users. We will follow these steps:

1. Create Cluster 1 and Cluster 2 without Kerberos configuration.

2. Run a TeraGen job and create the output file in Cluster 1's HDFS location.

3. Run a TeraSort job that reads the input file from Cluster 1 and stores the output in Cluster 2.

Ideally, you would expect the TeraSort job to fail because it is trying to access data in Cluster 1.

The gcloud command for creating Cluster 1 and Cluster 2 without Kerberos authentication is as follows:

```
gcloud dataproc clusters create cluster-1 \
  --region us-central1 --enable-component-gateway

gcloud dataproc clusters create cluster-2 \
  --region us-central1  --enable-component-gateway
```

Run a TeraGen job to generate data in Cluster 1's HDFS:

```
gcloud dataproc jobs submit hadoop \
  --cluster=cluster-1
  --region=us-central1 \
  --jar=file:///usr/lib/hadoop-mapreduce/hadoop-mapreduce-examples.jar \
  -- teragen 5000 /tmp/teragen_input
```

When specifying a local path (not a GCS path) as input or output, Dataproc stores the files in HDFS using the provided path location. To access the HDFS service in Cluster 1 from Cluster 2, prefix the path with the service name, hostname, and port number where the HDFS service is running. For example, the path *hdfs://cluster-1-m:8020/tmp/* points to the */tmp* folder in Cluster 1's HDFS.

Run a TeraSort job that takes input as an HDFS file present in Cluster 1 and generates output in Cluster 2:

```
gcloud dataproc jobs submit hadoop \
  --cluster=cluster-2 \
  --region=us-central1 \
  --jar=file:///usr/lib/hadoop-mapreduce/hadoop-mapreduce-examples.jar \
  -- terasort  hdfs://cluster-1-m:8020/tmp/teragen_input /tmp/terasort_output
```

You can see the TeraSort job running on Cluster 2 successfully reading input files from Cluster 1 and storing output in Cluster 2. This highlights a potential issue: isolating resources between clusters. Let's explore how enabling Kerberos can help.

Create kerberos-cluster-1 and kerberos-cluster-2 with Kerberos enabled:

```
gcloud dataproc clusters create kerberos-cluster-1 \
  --region us-central1 --enable-component-gateway --enable-kerberos

gcloud dataproc clusters create kerberos-cluster-2 \
  --region us-central1 --enable-component-gateway --enable-kerberos
```

Run a TeraGen job to generate output in kerberos-cluster-1's HDFS:

```
gcloud dataproc jobs submit hadoop \
  --cluster=kerberos-cluster-1 \
  --region=us-central1 \
  --jar=file:///usr/lib/hadoop-mapreduce/hadoop-mapreduce-examples.jar \
  -- teragen 5000 /tmp/teragen_input
```

Run a TeraSort job that takes input as an HDFS file on kerberos-cluster-1 and stores output in kerberos-cluster-2:

```
gcloud dataproc jobs submit hadoop \
  --cluster=kerberos-cluster-2 \
  --region=us-central1 \
  --jar=file:///usr/lib/hadoop-mapreduce/hadoop-mapreduce-examples.jar \
  -- terasort  hdfs://kerberos-cluster-1-m:8020/tmp/teragen_input \
  /tmp/terasort_output
```

The TeraSort job will fail with the following error messages since the HDFS on kerberos-cluster-1 now requires a valid Kerberos ticket to access Cluster 1 resources:

```
Caused by: javax.security.sasl.SaslException: DestHost:destPort kerberos-cluster-
1-m.us-central1-c.c.domain.internal.:8020 , LocalHost:localPort kerberos-cluster-
2-m/10.128.0.13:0. Failed on local exception: javax.security.sasl.SaslException:
GSS initiate failed [Caused by GSSException: No valid credentials provided
(Mechanism level: Server not found in Kerberos database (7) - LOOKING_UP_SERVER)]
[Caused by javax.security.sasl.SaslException: GSS initiate failed [Caused by
GSSException: No valid credentials provided (Mechanism level: Server not found in
Kerberos database (7) - LOOKING_UP_SERVER)]]

Caused by: GSSException: No valid credentials provided (Mechanism level: Server
not found in Kerberos database (7) - LOOKING_UP_SERVER)

Caused by: KrbException: Server not found in Kerberos database (7) -
LOOKING_UP_SERVER

Caused by: KrbException: Identifier doesn't match expected value (906)
```

Kerberos effectively isolates resources within a cluster. To enable Cluster 2 to read data from Cluster 1, you must explicitly configure cross-realm trust between the key distribution center servers.

10.4 Installing Ranger

Problem

You'd like to Install the Apache Ranger service to manage authorization of cluster resources.

Solution

Apache Ranger provides authorization control over cluster resources (HDFS, Hive tables, GCS buckets, etc.). To install it, follow these steps:

1. Create an encryption/decryption key.

2. Choose a password for the Ranger service and store it in an encrypted file.

3. Create a cluster with Ranger/Solr components with user attribution enabled.

Discussion

Apache Ranger is a centralized security framework designed to manage and enforce access control policies across cluster services like Hadoop, HDFS, Hive, Kafka, and more (illustrated in Figure 10-15). It provides fine-grained authorization by allowing administrators to define role-based access control (RBAC) and attribute-based access control (ABAC) policies.

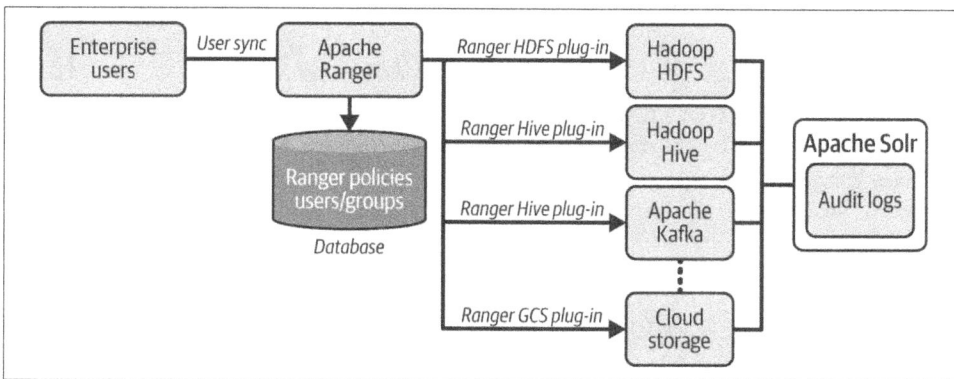

Figure 10-15. High-level architecture of Apache Ranger

Ranger works by deploying lightweight plug-ins within these data services, which intercept requests and enforce the security policies defined in the Ranger admin UI. The system also supports audit logging, enabling organizations to track and monitor access attempts. Ranger integrates with external systems like LDAP and Active Directory for user management and includes synchronization tools to keep its user database updated. By centralizing security policies, Ranger helps improve data governance and compliance and secure multitenancy in big data environments.

The prerequisites for including a Ranger component in a Dataproc cluster are:

- A key for encrypting and decrypting the password
- A password provided in an encrypted file

Let's explore these in more detail.

Creating an encrypt/decrypt key

Google Cloud KMS is a cloud service that allows you to securely manage cryptographic keys for your applications and services. It is used for data encryption, signing data, and managing the lifecycle of encryption keys in a centralized, secure manner.

A key ring in Google Cloud KMS is a logical grouping of cryptographic keys that provides an organizational structure for managing and controlling access to related keys within the same project and location. Create a key ring and key using the following commands:

```
gcloud kms keyrings create ranger-keyring --location=global

gcloud kms keys create ranger-key \
    --location=global \
    --keyring=ranger-keyring \
    --purpose=encryption
```

Encrypting a password using the key

After creating your encryption key, you'll need to encrypt your Ranger password using Google Cloud KMS to ensure secure storage. Google Cloud KMS requires that the password must be alphanumeric and contain at least one uppercase letter. Failure to comply with these requirements will result in a cluster creation error.

To encrypt your password using a key and store the output in a file, use the following command:

```
echo 'Rangerpassword123' | \
    gcloud kms encrypt \
        --location=global \
        --keyring=ranger-keyring \
        --key=ranger-key \
        --plaintext-file=- \
        --ciphertext-file=admin-password.encrypted
```

Copying the password to the GCS location

To copy the encrypted admin password to your GCS bucket, use the following gcloud command:

```
gcloud storage cp admin-password.encrypted gs://<bucket_name_here>/
```

Create a cluster with a Ranger component and user attribution enabled:

```
gcloud dataproc clusters create ranger-test \
  --optional-components=SOLR,RANGER \
  --region=us-central1 \
  --enable-component-gateway \
  --properties="dataproc:user-attribution.enabled=true,dataproc:ranger.kms.key.
uri=projects/<project_name>/locations/global/keyRings/ranger-keyring/
cryptoKeys/ranger-key,dataproc:ranger.admin.password.uri=gs://nstestb/
admin-password.encrypted"
```

With this command, we are passing three Dataproc properties:

`user-attribution.enabled`
: Jobs will be submitted as the user instead of the root account.

`dataproc:ranger.kms.key.uri`
: This is the resource name for the KMS key. This key is used for decrypting the Ranger admin password that has been encrypted and stored in a file.

`dataproc:ranger.admin.password.uri`
: This is the GCS location of the encrypted password.

Why Should We Enable User Attribution Mapping?

The `user-attribution.enabled` property in Dataproc is crucial for managing user identities when submitting jobs to a cluster. When this property is disabled (the default setting), all jobs are attributed to the root user, regardless of who actually submitted them. This can be problematic when you're using Ranger to implement fine-grained access control based on individual user IDs. To ensure that Ranger policies can accurately authorize users based on their actual identities, you must enable the `user-attribution.enabled` property. This allows Dataproc to correctly associate each job with the user who submitted it, enabling Ranger to enforce the appropriate permissions.

Once the cluster is created, log in to the Ranger web UI from the component gateway, as shown in Figure 10-16.

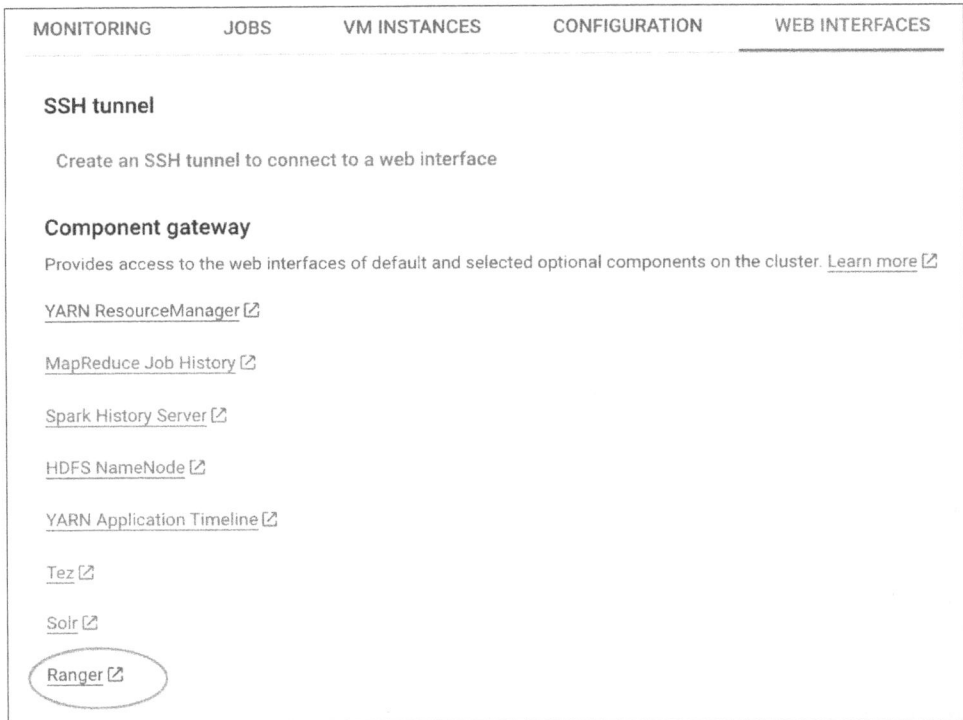

Figure 10-16. Cluster Web Interfaces page with access link to the Ranger web UI

Enter the username as admin and the password as the same one you encrypted using the key, as shown in Figure 10-17. If you followed the previous examples, the password will be Rangerpassword123.

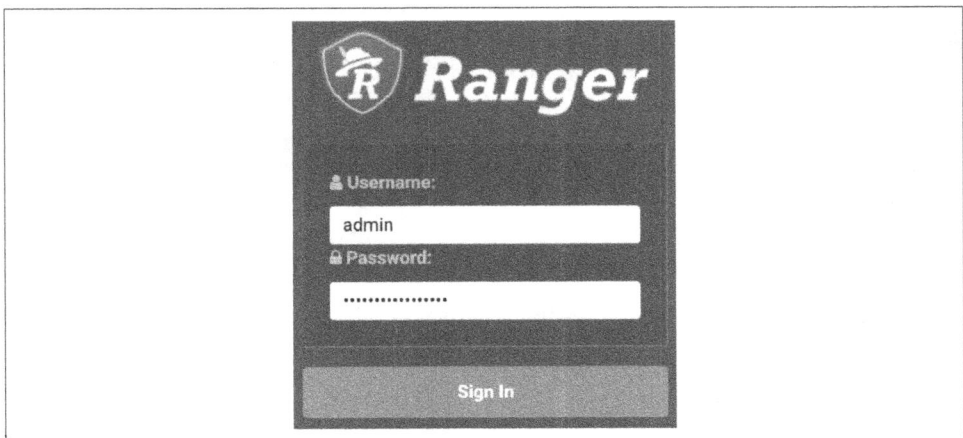

Figure 10-17. Ranger web UI log-in page

In an enterprise environment, Ranger is typically deployed on a Dataproc cluster with a focus on resilience and high availability. This involves using an external database, such as Cloud SQL, to store Ranger's metadata and a separate GCS bucket to house Solr data. This approach ensures that even if the Dataproc cluster fails, Ranger's critical information remains intact and can be easily recovered. By decoupling these components from the cluster itself, organizations can minimize downtime and ensure consistent enforcement of security policies across their data lake.

10.5 Securing Cluster Resources Using Ranger

Problem

You have a Dataproc cluster that is shared by multiple teams. To ensure data security and isolation, you need to consider the following:

- Restricting each team's access to only their designated HDFS folders
- Preventing unauthorized access to sensitive data belonging to other teams

How do you achieve these objectives in a shared cluster?

Solution

Apache Ranger provides authorization control over cluster resources (HDFS, Hive tables, GCS buckets, etc.). To secure resource access by a team, follow these steps:

1. Install the Ranger component in the cluster.
2. Configure Ranger policies and assign allow/deny permissions to users or groups.

Discussion

Consider a scenario of three distinct user groups (Finance, HR, and IT) who all want to securely store their data within a shared HDFS cluster. Each group requires exclusive access to their designated folder, as shown in Table 10-1.

Table 10-1. User groups and their corresponding folder locations

User group	HDFS folder location
Finance	/finance
HR	/hr
IT	/it

> In an enterprise setup, you can also configure Ranger to synchronize users from Active Directory or IAM. When synchronizing, choose the user ID to match the GCP identity username.

Creating users in Ranger

Navigate to the Ranger home page by logging in with your credentials, as shown in Figure 10-18.

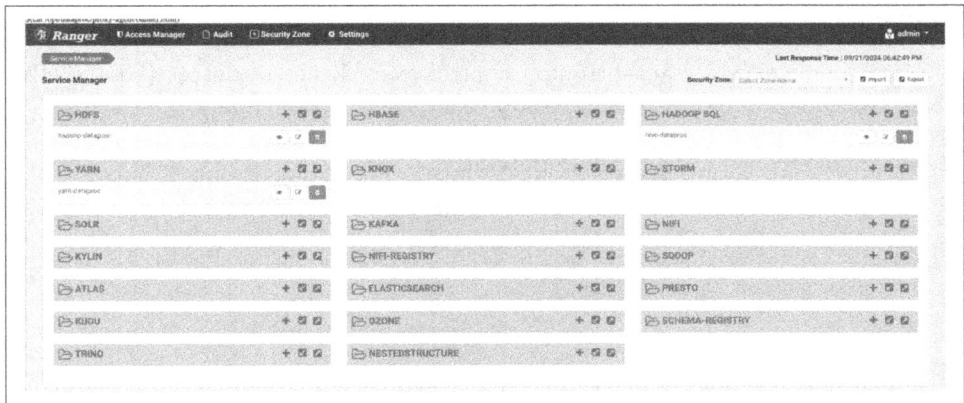

Figure 10-18. Ranger web UI home page

Create a user group, as shown in Figure 10-19. We will be adding all of the users to this group.

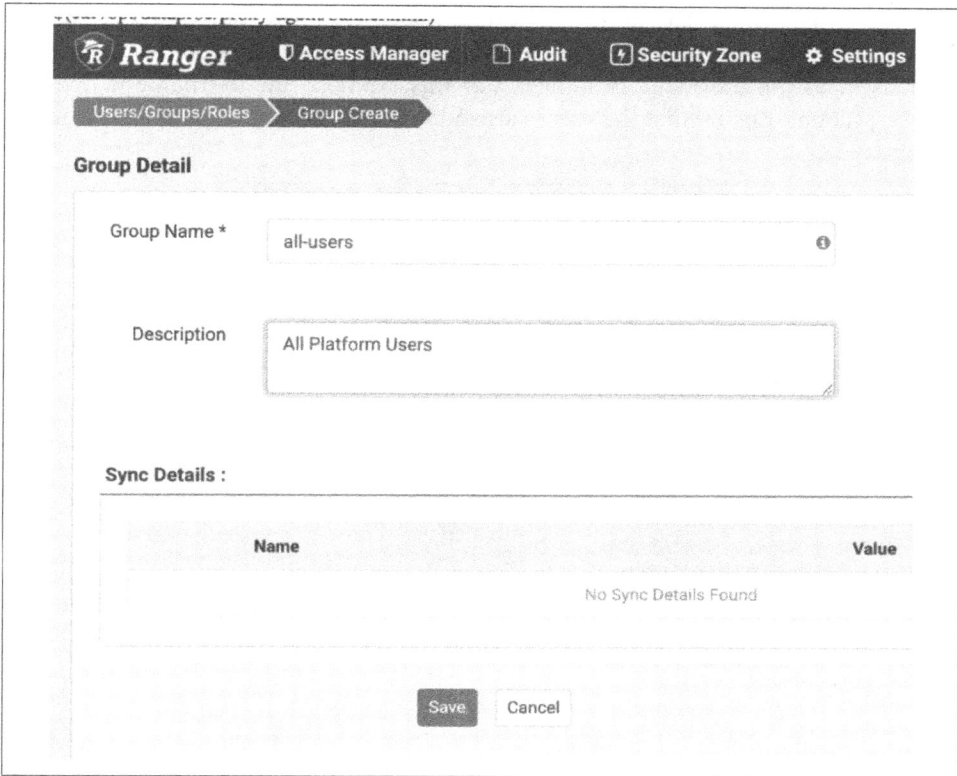

Figure 10-19. Creating a user group in the Ranger web UI

From the Settings tab, click Users/Groups/Roles, as shown in Figure 10-20.

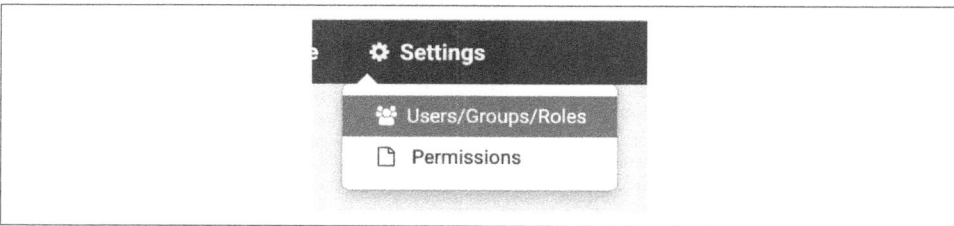

Figure 10-20. Navigating to view the users from the Ranger web UI

Select Add New User, as shown in Figure 10-21.

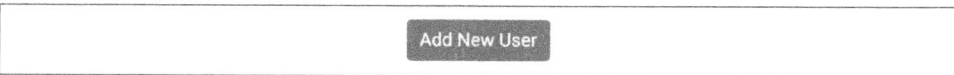

Figure 10-21. Click this button to add a new user

Fill in the user details. The username should match the IAM username but exclude the domain name. For example, if the Google IAM name is *user1@domainname.com*, enter user1 as the username in Ranger. For this example, the username is finance-service-account. Ensure that the user is also added to the all-users group, as shown in Figure 10-22.

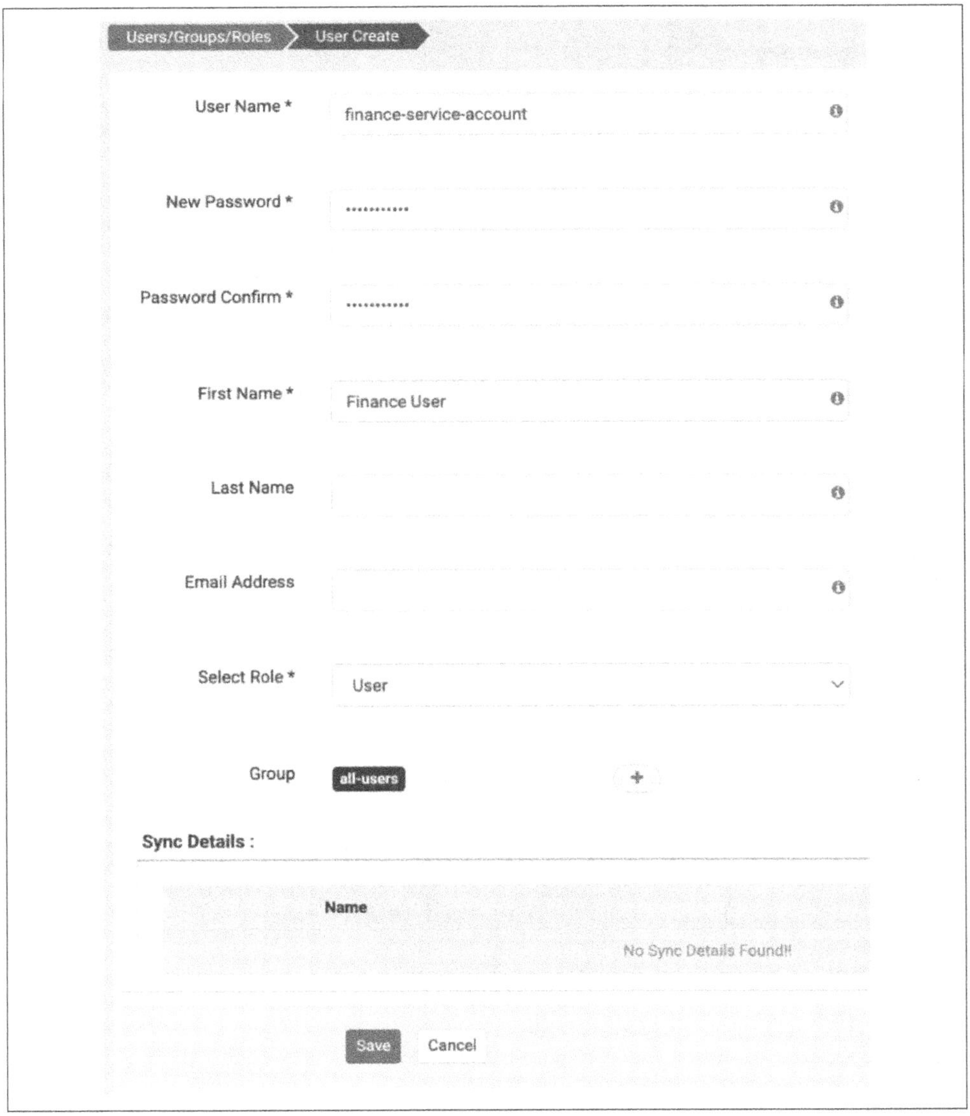

Figure 10-22. New user creation form

Repeat this step to create two more users: it-service-account and hr-service-account. Once all three users have been created, you should be able to see them in Ranger, as in Figure 10-23.

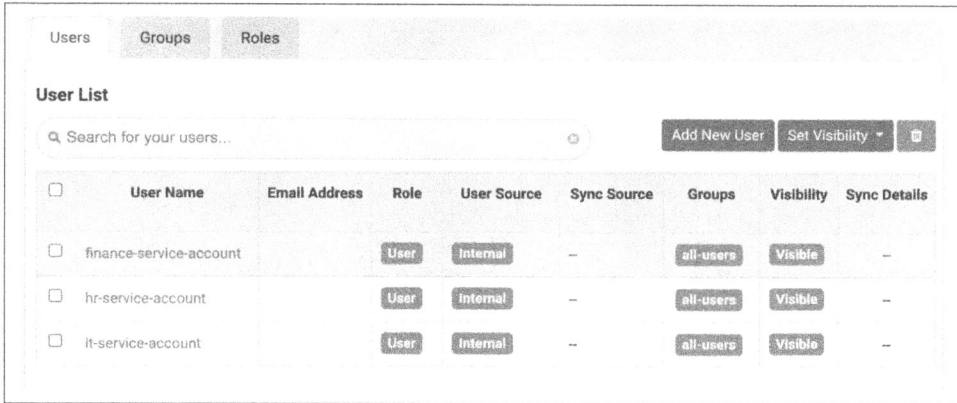

Figure 10-23. Ranger web UI showing the three users accounts that have been created

Creating Ranger policies to secure access to folders

To enforce policies on HDFS folder access, we'll create a Ranger policy within the HDFS service. On the Ranger home page, under the HDFS folder, click the "hadoop-dataproc" link, as indicated in Figure 10-24.

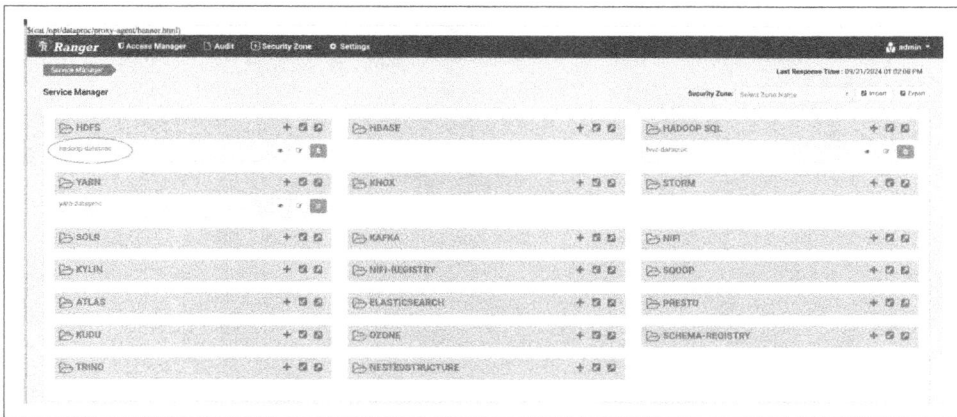

Figure 10-24. Accessing policies from the Ranger UI

This will show you the default existing policies. Click the Add New Policy button to create a policy, as shown in Figure 10-25.

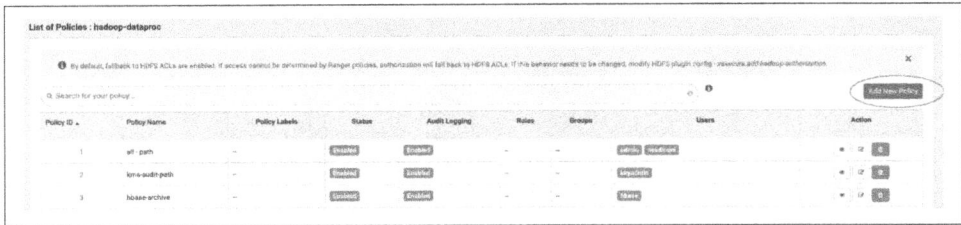

Figure 10-25. Default access policies for an HDFS service

Creating a Ranger policy

Once you've selected Add New Policy, a new page will populate. Fill in the following fields:

- Name of the policy
- Filepath where the policy should be applied
- Allow/deny condition

In our case, we want to block access to all users except the owner of the group. Let's walk through how we would create a policy for the Finance group. First, enter the policy name and the resource path on which the policy should be applied, as shown in Figure 10-26.

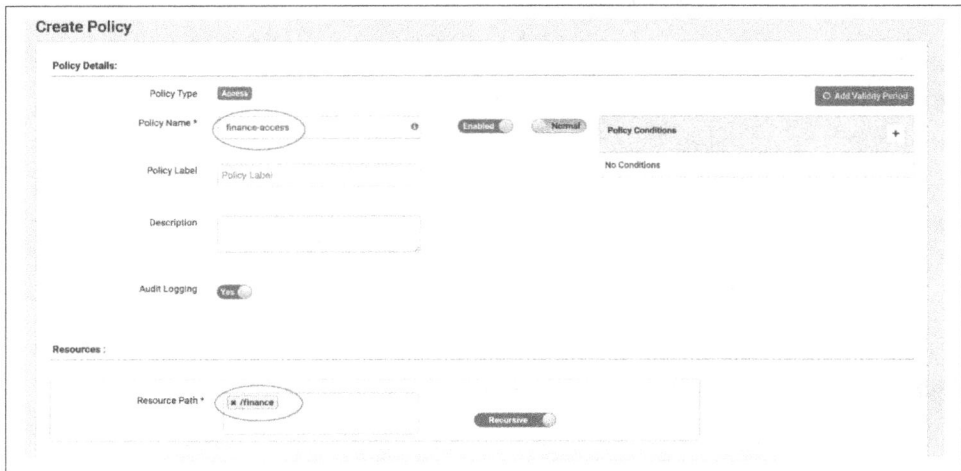

Figure 10-26. Naming the Ranger policy and choosing the resource path

Leave the Allow Conditions section empty, as shown in Figure 10-27.

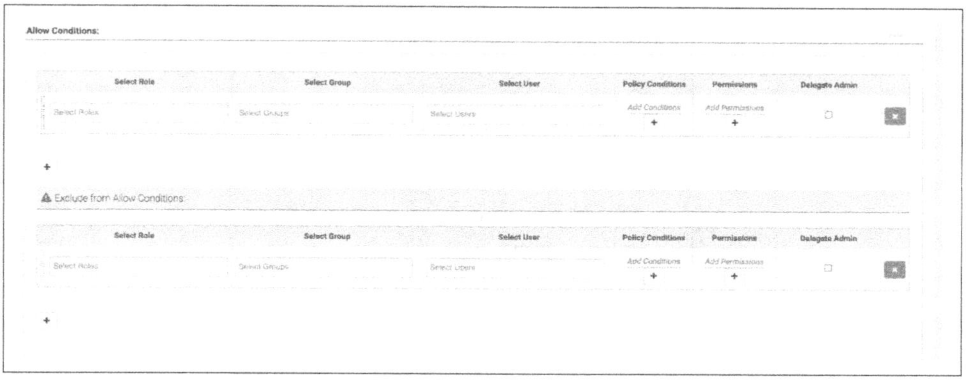

Figure 10-27. Allow conditions for the access policy

Configure a deny condition to deny all users except the finance-service-account, which should be the owner of this group, as shown in Figure 10-28.

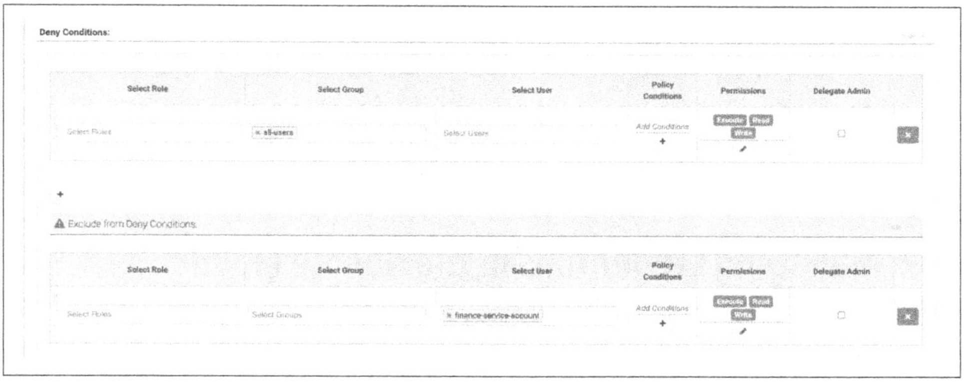

Figure 10-28. Deny conditions for the access policy

Click Add Policy for these changes to take effect.

Repeat this step for the IT and HR folders by creating a new Ranger policy for each group. Once you have created all three policies, you should be able to see policies like those shown in Figure 10-29 along with the default ones.

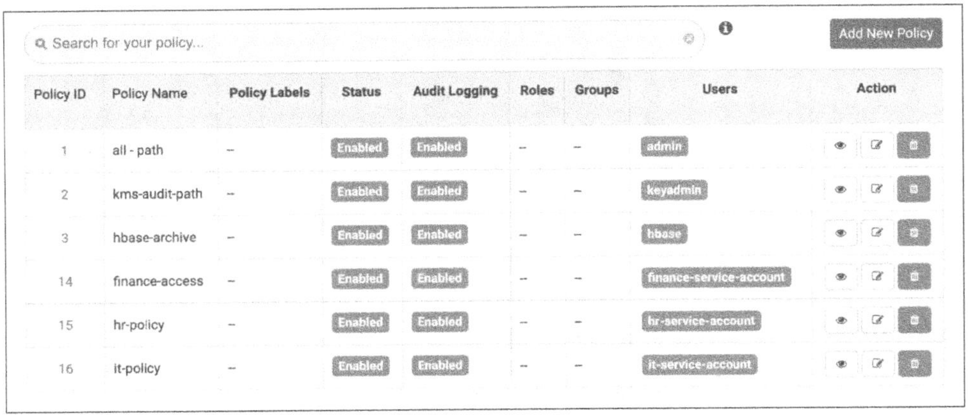

Policy ID	Policy Name	Policy Labels	Status	Audit Logging	Roles	Groups	Users	Action
1	all - path	--	Enabled	Enabled	--	--	admin	
2	kms-audit-path	--	Enabled	Enabled	--	--	keyadmin	
3	hbase-archive	--	Enabled	Enabled	--	--	hbase	
14	finance-access	--	Enabled	Enabled	--	--	finance-service-account	
15	hr-policy	--	Enabled	Enabled	--	--	hr-service-account	
16	it-policy	--	Enabled	Enabled	--	--	it-service-account	

Figure 10-29. Default and newly created access policies for an HDFS service

Now, if user-finance tries to access the HR folder, they will get an access error like this:

```
Caused by: org.apache.hadoop.ipc.RemoteException(org.apache.hadoop.security.
AccessControlException): org.apache.ranger.authorization.hadoop.exceptions.
RangerAccessControlException: Permission denied: user=finance-service-account,
access=EXECUTE, inode="/hr/teragen2/_temporary/1"
```

10.6 Managing Credentials in the Google Cloud Environment

Problem

The Spark job running on Dataproc needs to securely read data from MySQL by using a username and password. The password must be stored in a secure location to ensure proper authentication without exposing sensitive credentials.

Solution

GCP's Secret Manager provides a secure, centralized way to store sensitive information like passwords. By storing the MySQL password in Secret Manager, you can ensure that credentials are encrypted and managed securely.

Once the password is stored in Secret Manager, there are two options for accessing it in the Spark job:

- Use the Secret Manager API to programmatically retrieve the password at runtime, ensuring secure access without hardcoding credentials.

- Integrate Hadoop credentials with Secret Manager, allowing the Spark job to securely fetch the password during execution without modifying the job's code. This method simplifies the process by using built-in Hadoop credential mechanisms.

Discussion

GCP's Secret Manager provides a secure way to store sensitive information like passwords. By leveraging Secret Manager, you ensure that credentials are encrypted and securely managed. Let's start with creating a secret in Secret Manager.

Creating a secret

The user needs a Secret Manager role of admin or higher to create a secret.

If Secret Manager is not already enabled, issue the following gcloud command to enable it in your project:

```
gcloud services enable secretmanager.googleapis.com
```

Create a secret using the following gcloud command:

```
echo -n "your-password" | gcloud secrets create my-sql-password --data-file=-
```

Alternatively, you can use the Google Cloud console to create a secret.

Grant the secret accessor role to users who need to access the secret. In this case, jobs running in the Dataproc cluster will use a service account. Therefore, we'll grant the role to the service account:

```
gcloud projects add-iam-policy-binding YOUR_PROJECT_ID \
  --member=serviceAccount:YOUR_SERVICE_ACCOUNT_EMAIL \
  --role=roles/secretmanager.secretAccessor
```

We have now stored the password in Secret Manager.

Accessing a secret in a Spark job (PySpark) and using it to connect to a SQL server

First, install the Python package of Google Cloud Secret Manager. Here is the shell command for installing the google-cloud-secret-manager Python package:

```
pip install google-cloud-secret-manager
```

Here is the Python function that takes arguments as a secret name and returns a password string:

```scala
import com.google.cloud.secretmanager.v1.SecretManagerServiceClient

def getSecret(secret_id, project_id): String = {
  val client = SecretManagerServiceClient.create()
  val secret_name = f"projects/{project_id}/secrets/{secret_id}/versions/latest"

  val response = client.accessSecretVersion(name=secret_name)
  // Get the secret payload
  val secretData = response.getPayload.getData.toStringUtf8
  client.close()

  secretData
}
```

Next, use PySpark code to access the SQL password using Secret Manager and connect to MySQL:

```python
from pyspark.sql import SparkSession
import mysql.connector

# Fetch the MySQL password from Google Secret Manager
project_id = "your-gcp-project-id"
secret_id = "my-sql-password"
mysql_password = get_secret(secret_id, project_id)


# MySQL connection properties
jdbc_url = "jdbc:mysql://<mysql-host>:3306/<db-name>"
connection_properties = {
    "user": "your-mysql-username",
    "password": mysql_password,
    "driver": "com.mysql.cj.jdbc.Driver"
}

# Reading data from MySQL using the password fetched from Secret Manager
df = spark.read.jdbc(
    url=jdbc_url,
    table="<your-table-name>",
    properties=connection_properties
)
```

Hadoop credentials approach

Hadoop credentials are a secure way to store and manage sensitive information like passwords and tokens that are needed for accessing external resources such as databases or other services within a Hadoop cluster. They provide a centralized, protected way to handle credentials, ensuring that they are not stored in plain text and are accessible only to authorized components.

When working with on-premises versions of Hadoop, credentials are typically stored in file formats such as *.credentials* files or Java KeyStore (JKS) formats, which are then referenced by applications for secure access. However, in a Google Cloud environment, Secret Manager provides a centralized and more secure way to store and manage sensitive credentials. Instead of relying on local files, Hadoop credentials can be integrated with Secret Manager, allowing secrets to be securely stored, retrieved, and managed using the centralized service. This integration enhances security by leveraging GCP's IAM roles and policies for access control, reducing the risk of exposing sensitive data in plain-text files while simplifying secret management across distributed cloud applications.

The following Hadoop credential command creates a new secret and stores it in Secret Manager (note: this command has to run from a cluster VM machine):

```
hadoop credential create mysql-password -provider gsm://projects/PROJECT_ID -v \
   <value-of-pass-word-here>
```

The -provider field specifies that Google Secret Manager (*gsm://*) within the given PROJECT_ID should be used to store or retrieve the actual secret value.

Here's a PySpark code snippet for accessing the password runtime in a Spark job:

```
from pyspark.sql import SparkSession
from org.apache.hadoop.conf import Configuration
from org.apache.hadoop.security import Credentials

# Initialize Spark session
spark = SparkSession.builder \
    .appName("PySpark Hadoop Credentials") \
    .getOrCreate()

# Hadoop Configuration to access credentials
hadoop_conf = spark._jsc.hadoopConfiguration()

# Get credentials provider path
provider_path = "gsm://projects/YOUR_PROJECT_ID"

# Set the provider path in the Hadoop configuration
hadoop_conf.set("hadoop.security.credential.provider.path", provider_path)

# Use Hadoop Credentials API to access the stored password
credentials = Credentials.readTokenStorageFile(None, hadoop_conf)
password_token = credentials.getSecretKey("mysql.password")


# MySQL connection properties
jdbc_url = "jdbc:mysql://<mysql-host>:3306/<db-name>"
connection_properties = {
    "user": "your-mysql-username",
    "password": password,  # Use the password fetched from Hadoop credentials
    "driver": "com.mysql.cj.jdbc.Driver"
}
```

```
# Read data from MySQL
df = spark.read.jdbc(
    url=jdbc_url,
    table="<your-table-name>",
    properties=connection_properties
)

# Show the data
df.show()

# Stop the Spark session
spark.stop()
```

When submitting jobs using the Dataproc API, you can pass the `hadoop.secu rity.credential.provider.path` as a configuration option when submitting the job.

Here is a gcloud command to run a job with Hadoop security credentials configuration:

```
gcloud dataproc jobs submit pyspark your_pyspark_job.py \
    --cluster=your-cluster-name \
    --region=your-region \
    --properties=spark.hadoop.hadoop.security.credential.provider.path=gsm://
projects/YOUR_PROJECT_ID
```

When Should You Access the Secret Manager API Versus Integrating Hadoop Credentials with Secret Manager?

If your existing Spark jobs already use native Hadoop credentials for accessing secrets or if you aim for a cloud-agnostic, abstract approach, integrating with Hadoop credentials might be a suitable option. This approach leverages the familiar Hadoop mechanisms and can simplify the integration process for existing workloads.

If you need fine-grained control over secrets, such as rotating credentials, logging access, or retrieving multiple secrets during the job's execution, using the Secret Manager API directly provides greater flexibility. This method is best suited for applications that require dynamic access to secrets or need to manage different versions of credentials.

10.7 Enforcing Restrictions Across All Clusters

Problem

Your organization needs to enforce consistent policies for Dataproc clusters by restricting machine types to N2D and C2D series and requiring predefined cost center labels for cost tracking and compliance. How can you apply these constraints effectively across all clusters?

Solution

To enforce the rules to use specific machine types and labels, create the following organization constraints:

Create an organization policy for machine types
Define an organization policy that restricts the allowed machine types for Dataproc clusters to only N2D and C2D series. This policy will apply to all projects and folders within your organization, ensuring that no clusters can be created with noncompliant machine types.

Create an organization policy for cost center labels
Define another organization policy that enforces the presence of predefined cost center labels on all Dataproc clusters. This policy will require users to provide the necessary labels during cluster creation, enabling accurate cost tracking and compliance.

Discussion

Google Cloud organization constraints are essentially rules that you can set at the organizational level to govern how resources are used across all the projects under your organization. Think of them as guardrails that ensure your cloud environment stays within predefined boundaries, helping you maintain control, security, and compliance.

In large enterprises, architecture and security teams come up with a set of restrictions to be enforced across all projects. Here are examples of some constraints you may see in enterprises:

- Limits on the regions where resources can be created
- Restrictions on the creation of VMs to specific machine types
- Enforcements on the use of shielded VMs for enhanced security
- Limits on the operation systems or versions allowed in compute engines
- Mandates on the use of specific labels for cost tracking and allocation

These rules can be enforced at multiple levels, typically in a hierarchical way. Refer to Figure 1-2 to understand how resources are organized in a Google Cloud environment. Here's a brief recap:

Organization level
> Enforces rules across all resources within the organization

Folder level
> Applies rules to resources under a specific folder; useful for grouping departments or teams

Project level
> Enforces rules on resources within a specific project

Resource level
> Applies constraints directly on specific resources, such as VMs, networks, or Dataproc clusters

Enforcing specific machine types in a Dataproc cluster

Create a YAML file with the details of a custom constraint to enforce that Dataproc master, worker, and secondary worker nodes must use machine types from the N2D and C2D series. Save the file with the name *machine_type_constraint.yaml*:

```
name: organizations/1099439180993/customConstraints/custom.dataprocMasterMachine
Type
resource_types: dataproc.googleapis.com/Cluster
method_types:
  - CREATE
  - UPDATE
condition: |-
  !(
    resource.config.masterConfig.machineTypeUri.matches('n2d-standard.*') ||
    resource.config.masterConfig.machineTypeUri.matches('c2-standard.*') ||
    resource.config.workerConfig.machineTypeUri.matches('n2d-standard.*') ||
    resource.config.workerConfig.machineTypeUri.matches('c2-standard.*') ||
    resource.config.secondaryWorkerConfig.machineTypeUri.matches(
      'n2d-standard.*') ||
    resource.config.secondaryWorkerConfig.machineTypeUri.matches('c2-standard.*')
  )
action_type: DENY
display_name: dataprocmachinetypes
description: ''
```

Running the following gcloud command will apply the custom constraint at the organization level:

```
gcloud org-policies set-custom-constraint machine_type_constraint.yaml
```

Enforcing the cost_center label for Dataproc clusters

Create a YAML file with the details of a constraint to enforce the presence of a label named "cost_center." Save the file with the name *cost_label_constraint.yaml*:

```
name: organizations/1099439180993/customConstraints/custom.dataprocEnvLabel
resource_types: dataproc.googleapis.com/Cluster
method_types:
  - CREATE
  - UPDATE
condition: '!has(resource.labels.cost_center)'
action_type: DENY
display_name: dataproclabel
description: ''
```

Here is the gcloud command for creating a cost label constraint:

```
gcloud org-policies set-custom-constraint cost_label_constraint.yaml
```

Testing the organization policies

If you create a cluster with a machine type as E2 series that is missing the cost center label, an error will be thrown. This gcloud command tests cluster creation and should produce an error due to the machine type restriction:

```
gcloud dataproc clusters create <CLUSTER_NAME> \
  --region <REGION> \
  --master-machine-type e2-standard-2 \
  --worker-machine-type e2-standard-2 \
  --num-workers 2
```

10.8 Tokenizing Sensitive Data

Problem

A data team handling sensitive information on Dataproc needs to implement measures to protect personally identifiable information (PII) stored in Hive tables and Parquet files. They face the challenge of ensuring that this data is not available as plain text while still allowing access for users performing exploratory analyses. How can they effectively enforce data-protection policies and manage access controls to address these concerns?

Solution

To protect sensitive data while allowing data exploration, the team can use the Cloud DLP API to deidentify sensitive information within their Hive tables and Parquet files. This involves creating a deidentification template that specifies the types of PII to mask and the techniques to use (e.g., redaction, masking). Then, they need to modify their Spark jobs to apply this template before making the data available for analysis.

Here are the steps:

1. Create a deidentification template in Cloud DLP. Define the types of PII (names, addresses, etc.) and select the deidentification techniques (redaction, masking, tokenization, etc.) that you want to apply.
2. Modify Spark jobs to integrate with the Cloud DLP API. Ensure that the deidentification template is applied before data is accessed from Hive tables or Parquet files.

Discussion

Let's walk through an example where we have a dataset in Hive or Parquet with the fields shown in Table 10-2.

Table 10-2. Sample customer table data

customer_id	first_name	last_name	zipcode	upc	purchase_amount	purchase_date
1	John	Doe	76052	22423	50.00	2024-09-01
2	Jane	Smith	75063	29231	75.00	2024-09-02

From this data, we can categorize first_name and last_name as PII. To protect customers' privacy, we want to avoid displaying these columns in plain text. However, we still want to provide access to this data in a secure way, enabling users to perform exploratory analysis. For example, even if the data is tokenized, users can still join two tables based on the tokenized values in these columns.

Creating a deidentified template in DLP

Cloud DLP helps organizations discover, classify, and protect sensitive data stored in Google Cloud. It allows you to apply deidentification techniques like masking, redaction, and tokenization to safeguard data while maintaining its utility for analysis.

First, enable a Cloud DLP API if you have not done so already. Here is the gcloud command for enabling the Cloud DLP API:

```
gcloud services enable dlp.googleapis.com
```

We will be using a tokenization technique called *pseudonymization*. This technique requires a wrapped key for secure data handling. Follow the documentation instructions (*https://oreil.ly/9GRAI*) to create a wrapped key.

Creating a KMS key

To ensure secure storage and management of sensitive information like the Ranger admin password, we'll leverage Google Cloud KMS. This involves creating a KMS key to encrypt the password, which adds an extra layer of security. To create a KMS key, first create a key ring using this gcloud command:

```
gcloud kms keyrings create key-ring-for-customer-deid --location global
```

Then, create a KMS key using the following gcloud command:

```
gcloud kms keys create customer-deid-key \
  --location global \
  --keyring key-ring-for-customer-deid \
  --purpose encryption
```

Grant IAM permission to access the key with the following gcloud command:

```
gcloud kms keys add-iam-policy-binding customer-deid-key \
  --location global \
  --keyring my-key-ring \
  --member "serviceAccount:dlp-api@google.com" \
  --role "roles/cloudkms.cryptoKeyEncrypterDecrypter"
```

Create a Base64-encoded Advanced Encryption Standard (AES) key:

```
openssl rand -out "./aes_key.bin" 32
base64 -i ./aes_key.bin
```

Finally, wrap the AES key using the following Cloud KMS key:

```
curl "https://cloudkms.googleapis.com/v1/projects/PROJECT_ID/locations/global/
keyRings/dlp-keyring/cryptoKeys/dlp-key:encrypt" \
  --request "POST" \
  --header "Authorization:Bearer $(gcloud auth application-default \
  print-access-token)" \
  --header "content-type: application/json" \
  --data "{\"plaintext\": \"BASE64_ENCODED_AES_KEY\"}"
```

The result of this command will provide cipher text to be used in later steps to create a deidentified template.

Creating a deidentification template

Search for DLP/Sensitive Data Protection and select the service, as shown in Figure 10-30.

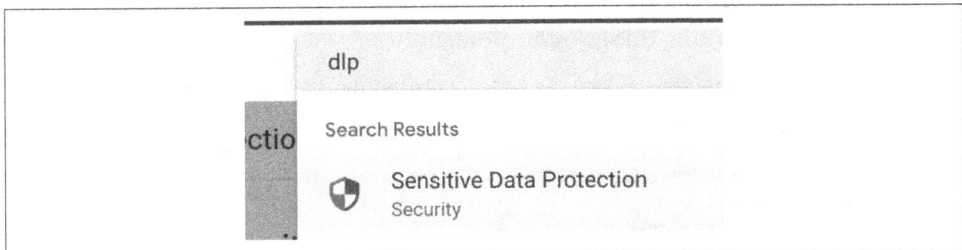

Figure 10-30. Searching for the DLP service

From the Configuration tab, select Templates and De-Identify. Click +Create Template, as highlighted in Figure 10-31.

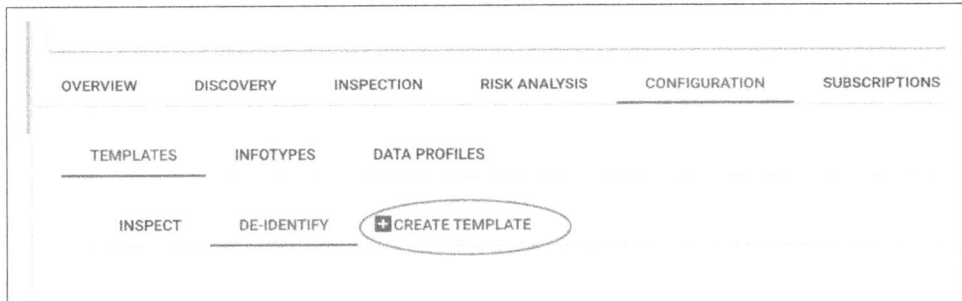

Figure 10-31. Creating a new deidentification template

Complete the template fields as follows:

Template type
    De-identify (remove sensitive data)

Data transformation type
    Record

Name
    First_And_Last_Name_Tokenization_Template

Display name
    Template for tokenizing first and last name of customer

Figure 10-32 shows the completed fields.

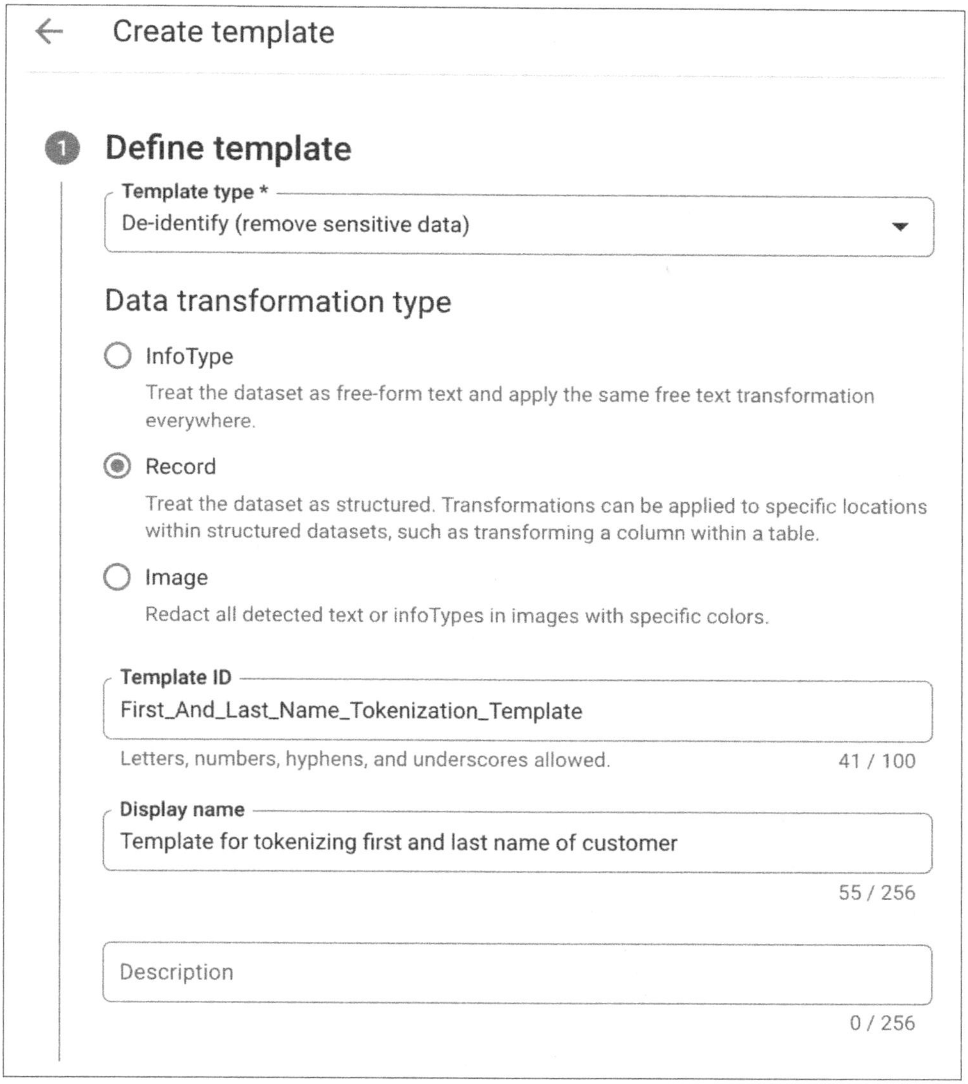

Figure 10-32. Specifying the details of the deidentification template

Choose the location as global or select a specific region and click Continue, as shown in Figure 10-33.

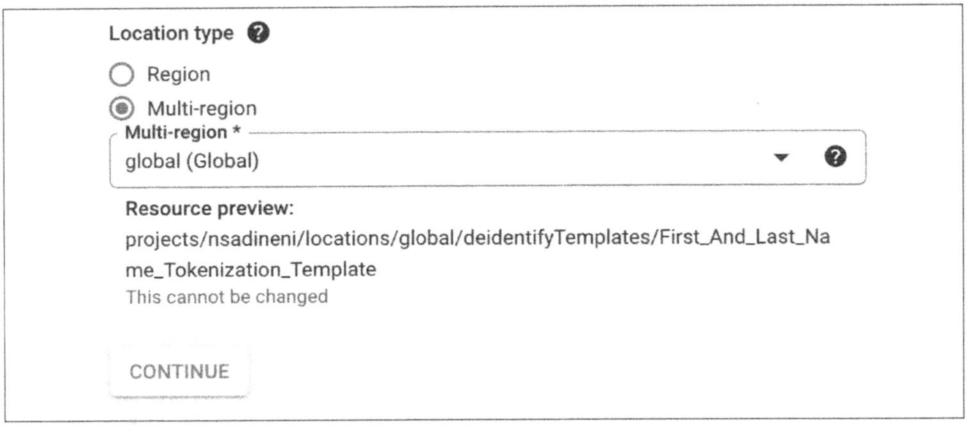

Figure 10-33. Choosing the location as global

Figure 10-34 shows which fields need to be entered for the transformation rule. In this example, you'll want to select "Primitive field transformation" as your transformation type.

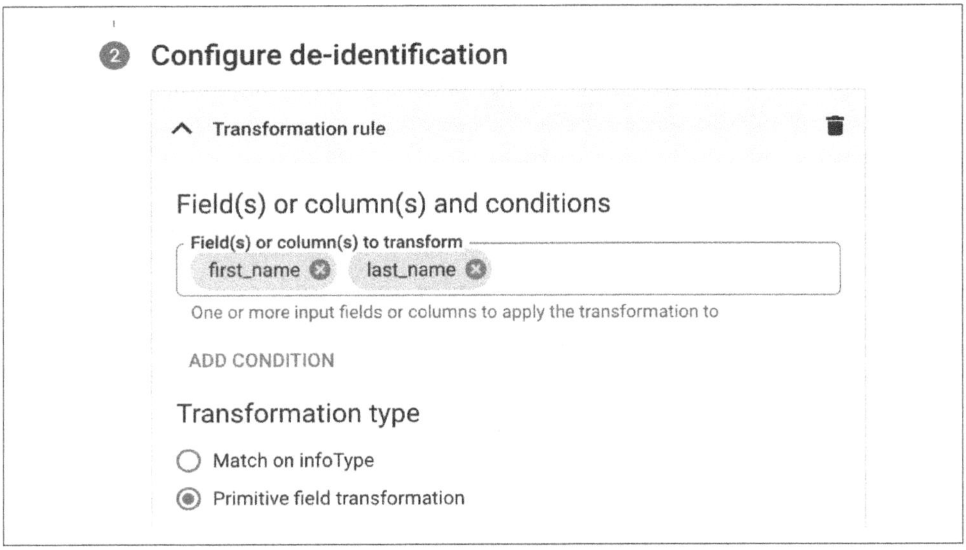

Figure 10-34. Configuring the field names for the transformation rule

Next, select "Pseudonymize (cryptographic deterministic token)" as your transformation method, as shown in Figure 10-35. The crypto key resource name should be the same as the one you previously created. Also provide the wrapped key.

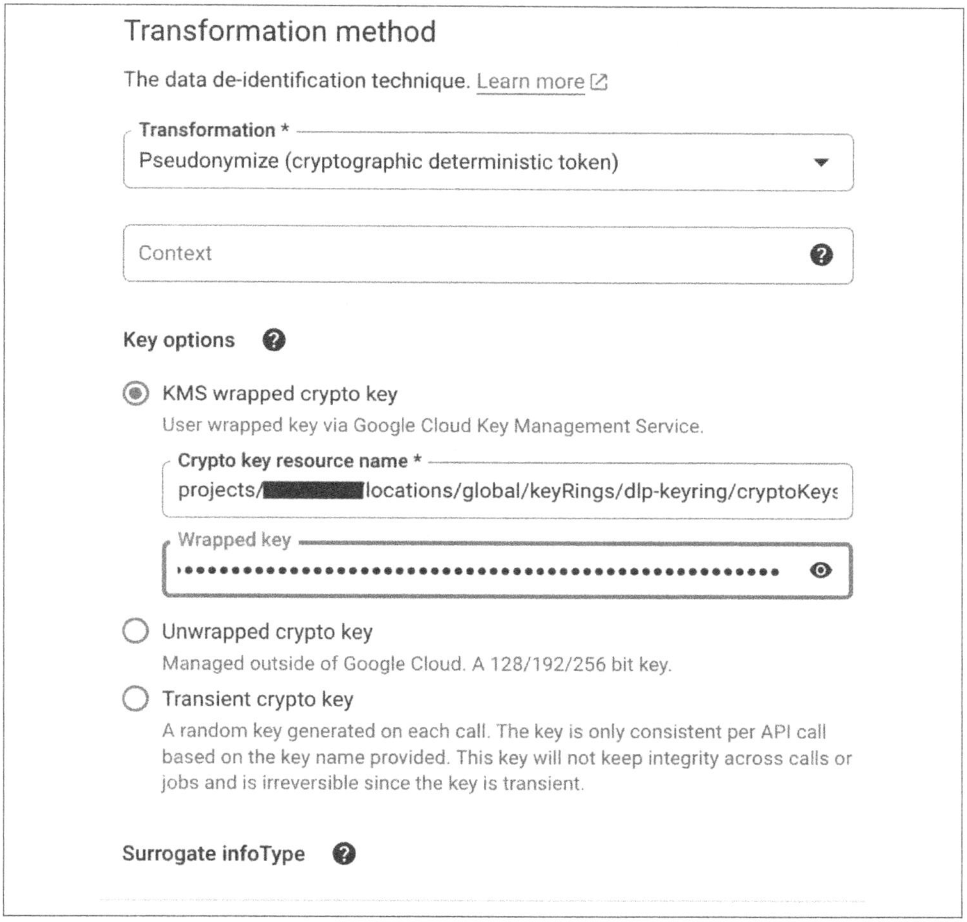

Transformation method

The data de-identification technique. Learn more ☑

Transformation *
Pseudonymize (cryptographic deterministic token)   ▼

Context   ❷

Key options   ❷

⦿ KMS wrapped crypto key
User wrapped key via Google Cloud Key Management Service.

Crypto key resource name *
projects/██████locations/global/keyRings/dlp-keyring/cryptoKeys

Wrapped key
••• ◉

◯ Unwrapped crypto key
Managed outside of Google Cloud. A 128/192/256 bit key.

◯ Transient crypto key
A random key generated on each call. The key is only consistent per API call
based on the key name provided. This key will not keep integrity across calls or
jobs and is irreversible since the key is transient.

Surrogate infoType   ❷

Figure 10-35. Configuring the transformation rules

Finally, click the Create button. Now you have successfully created the template.

Tokenizing first name and last name using a Spark job

First, install the google-cloud-dlp Python package to interact with the Cloud DLP API. Here is the shell command for installing the google-cloud-dlp python package:

```
pip install google-cloud-dlp
```

Here is the Python code you'll need to read data from a Parquet file and tokenize first_name and last_name using the DLP template:

```
from google.cloud import dlp_v2
from pyspark.sql import SparkSession
from pyspark.sql.functions import udf
from pyspark.sql.types import StringType
from pyspark.sql.types import StructType, StructField, StringType
from pyspark.sql.functions import udf, col

# Initialize Spark session
spark = SparkSession.builder.appName("DLPDeID").getOrCreate()

df = spark.read.parquet("gs://nstestb/customer_data/")
df.show()


# Function to de-identify first_name and last_name using DLP template
def deidentify_names(record, project_id, template_id):
    dlp_client = initialize_dlp_client()

    # Construct template name manually
    template_name = f"projects/{project_id}/deidentifyTemplates/{template_id}"

    # Create the content item with first_name and last_name for de-identification
    item = {"table": {
        "headers": [{"name": "first_name"}, {"name": "last_name"}],
        "rows": [{
            "values": [{
                "string_value": record['first_name']},
                {"string_value": record['last_name']
            }]
        }]
    }}

    # Prepare the Deidentify request
    request = {
        "parent": f"projects/{project_id}",
        "deidentify_template_name": template_name,
        "item": item,
    }

    # Call the DLP API
    response = dlp_client.deidentify_content(request)

    # Extract pseudonymized values from the response
    pseudonymized_first_name = response.item.table.rows[0].values[0].string_value
    pseudonymized_last_name = response.item.table.rows[0].values[1].string_value

    return pseudonymized_first_name, pseudonymized_last_name
```

```
# Define a UDF to call DLP API
def pseudonymize(first_name, last_name):
    pseudonymized_fn, pseudonymized_ln = deidentify_names(
        {
            "first_name": first_name,
            "last_name": last_name
        },
        project_id,
        template_id
    )
    return pseudonymized_fn , pseudonymized_ln

# Register the UDF
pseudonymize_udf = udf(pseudonymize, StructType([
    StructField("first_name", StringType()),
    StructField("last_name", StringType())
]))


# Apply the UDF to your DataFrame and create separate columns
df = df.withColumn(
    "pseudonymized_names",
    pseudonymize_udf(col("first_name"), col("last_name"))
).select(
    "*",  # Select all existing columns
    col("pseudonymized_names.first_name").alias("pseudonymized_first_name"),
    col("pseudonymized_names.last_name").alias("pseudonymized_last_name")
).drop(
    "pseudonymized_names"  # Optionally drop the struct column
)

# Show the results
df.show(truncate=False)
```

Once these steps are complete, your output should look similar to the following examples: the DataFrame prior to tokenization in Figure 10-36 and after tokenization in Figure 10-37.

```
: df = spark.read.parquet("gs://nstestb/customer_data/")
  df.show()


+-----------+----------+---------+-------+-----+---------------+-------------+
|customer_id|first_name|last_name|zipcode|  upc|purchase_amount|purchase_date|
+-----------+----------+---------+-------+-----+---------------+-------------+
|          1|      John|      Doe|  76052|22423|             50|   2024-09-01|
|          2|      Jane|    Smith|  75063|29231|             75|   2024-09-02|
+-----------+----------+---------+-------+-----+---------------+-------------+
```

Figure 10-36. DataFrame output with first_name and last_name showing as plain text

```
# Show the results
df.show(truncate=False)

[Stage 2:>                                              (0 + 1) / 1]
+-----------+----------+---------+-------+------+---------------+-------------+------------------------------------+------------------------------------+
|customer_id|first_name|last_name|zipcode|upc   |purchase_amount|purchase_date|pseudonymized_first_name            |pseudonymized_last_name             |
+-----------+----------+---------+-------+------+---------------+-------------+------------------------------------+------------------------------------+
|1          |John      |Doe      |76052  |22423 |50             |2024-09-01   |AUKE4V7siF32U59ArcwtTfKRhMBO        |AXMFe6pKhLiGc5FxiwC0r8acc2Y=        |
|2          |Jane      |Smith    |75063  |29231 |75             |2024-09-02   |ARA1bMzrvu6jyiwWfcVEOfYmc8Vq        |AaecJ6fpWd4pbQlVtZMtaZiczC7ELA==    |
+-----------+----------+---------+-------+------+---------------+-------------+------------------------------------+------------------------------------+
```

Figure 10-37. DataFrame output with first_name and last_name tokenized in newly added columns

## CHAPTER 11

Performance Tuning and Cost Optimization

In the landscape of big data engineering, optimizing both performance and costs is important. This requires a deep understanding of and well-designed big data applications as well as careful benchmarking to ensure optimal performance. Profiling and benchmarking are usually carried out during or after the application-development phase, which helps to optimize both performance and cost.

This chapter delves into the crucial aspects of performance tuning and cost optimization within Dataproc. You'll learn how to size Dataproc clusters, benchmark them, choose appropriate disks, utilize Spark and YARN UIs, optimize Spark jobs, profile them using Sparklens, identify errors, and calculate and optimize the cost of your Dataproc clusters.

11.1 Sizing a Dataproc Cluster

Problem

Your team is onboarding a new application in Dataproc and wants to estimate the size of the cluster.

Solution

Create a list to capture information about the storage and compute requirements and plan the cluster accordingly.

Discussion

To effectively size a Dataproc cluster, you should choose the right set of components in the cluster, including the following:

- Nature of the cluster (static, ephemeral, or serverless)
- Master and worker machine type
- Number of primary workers
- Number and type of secondary workers
- Disk type and size to be attached to each worker node
- Autoscaling configuration

To choose the proper components, start by asking the following questions about the nature of the applications or workloads that are going to run in the cluster:

Application concurrency and compute needs
- How many applications will run simultaneously?
- What type of applications (e.g., Spark, MapReduce, Hive, Pig)?
- What level of computer resources and parallelism is required to meet performance goals?
- What is the core-to-memory ratio requirement?

Data storage and location
- Where is the input data sourced from (e.g., on disk, GCS)?
- Where will the output data be stored?

Data transformation characteristics
- What types of transformations will be performed on the data?
- Do the transformations involve significant data shuffling (e.g., large joins, aggregations)?
- Are the transformations computationally intensive (e.g., Cartesian joins)?

Data volume
- What is the total size of the data to be processed?
- What is the total size of the data to be read from the source and written to the targets?

Workload capacity range
- What are the anticipated minimum and maximum workload capacities required for the cluster?

Let's consider some example scenarios where you are given a requirement and have to recommend the right sizing of the cluster.

Scenario 1: Configuring a cluster for batch requirements

We need to configure a cluster size for a nightly batch application that runs from 5 P.M. to 3 A.M. (10 hours). The application processes one hundred jobs, with a maximum of five jobs running in parallel. Data sizes for each job range from 100 MB to 100 GB, with an average size of 50 GB. The transformations range from simple to medium complexity, avoiding highly compute-intensive operations such as Cartesian joins or window functions. The data is stored in GCS.

From these requirements, we can establish the following facts:

Max parallel jobs
> The cluster must handle up to five jobs running in parallel at any given time.

Data size per job
> Each job processes 50 GB of data on average, with sizes ranging from 100 MB to 100 GB. A large portion of jobs may fall closer to the 50 GB average, but a notable proportion could approach 100 GB.

Core utilization
> The average processing capacity for each core is assumed to be 250 MB. For an average job size of 50 GB, the total core requirement is calculated as:

$$\text{Cores per job} = 50 \text{ GB} \div 250 \text{ MB per core} = 200 \text{ cores}$$

> For five parallel jobs, the total number of cores needed would be:

$$\text{Total cores} = 5 \times 200 = 1{,}000 \text{ cores}$$

Cluster lifespan
> The cluster is ephemeral, meaning it will only be active during the nightly processing window (5 P.M. to 3 A.M.).

Table 11-1 lists the recommended configuration.

Table 11-1. Dataproc cluster configuration for Scenario 1

Component	Configuration
Nature of the cluster	Ephemeral
Master machine type	n2d-standard-8
Worker machine type	n2d-standard-32
Number of primary workers	Two
Number of secondary workers	Two
Disk size to be attached to each worker node	Local SSD 375 GB
Autoscaling configuration	Enabled for secondary workers Min 5 to max 25

Rationales:

- Since the jobs only run nightly, an ephemeral cluster allows for cost-efficient use of resources.

- The master node is primarily responsible for managing the cluster and coordinating job execution, not for processing heavy workloads. Therefore, a machine type with eight vCPUs (n2d-standard-8) is sufficient to handle these administrative tasks.

- The worker node is where the actual processing happens. With the estimated need of two hundred cores per job, a worker machine with 32 vCPUs (n2d-standard-32) strikes a good balance of power and efficiency for processing 50 GB of data per job on average.

- Starting with two primary workers is adequate as these nodes will mainly handle cluster management, so minimal resources are needed initially. The setup can scale as needed based on job requirements and autoscaling triggers.

- Enabling autoscaling for secondary workers allows the cluster to dynamically adjust based on the number of parallel jobs running. This prevents resource wastage when fewer jobs are processed and ensures that the cluster can handle more jobs if needed.

- Autoscaling from 2 to 25 secondary workers provides flexibility. During off-peak hours, fewer workers are needed, but during peak job-execution periods (e.g., when five jobs run in parallel), the cluster can scale to meet the demand without unnecessary manual intervention.

- A good size for each worker is 375 GB of local SSD storage. This ensures that there is enough temporary storage for intermediate data and processing files during the batch jobs without running into disk-space issues.

- For 50 GB jobs, an estimate of two hundred cores is a practical and reasonable configuration. This ensures that jobs are processed within an acceptable time frame.

Scenario 2: Designing a streaming application

We need to design a Spark-based streaming application for continuous data ingestion, real-time aggregations, and timely output to BigQuery. The application will consume data from a 60-partition Kafka topic with a constant volume throughout the day. The application must adhere to strict SLAs to ensure reliable, low-latency processing. The cluster must handle this continuous workload with high reliability and minimal latency.

From these requirements, we can establish the following facts:

- The streaming application consumes data from a 60-partition Kafka topic. Each Kafka partition is processed by one consumer thread in Spark, meaning there are 60 threads to handle the data.

- Each n2d-standard-32 worker node has 32 vCPUs. With two worker nodes, there are a total of 64 vCPUs, which is sufficient to handle 60 threads (one per Kafka partition) with some overhead for system processes.

- The application requires continuous operation, which means the cluster must run all day without interruption.

- The workload is constant throughout the day, so there is no need for dynamic scaling. The resources are predictable and stable, and the application does not need to adapt to fluctuating loads.

- The application requires 1 TB of PD SSD storage per worker node for fast data access and temporary storage of streaming data.

Table 11-2 lists the recommended configuration.

Table 11-2. Recommended cluster configuration

Component	Configuration
Nature of the cluster	Static
Master machine type	n2d-standard-8
Worker machine type	n2d-standard-32
Number of primary workers	Two
Number of secondary workers	Zero
Disk size to be attached to each worker node	PD SSD 1 TB
Autoscaling configuration	Disabled

Rationales:

- With 60 partitions in Kafka and two worker nodes with 32 vCPUs each, the cluster can easily support 60 threads, one per partition, and handle the constant stream of data. There's no need for three primary workers unless you have additional tasks or require fault tolerance, which would justify adding another worker for redundancy or capacity.

- Allowing 1 TB of PD SSD per worker node ensures ample storage space for processing streaming data, minimizing latency in disk I/O operations. SSD is ideal for the high-speed, low-latency requirements of Spark streaming.

- Since the workload is constant and the system requires predictable, stable resource allocation for low-latency processing, autoscaling is not necessary. The cluster size remains fixed to meet the steady demand.

11.2 Choosing the Right Disks for Big Data Workloads on Dataproc

Problem

You're remigrating or deploying a new big data application to Dataproc. You need to select and configure the appropriate disks for optimal performance and cost efficiency.

Solution

The optimal disk configuration depends on your workload type and cost and your performance priorities. Based on these criteria, we would make the following recommendations:

MapReduce or Hive workloads
- Choose PD SSD disks for high throughput and fast reads/writes.

- Consider PD balanced or local SSDs or a combination of both, which are more performant than PD standard and have a lower cost compared to PD SSDs.

- If there are no strict SLAs, consider standard PDs backed by hard disk drives (HDDs) for a lower cost, but be aware of potential performance limitations.

Spark workloads
- If your job has strict SLAs and expensive shuffles that require high IOPS, use a combination of local SSD and PD SSD disks.

- If you have cost and performance priorities, use a combination of local SSDs and PD balanced for greater performance and optimized cost.

- If it's a simple batch job with no strict SLAs, leverage PD-standard disks.

Discussion

There are four possible disk options: PD standard, PD SSD, PD balanced, and local SSDs. PD standard, PD SSD, and PD balanced are all storage devices but have significant differences in how data is stored and accessed. Understanding these key differences is crucial for cost and performance optimization.

PD standards are HDDs that use mechanical spinning disks for data writes and reads. PD SSD, PD balanced, and local SSDs use flash memory chips, making them nonmechanical, and they offer faster data access and lower latency.

If you have a long-running Spark job or a MapReduce job with high IOPS requirements that demands better performance and low-latency shuffles, consider configuring local SSDs along with persistent boot disk backed by SSDs. Leverage PD-balanced disks and local SSDs as a compromise between the speed of PD SSDs and the affordability of HDDs.

> When you add local SSDs, Dataproc automatically leverages them for writing intermediate shuffle output and will use PDs for the operating system.

Table 11-3 lists the features of PD SSDs and local SSDs and their differences.

Table 11-3. Summary of the key differences between PD SSDs and local SSDs

Feature	PD SSDs	Local SSDs
Persistence	Yes	No (temporary storage)
IOPS	High	Very high
Latency	Moderate	Very low
Attachment	Network	Direct to VM
Cost	Higher	Lower
Capacity	Varies	375 GB each (up to eight per node)

Deciding which SSD to use depends on your desired speed, throughput, and cost as well as the data-persistence requirements of your Dataproc workload. Here are a few things to consider:

- PD SSD is a network-attached persistent disk. Local SSDs are attached directly to the VMs and are highly performant.
- If the Spark job requires caching, low-latency shuffles, and intermediate data storage at a lower cost, local SSDs with the NVMe interface are a cost-effective

way to enhance performance alongside PDs backed by SSDs (PD SSD or PD balanced) for caching and shuffle operations.

- Even with the same total storage, adding more local SSDs increases IOPS, improving Dataproc's performance for workloads that need fast shuffle and caching operations.

- Consider using local SSDs and PD balanced for better performance and optimized cost when running Spark jobs.

- Consider using local SSDs and PD SSDs if performance takes more precedence than cost when running Spark jobs.

11.3 Benchmarking Clusters with Performance Tuning

Problem

You want to run Benchmark tests to measure the cluster performance of disk, compute, and network aspects.

Solution

Run the benchmarking script to capture the performance of compute, disk, and network.

Discussion

Benchmarking processes run specific jobs to measure performance in a given environment. These jobs can be industry standards like TeraGen, TeraSort, or TPC-DS or custom jobs representative of your application's workload.

Let's consider some important questions:

What should we benchmark?
> You'll want to benchmark computing performance of the machine and read/write performance of all your I/Os, including disk, GCS, network, database connection, and so on. There could be additional application-specific requirements, such as calling a third-party API, encrypting data, and the like.

Which tools should I use?
> When benchmarking, it is beneficial to use real workloads if possible. By doing so, you can ensure that the results of your benchmarking are accurate and reflect real-world usage. If you cannot use a single workload that sufficiently reflects your workload environment, you should run multiple workloads. This will help ensure that your testing covers a wider range of scenarios and use cases.

If real workloads can't be run, then you can use open source tools like TeraGen, Tera-Sort, DFSIO, and the like. Each of these tools covers a unique set of performances. Let's take a closer look at each of these open source tools:

TeraGen
> Generates data with each record or line counting as 100 bytes

TeraSort
> Used to sort the input data; involves shuffling data across nodes, categorized as network-intensive operations

DFSIO
> Creates files with a specific size; involves I/O-intensive operations and reading/writing from disk

Creating a cluster to run benchmarking jobs

Use the following command to create a Dataproc cluster for running the benchmark jobs:

```
gcloud dataproc clusters create benchmarkn2d \
  --region us-central1 \
  --master-machine-type n2-standard-4 \
  --master-boot-disk-size 500 \
  --num-workers 5 --worker-machine-type n2d-standard-8 \
  --worker-boot-disk-size 500 \
  --image-version 2.1-debian11 \
  --enable-component-gateway \
  --max-idle=30m
```

> To ensure accurate and reliable benchmarking results, it is essential to minimize external factors that may interfere with the job's execution. This includes ensuring that no other jobs are running concurrently with the benchmark jobs since they can compete for resources and affect the performance measurements.

Running a TeraGen job to generate 500 GB of data

TeraGen generates random data, and each record created is 100 bytes in size. This allows users to configure the number of records/lines to be generated for the file. For example, passing 1000 as a parameter generates $10,000 \times 100$ bytes = 1 MB of data. To generate 500 GB, we pass the value of 5000000000 (five billion records). Each record, with a size of 100 bytes, generates a total of 500 billion bytes, equivalent to 500 GB of data (shown in Figure 11-1).

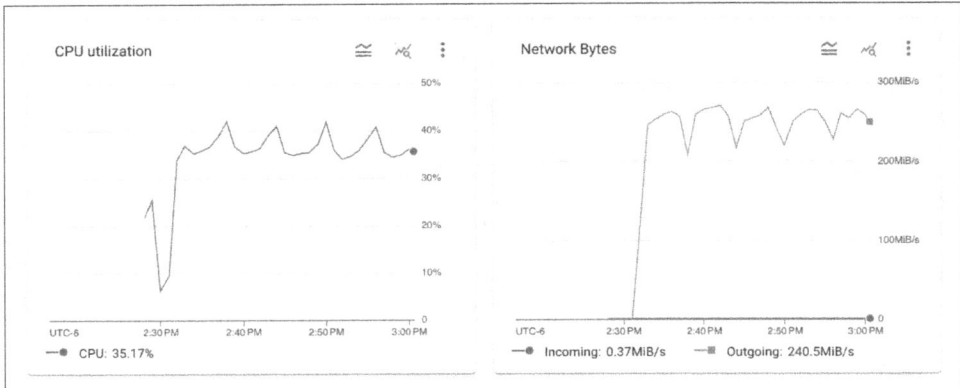

Figure 11-1. Capturing CPU utilization and network bytes in a cluster

Use the following command to run a TeraGen job:

```
time gcloud dataproc jobs submit hadoop \
  --cluster=benchmarkn2d \
  --properties mapreduce.job.maps=39 \
  --region=us-central1 \
  --jar=file:///usr/lib/hadoop-mapreduce/hadoop-mapreduce-examples.jar \
  -- teragen 5000000000 gs://<bucket-name-here>/terasort/input
```

> To print the time duration at the end of the job, add the `time` command as a prefix to the command. Using the `time` command before the command in the terminal will display the time taken to complete the job when it's finished.
>
> The code needed for running TeraSuite (TeraGen, TeraSort, etc.) jobs is present in the JAR file named *hadoop-mapreduce-examples.jar*. During the cluster-creation process, this JAR file is included by default.

Running TeraSort to sort 500 GB of data generated by TeraGen

TeraSort takes the input and output files as parameters. It performs a sorting operation on the input data and stores the output in the output folder. Here's an example command for running a TeraSort job:

```
time gcloud dataproc jobs submit hadoop \
  --cluster=benchmarkn2d \
  --region=us-central1  \
  --jar=file:///usr/lib/hadoop-mapreduce/hadoop-mapreduce-examples.jar \
  -- terasort gs://<bucket-name-here>/terasort/input gs://<bucket-name-here>/
terasort/output
```

Capturing benchmarking results

When you are conducting benchmarking tests using TeraSuite or real-world applications, it is crucial to capture various metrics to evaluate the performance and resource utilization of the system. Here are the metrics that should be captured:

Job duration and runtime

Measure the total elapsed time from the start of the job until its completion. This metric provides an overall indication of the job's performance and can be used to compare the efficiency of different configurations or optimizations.

Resource utilization metrics

To understand how your system handles resource demands, it's essential to monitor the following key metrics:

CPU utilization

Monitor the percentage of CPU resources utilized during the job's execution. High CPU utilization indicates that the job is making efficient use of the available processing power. Low CPU utilization, on the other hand, may suggest underutilization of resources or potential bottlenecks.

Network I/O bandwidth

Measure the amount of data transferred over the network during the job's execution, in terms of both inbound and outbound traffic. This metric helps assess the network's capacity and performance in supporting the job's requirements.

Disk read/write speeds

Monitor the read/write speeds of the storage devices used by the job. These metrics provide insights into the storage subsystem's performance and can identify any potential I/O bottlenecks that may affect the job's execution time.

> These metrics are automatically captured when using the Google Cloud Monitoring API. For instructions on how to resource metrics, see Chapter 12.

By capturing and analyzing these metrics, you can gain valuable insights into the performance characteristics of the system being tested. This information can help identify areas for improvement, optimize resource utilization, and ensure that the system meets the desired performance requirements.

> ## How Often Should I Run Benchmarking?
>
> The benchmarking scripts should be executed whenever critical platform or application changes are made to ensure that no negative impact occurs. Each time benchmarking is performed, the results should be compared to previous results to identify any deviations.

11.4 Navigating the Spark UI

Problem

As a user, you want to monitor the Spark UI to view jobs, stages, tasks, and all the metrics. This will allow you to identify any potential issues and optimize Spark applications.

Solution

Navigate to Web Interfaces in the Dataproc history server or the existing cluster and then access the Spark history server, as shown in Figure 11-2.

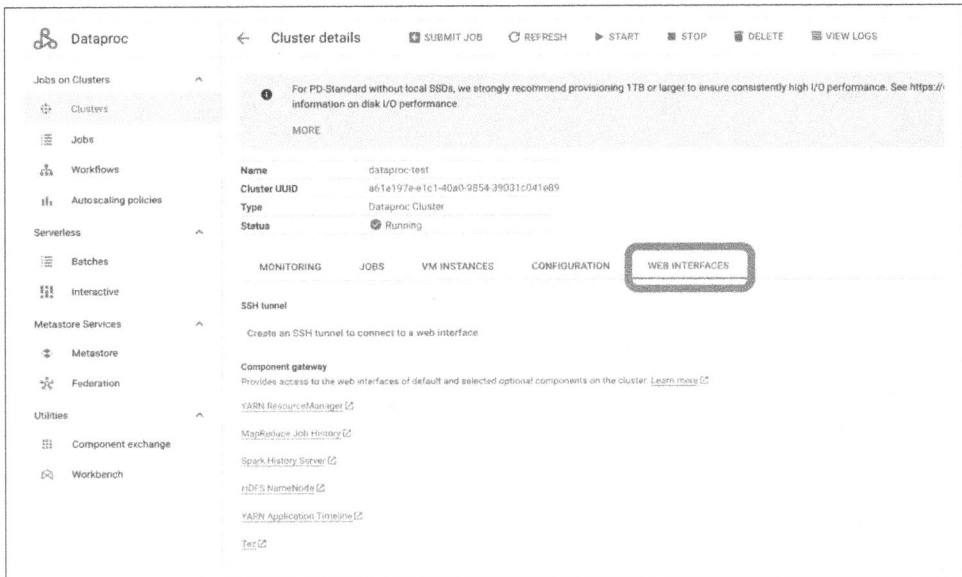

Figure 11-2. Navigating to Web Interfaces in your existing cluster

The Spark history server allows you to monitor the Spark applications in the Spark UI, as shown in Figure 11-3.

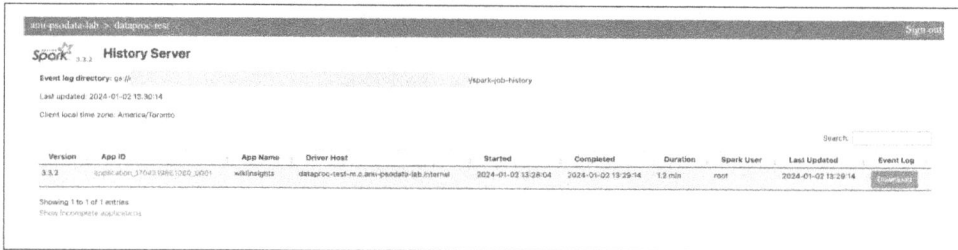

Figure 11-3. Accessing the Spark UI

Discussion

The Spark history server renders the event logs. It is very important to understand the Spark UI to optimize the performance of Spark jobs.

You can navigate to the Spark history server of the existing cluster, as shown in Figure 11-2. If the cluster is deleted and you have set up a Dataproc Persistent History Server (PHS), you can navigate to the Spark UI from that. If you haven't set up a PHS, you can download the event logs from GCS and either configure Spark history server locally or configure PHS to view the Spark UI. This discussion assumes you have access to the Spark UI.

Let's consider an example. We will use a sample Spark job that reads CSV input files, applies aggregations, and writes the results in Parquet format:

```
import sys
from pyspark.sql import SparkSession
from pyspark.sql.functions import *
from pyspark.sql.types import *

spark = SparkSession.builder.appName("wikiinsights") \
    .master("yarn") \
    .getOrCreate()

#Get the input path and output path
inputPath = sys.argv[1]
outputPath = sys.argv[2]

print("Reading the CSV input file")

df = spark.read.csv(inputPath,inferSchema=True,header=True)

df.groupBy("language").count().write.mode("overwrite").parquet(outputPath)

print("Application Completed!!!")

# Closing the Spark session
spark.stop()
```

In this example, the Spark application is reading sample Wikipedia page views from January 2008 with monthly data stored as CSV file partitions in GCS. This Spark application has two actions (read and write), and the transformations are `groupBy` and `count` on grouped data.

Spark actions trigger one or more Spark jobs. The jobs are further broken down into stages and then tasks that run on executors. The results are returned to the driver.

There are multiple tabs in the Spark UI that provide useful information: Jobs, Stages, Storage, Environment, Executors, and SQL.

The Jobs tab displays a list of jobs, as shown in Figure 11-4. Each job is broken down into stages, and one job might have multiple stages. Clicking the job description takes you to the corresponding Stages section, which shows a list of stages and execution times. Stages are broken down into tasks, which are then assigned to executors.

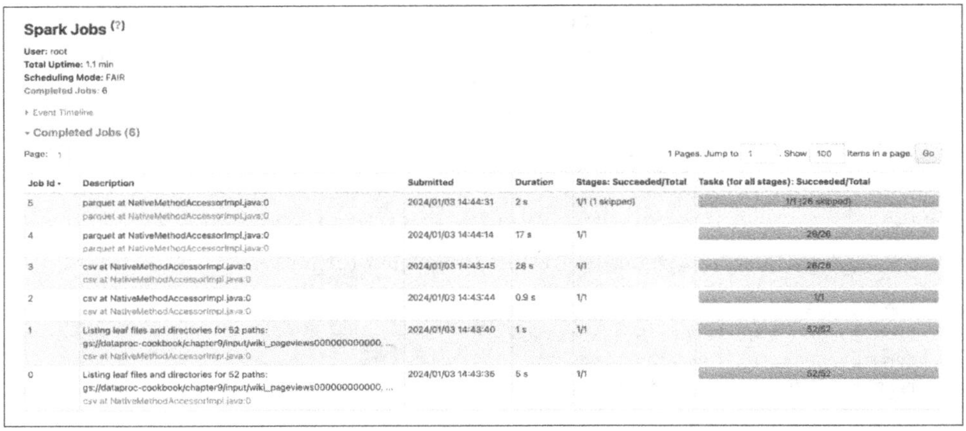

Spark Jobs (?)

User: root
Total Uptime: 1.1 min
Scheduling Mode: FAIR
Completed Jobs: 6

▸ Event Timeline

▾ Completed Jobs (6)

Page: 1

1 Pages. Jump to 1   . Show 100   items in a page. Go

Job Id ▾	Description	Submitted	Duration	Stages: Succeeded/Total	Tasks (for all stages): Succeeded/Total
5	parquet at NativeMethodAccessorImpl.java:0 parquet at NativeMethodAccessorImpl.java:0	2024/01/03 14:44:31	2 s	1/1 (1 skipped)	1/1 (26 skipped)
4	parquet at NativeMethodAccessorImpl.java:0 parquet at NativeMethodAccessorImpl.java:0	2024/01/03 14:44:14	17 s	1/1	26/26
3	csv at NativeMethodAccessorImpl.java:0 csv at NativeMethodAccessorImpl.java:0	2024/01/03 14:43:45	28 s	1/1	26/26
2	csv at NativeMethodAccessorImpl.java:0 csv at NativeMethodAccessorImpl.java:0	2024/01/03 14:43:44	0.9 s	1/1	1/1
1	Listing leaf files and directories for 52 paths: gs://dataproc-cookbook/chapter9/input/wiki_pageviews000000000000, ... csv at NativeMethodAccessorImpl.java:0	2024/01/03 14:43:40	1 s	1/1	52/52
0	Listing leaf files and directories for 52 paths: gs://dataproc-cookbook/chapter9/input/wiki_pageviews000000000000, ... csv at NativeMethodAccessorImpl.java:0	2024/01/03 14:43:36	5 s	1/1	52/52

Figure 11-4. The Jobs tab in Spark

Clicking the Stages tab displays the tasks summary metrics and aggregated metrics by executors, as shown in Figure 11-5. This section has task aggregation metrics that help with monitoring for data skew, spills to disk, or straggled tasks. For example, if there is a big gap between median and maximum duration and one task is significantly longer than the others, then there could be skew. In that case, you can consider techniques like repartitioning or salting to reduce the skews. If the skew is less but each task is handling a large amount of data, then it will take more time to complete. In those instances, you can tweak Spark shuffle partitions.

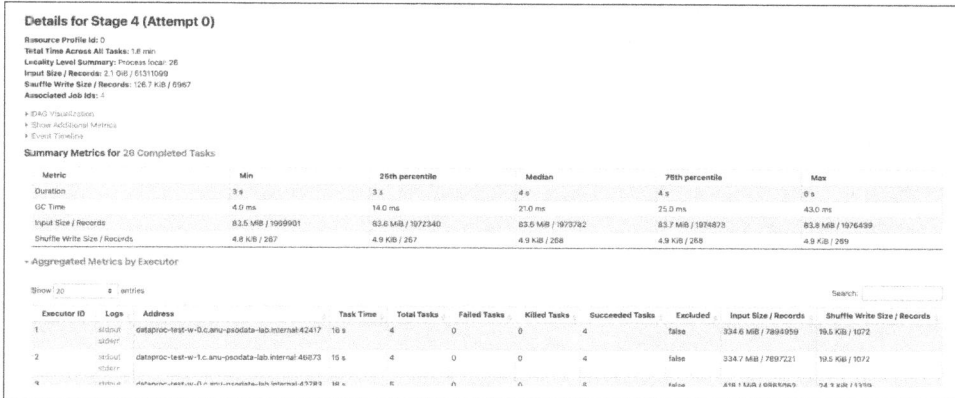

Figure 11-5. Details of the Tasks tab in Spark

Selecting the Executors tab displays the number of executors that process the data and where they ran, as shown in Figure 11-6. In this tab, you can expand the window to show additional metrics and look at peak execution and storage execution memory to see if the executor compute size is optimal for this job.

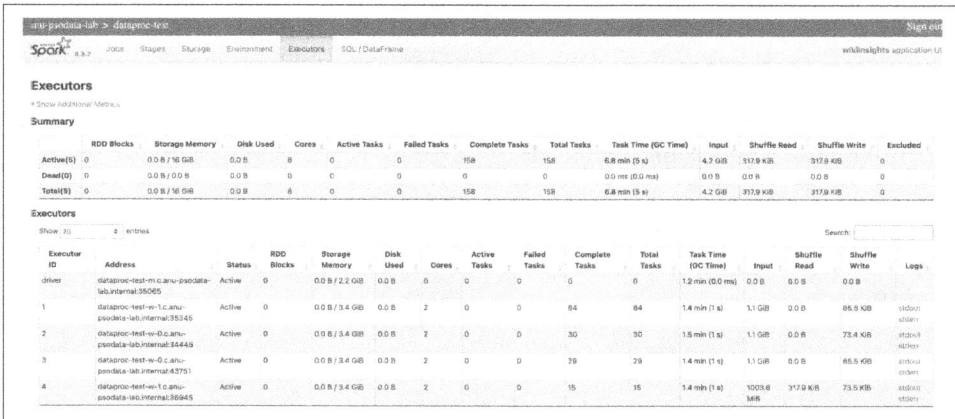

Figure 11-6. The Executors tab in Spark

The Environment tab shows the runtime details of the application. The Storage tab displays details of persisted RDDs and DataFrames.

Spark SQL is one of the most important tabs to monitor when optimizing Spark queries. Figure 11-7 shows a list of the Spark queries that are executed as well as the execution time of the queries.

SQL / DataFrame				
Completed Queries: 2				
▾ Completed Queries (2)				
Page: 1	1 Pages. Jump to 1 . Show 100 items in a page. Go			
ID ▾	Description	Submitted	Duration	Job IDs
1	parquet at NativeMethodAccessorImpl.java:0 +details	2024/01/02 18:28:52	21 s	[4][5]
0	csv at NativeMethodAccessorImpl.java:0 +details	2024/01/02 18:28:25	1 s	[2]

Figure 11-7. The SQL tab in Spark

When you submit the Spark application to the driver, the Catalyst Optimizer generates the optimized logical plan by applying rule-based and cost-based optimization. The SQL tab shows the physical plan of the queries. This tab presents the execution time of each operator, the time to get metadata like number of files, and shuffle exchange metrics, such as the number of bytes written by a shuffle. These are very important metrics when determining the appropriate disks to use and executor sizing and for optimizing the application code.

11.5 Optimizing Spark Jobs

Problem

Your Spark job on Dataproc is taking longer than expected. How can you improve its performance?

Solution

The most appropriate solution will depend on your workload:

- For compute-intensive workloads, leverage high-CPU machines.
- For memory-intensive workloads, leverage high-memory machines.
- For the disks, leverage local SSDs in addition to PDs based on SSDs.

Then set the appropriate executor memory, executor cores, driver memory, driver cores, and shuffle partition configurations.

Discussion

To identify the bottleneck that is affecting your speed, you'll want to take the following steps:

1. Monitor the Spark UI (see Recipe 11.4).
2. Profile by using tools like Sparklens to observe the wall-clock time, task skew, and critical path (see Recipe 11.6).

In the Sparklens profiling report, if the critical path shows that adding more resources will significantly improve the performance, then complete the following steps:

1. Benchmark with n2-highmem or n2d-highmem. Begin with N2D 16+ cores and benchmark jobs with increased executors in the development environment. N2D VMs powered by AMD EPYC processors support high-bandwidth networks.
2. Leverage primary workers and preemptible secondary workers. Instead of having 50 primary workers, divide it into 25 primary workers and 25 preemptible or spot secondary workers to optimize cost.
3. If it's I/O intensive, combine high-throughput VMs with high-performance local SSDs and PD balanced or PD SSDs.

Now that we have the basics established, let's consider optimizing the Spark configuration.

Dynamic allocation is enabled by default in Dataproc. This means that the executors are added and removed from the Spark application dynamically as needed, based on the resource requirement. If you have multiple Spark applications running on a cluster, you can configure `spark.dynamicAllocation.maxExecutors` accordingly to make sure a specific application doesn't take away all of the resources.

You can also configure the size (cores and memory) of individual executors. The total number of executors depends on the chosen executor size and your cluster's worker nodes. For example, considering three n2-highmem-16 worker nodes:

- Executors taking two cores and 16 GB of memory might have a parallelism of 36 (six executors per node).
- Executors taking three cores and 16 GB of memory might have a parallelism of 36 (four executors per node).

Each task operates on a single core (`spark.task.cpus` = 1 by default). For instance, if you set the executor core count to three, three tasks will execute on each executor, sharing the allocated executor memory.

By default, Dataproc reserves 80% of node manager memory (`yarn.nodeman ager.resource.memory-mb`) for the Spark application, leaving 20% for YARN daemons and the operating system. The data presented in Table 11-4 is based on the assumption that `yarn.nodemanager.resource.memory-mb` is set to 80%. You can use this as a template and build a reusable calculator to choose the executor and driver configuration for Spark applications on ephemeral clusters.

Table 11-4. Calculating driver and executor compute configuration based on the machine type

Input			
spark.executor.cores	2		
vCores	16	80% vCores	13
RAM per node	128	80% RAM	102
Number of nodes	3		
Machine type	n2-highmem-16		
Output			
Number of executors	18	Available vCores × number of nodes / `spark.executor.cores`	
spark.driver.cores	2		

It's also important to consider Spark configurations. `Spark.shuffle.partitions` is the number of shuffle sort partitions. This should be at least equal to the number of cores for large data to optimize resource utilization. When you don't right-size shuffle partitions, you will have each task handling huge amounts of data, leading to longer execution times. There are a couple of options to manage compression:

Shuffle spills
Compress spills with `spark.shuffle.spill.compress=true` to improve shuffle performance.

Broadcast variables
Compress broadcast variables with `spark.broadcast.compress=true` for faster transfers.

Here is a sample configuration for a long-running memory and I/O-intensive Spark job:

```
N2d-highmem-16 (15 Primary and 15 Secondary spot instances)

spark.executor.memory: 15G
spark.driver.memory: 15G
spark.driver.cores: 2
spark.executor.cores: 2
spark.task.maxFailures: 20

300GB pd-SSD and 4 Local SSDs
```

> This is a starting point, and optimal configurations will vary based on your specific workload and requirements. Benchmark and test different options to find the best fit.

11.6 Installing Sparklens for Profiling Spark Applications

Problem

You have developed a Spark application to read CSV files from GCS, perform aggregations, and write in Parquet format. You have run the Spark application on Dataproc, and you want to profile the Spark application to make sure the application code and resource utilization are optimal.

Solution

There are a few options to consider:

- Use Sparklens for profiling the Spark jobs.
- Download and copy the Sparklens JAR to GCS. The JAR can be compiled from the Sparklens GitHub page (*https://oreil.ly/ip9o2*) (the init action might help to automate), or you can download the JAR directly from the public GitHub repo (*https://oreil.ly/4OIVH*).
- SSH into a Dataproc cluster and submit the Sparklens application, passing the history server event log path as an argument:

```
spark-submit --jars {sparklens_gcs_path_here} --class com.qubole.sparklens.
app.ReporterApp qubole-report \
  {spark_event_log_gcs_path} source=history \
  --packages qubole:sparklens:0.3.2-s 2.11 \
  --conf spark.extraListeners=com.qubole.sparklens.QuboleJobListener
```

Discussion

Sparklens is an open source Spark profiling tool. Typical goals for profiling are:

- Reduce the application execution time
- Make the application performant with fewer resources

The following `spark-submit` command can be used to profile and submit the event log of a sample Spark job (shared in Recipe 11.5) to Sparklens. This application runs on Dataproc n2-highmem-4 (two workers):

```
spark-submit --jars gs://biglaketest8/sparklens/sparklens_2.12-1.9.3.jar \
  --class com.qubole.sparklens.app.ReporterApp qubole-report \
```

```
gs://dataproc-temp-us-west1-1072535324208-9swwvmwl/a61e197e-e1c1-40a0-9854-
39031c041e89/spark-job-history/application_1704219861060_0005 \
  source=history --packages qubole:sparklens:0.3.2-s 2.11 \
  --conf spark.extraListeners=com.qubole.sparklens.QuboleJobListener
```

Now, let's interpret the Sparklens profiling output. To do this, we'll use the sample Sparklens reports:

- Driver and executor wall-clock times
- Critical path
- Simulation of wall-clock time by adding or reducing executors
- Compute wastage and utilization
- Task skew

Driver and executor wall-clock times

Driver wall-clock time is the amount of time spent by a driver. In this example, it is 17.99%. Executor wall-clock time is 82.01%. This is a good start. The time spent by the driver should be significantly less than the executors because tasks run on executors.

Critical path

We usually think that if we have infinite resources, applications will run faster. This might not be true in all cases. Critical path is the minimal amount of time with infinite resources. For our case, adding more executors will improve the performance. There is room for improvement here to add more resources. Figure 11-8 shows that the total wall-clock time for the app is 1 minute and 7 seconds as well as the critical path and execution time when you just run on a single executor.

```
Time spent in Driver vs Executors
Driver WallClock Time      00m 12s    17.99%
Executor WallClock Time    00m 55s    82.01%
Total WallClock Time       01m 07s


Minimum possible time for the app based on the critical path (with infinite resources)  00m 34s
Minimum possible time for the app with same executors, perfect parallelism and zero skew 00m 55s
If we were to run this app with single executor and single core                         00h 04m
```

Figure 11-8. Driver versus executor wall-clock time, critical path, and single executor run time

Sparklens simulates executor wall-clock time and cluster utilization by adding more resources. The Spark application currently runs on six executors; the actual execution time is about 1 minute. With more executors (i.e., with infinite resources), it will

---

complete in 34 seconds, which is more than 50% performant. If the application has tight SLAs, you can measure the performance by adding more executors.

Let's consider the section in the report shown in Figure 11-9.

Stage-ID	WallClock Stage%	OneCore ComputeHours	Task Count	PRatio	------Task------		OIRatio	|* ShuffleWrite%	ReadFetch%	GC%	*|
					Skew	StageSkew					
0	4.81	00h 00m	52	8.67	42.80	0.41	0.00	|* 0.00	0.00	1.85	*|
1	2.14	00h 00m	52	8.67	9.30	0.18	0.00	|* 0.00	0.00	0.00	*|
2	2.08	00h 00m	1	0.17	1.00	0.75	0.00	|* 0.00	0.00	3.64	*|
3	54.63	00h 02m	26	4.33	2.54	0.42	0.00	|* 0.00	0.00	1.99	*|
4	32.12	00h 01m	26	4.33	1.33	0.30	0.00	|* 0.14	0.00	0.59	*|
6	4.22	00h 00m	1	0.17	1.00	0.96	0.03	|* 0.00	0.00	0.32	*|

Figure 11-9. Sparklens report detailing per-stage performance metrics

It shows three important metrics:

PRatio

Parallelism on stage. This is the number of tasks in a stage divided by the total number of executor cores. In this case, there are six executors, each having one core, so a total of six cores. This means that it can run six tasks in parallel. Let's observe the stages with a higher wall-clock stage percentage—in this case, they are stages 3 and 4. The PRatio is 4.33, which means it takes around four iterations to complete all the tasks in these two stages. This helps you analyze if increasing resources will improve the performance by reducing the PRatio.

Task skew

The degree of skew in the stage. Stages 3 and 4 have lower task skew, which is better.

OIRatio

Output-to-input ratio in each stage. This helps check if there are a lot of intermediate outputs. If this is high, consider running tests with local SSDs and PD SSDs.

One of the next actions from this Sparklens report is benchmarking this application with more executors to reduce execution time.

See Also

- Tutorial on Qubole's Spark Tuning Tool (*https://oreil.ly/lFdZK*)
- "SparkLens: A Profiling Tool for Spark Optimization" (*https://oreil.ly/h-s84*) (Medium article)
- "How to Optimize Spark Applications for Performance Using Sparklens (*https://oreil.ly/Jwl3c*)" (LinkedIn post)

11.7 Identifying Spark Job Errors and Bottlenecks

Problem

You want to understand the methods for identifying specific error causes and performance bottlenecks in Spark applications.

Solution

There are multiple methods for identifying specific errors and performance bottlenecks in Spark applications:

Driver out of memory (OOM)
Increase driver memory or reduce data by using an efficient algorithm

Executor OOM
Increase executor memory allocation to reduce the amount of data processed by each executor

Spill to disk/memory
Increase Spark memory and reduce the amount of data that gets sent to each task by repartitioning the data

Garbage collection (GC) time
Increase memory allocation to reduce GC pressure

Slow-running tasks
Analyze task execution times to identify outliers and investigate the cause of slowdowns in specific tasks (e.g., data skew, resource contention)

Discussion

Understanding and resolving Spark errors, such as out of memory, involves first replicating them with test data. We'll begin by generating this data, demonstrate triggering common errors, and then discuss their potential solutions.

Setting up test data

The following example program will generate sample data that can be used to replicate each of the scenarios. It will generate a file of around 1.8 GB in size:

```
import org.apache.spark.sql.SparkSession
import org.apache.spark.sql.functions._
import scala.util.Random

object EmployeeDataGenerator {

  def main(args: Array[String]): Unit = {
```

```scala
val spark = SparkSession
  .builder()
  .appName("EmployeeDataGenerator")
  .master("local[*]") // Adjust for your cluster environment
  .getOrCreate()

import spark.implicits._

// Data Generation Functions
def randomName(): String =
  Random.shuffle(
    Seq("JohnMary", "MichaelSarah", "DavidEmily", "JenniferRobert")
  ).head
def randomDepartment(): String =
  Random.shuffle(
    Seq("Sales", "InformationTechnology", "HR", "Marketing", "Finance")
  ).head
def randomSalary(): Int = Random.nextInt(150000) + 30000
def randomCity(): String =
  Random.shuffle(
    Seq("New York", "Los Angeles", "Chicago", "Dallas", "Austin")
  ).head
def randomCountry(): String =
  Random.shuffle(
    Seq("USA", "Canada", "Mexico")
  ).head
def randomState(): String =
  Random.shuffle(
    Seq(
      "New York", "California", "Los Angeles",
      "Chicago", "Dallas", "Austin"
      )
  ).head

// Schema
val schema =
  Seq("id", "name", "department", "salary", "city", "country", "state")


// Generate Rows (Tune 'numRecords' for dataset size)
val numRecords = 200000000 // Approximately 1 GB, adjust as needed
val numPartitions = 100

// Parallelized Data Generation
val dataRDD = spark.sparkContext
  .parallelize(1 to numRecords, numPartitions)
  .mapPartitions { partition =>
    partition.map(_ =>
      (
        Random.nextInt(1000000),
        randomName(),
```

```
              randomDepartment(),
              randomSalary(),
              randomCity(),
              randomCountry(),
              randomState()
          )
        )
      }

    // Create DataFrame
    val df = dataRDD.toDF(schema: _*)

    // Output (Save in your desired format)
    df.write.mode("overwrite").parquet(
      "gs://<bucket-name-here>/sparktest/empout"
    )
  }
}
```

Replicating driver OOM

The driver OOM error in Apache Spark occurs when the driver cannot accommodate the amount of data that needs to be processed in its memory. For instance, if the driver memory is 1 GB and an attempt is made to process more than 1 GB in the driver itself, an OOM error may result.

The following sample code snippet generates an OOM error. Save the file as *driver-oom.py*:

```
from pyspark.sql import SparkSession

spark = SparkSession \
    .builder \
    .appName("PySparkCollectExample") \
    .getOrCreate()

# Read from parquet file present in GCS
df=spark.read.parquet("gs://<bucket-name-here>/sparktest/empout")

# Collect the DataFrame into the driver
collected_data = df.repartition(1).collect()
```

To run a Spark job, use the following `spark-submit` command:

```
spark-submit --conf spark.driver.maxResultSize=5G  --deploy-mode cluster \
  --executor-memory 1G test.py
```

The job will fail with error code 143 as the resource manager kills the driver container that tries to collect ~1.8 GB of data. This happens because the job is more than the allowed 1 GB limit (the executor memory was configured as 1 GB in the `spark-submit` command).

To prevent such scenarios, it is advisable to refrain from performing the collect operation on large datasets. The `collect` function in a DataFrame fetches the entire dataset to the driver program. It is acceptable to perform this operation on small datasets or, alternatively, to read a limited number of records using the `show` or `take` functions.

Replicating executor OOM

The executor OOM error occurs in Apache Spark when an executor runs out of memory while processing data. This can happen for several reasons:

Insufficient executor memory
The amount of memory allocated to each executor is insufficient to process the data assigned to it. This can be caused by a number of factors, including:

- The dataset is too large for the available memory.
- The executor is running multiple tasks that require a lot of memory.
- The executor is running tasks that are not efficient in their memory usage.

Memory leaks
A memory leak is a situation where a program allocates memory but does not release it when it is no longer needed. This can cause the executor to run out of memory even if the dataset is not particularly large.

The following code snippet will replicate the executor OOM error:

```
#Read input file from GCS
val df1=spark.read.parquet("gs://<bucket-name-here>/sparktest/empout")
#create a temporary table with name emp
df1.createOrReplaceTempView("emp");
#Forcing the join operation to happen on one single task
spark.conf.set("spark.sql.shuffle.partitions", "1")
#Self join with same table on name column match
val rdf=spark.sql("select t1.*,t2.* from emp t1,emp t2 where t1.name=t2.name")
#Show/print 1 record from join output
rdf.show(1,false)
```

This code will cause the executor (at the join stage) to consume a large amount of memory and get killed by YARN for exceeding the allocation memory threshold, as shown in Figure 11-10.

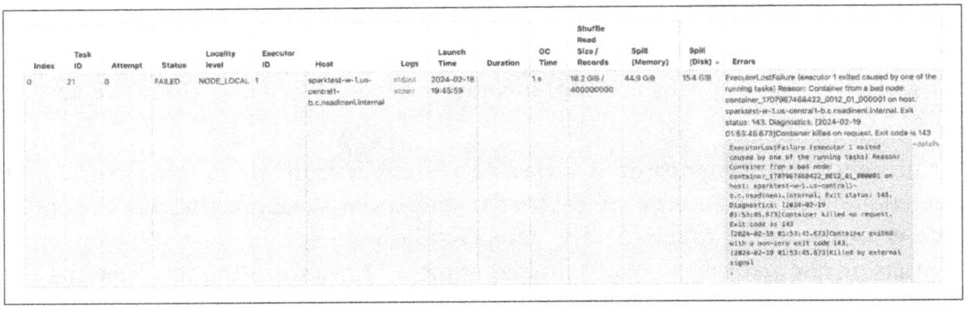

Figure 11-10. Executor killed by YARN ResourceManager for memory utilization

Replicating spill to memory (spill to disk)

Spill to memory or *spill to disk* in a Spark framework refers to the process of temporarily storing intermediate data on memory or disk when the in-memory storage of the Spark executor is insufficient to hold all the data. This typically occurs when the size of the intermediate data exceeds the executor's memory capacity.

Spill to memory or spill to disk can happen for various reasons, including:

Insufficient executor memory
> When the size of the intermediate data exceeds the executor's memory capacity, Spark is forced to spill the data to disk or memory.

Large intermediate datasets
> Some operations, such as joins or aggregations, can produce large intermediate datasets that cannot fit in memory.

Skewed data
> If the data is skewed, with a small number of keys receiving a large number of values, that can lead to memory pressure and spills.

Spill to memory or spill to disk can significantly affect performance since it introduces additional overhead in terms of data serialization, deserialization, and disk I/O. Spilling to disk is particularly expensive as it involves reading and writing data to physical storage devices, which is much slower than accessing data in memory.

Let's take a look at a code snippet that could trigger spill to memory and disk at the task level:

```
#Read input file from GCS
val df1=spark.read.parquet("gs://<bucket-name-here>/sparktest/empout")
#create a temporary table with name emp
df1.createOrReplaceTempView("emp");
#Self join with same table on name column match
val rdf=spark.sql("select t1.*,t2.* from emp t1,emp t2 where t1.name=t2.name")
```

```
#Show/print 1 record from join output
rdf.show(1,false)
```

This code reads from GCS and attempts to perform a join operation. The initial step of execution involves reading input files from Parquet files. Figure 11-11 shows a snapshot of one of the tasks when run with lower executor memory (1 GB). It took 2 minutes to complete and required spilling 1 GB of data to memory and 465 MB to disk. Due to the spill operation, the task ran longer than the typical execution time.

Show 20 entries

Search:

Index	Task ID	Attempt	Status	Locality level	Executor ID	Host	Logs	Launch Time	Duration	GC Time	Input Size / Records	Shuffle Write Size / Records	Spill (Memory)	Spill (Disk)	Errors
0	2	0	SUCCESS	PROCESS_LOCAL	2	sparktest-w-0.us-central1-b.c.nsadineni.internal	stdout stderr	2024-02-18 20:32:55	2.0 min	0.9 s	92.1 MiB / 10000000	465 MiB / 10000000	1 GiB	440.5 MiB	

Figure 11-11. Task spilling to memory and disk running for a 2-minute duration

The task when executed with higher executor memory (4 GB) was completed in 34 seconds, and no overflow to memory or disk was observed, as shown in Figure 11-12.

Index	Task ID	Attempt	Status	Locality level	Executor ID	Host	Logs	Launch Time	Duration	GC Time	Input Size / Records	Shuffle Write Size / Records	Errors
0	1	0	SUCCESS	PROCESS_LOCAL	2	sparktest-w-1.us-central1-b.c.nsadineni.internal	stdout stderr	2024-02-18 20:40:18	34 s	0.3 s	92.1 MiB / 10000000	465 MiB / 10000000	

Figure 11-12. Task with no spills completed in 34 seconds

There are several techniques you can employ to avoid spills:

Increase executor memory
: Increasing the executor memory can prevent spills by providing more memory for intermediate data. However, this may not be feasible in all cases, especially for large datasets.

Tune Spark parameters
: Tuning Spark parameters such as executor memory and memory management properties like memory fraction (`spark.memory.fraction`) and storage fraction (`spark.memory.storageFraction`) can help optimize memory usage and reduce the likelihood of spills.

Optimize data partitioning
: Optimizing data partitioning (`spark.sql.shuffle.partitions`) can help distribute data more evenly across executors, reducing the risk of memory pressure on individual executors.

It is possible to have a few tasks running longer than other tasks in a Spark program. This can be caused by several factors:

Data skew
  If the data is skewed, with a small number of keys receiving a large number of values, that can lead to memory pressure and spills. This can cause some tasks to run more slowly than others.

Slow tasks
  Some tasks may be inherently slower than others due to the nature of the computation. For example, tasks that involve complex calculations or I/O operations may take longer to complete.

Resource contention
  If tasks are competing for the same resources, such as memory or CPU, that can lead to slower execution times.

Here are some tips for preventing a few tasks from running longer than other tasks:

- Optimizing data partitioning can help distribute data evenly across executors, reducing the risk of memory pressure on individual executors.
- Increasing the executor memory can prevent spills by providing more memory for intermediate data.
- Tuning Spark parameters can optimize memory usage and reduce the likelihood of spills.

11.8 Understanding the YARN UI

Problem

You want to understand the YARN UI and the important metrics to monitor.

Solution

The YARN service offers three web UIs, each helping to perform specific tasks.

The YARN ResourceManager (YARN RM) is the central management UI for the YARN cluster. It offers the following options:

Cluster overview
  Provides information on resource utilization and running and completed applications

Nodes
  Indicates the node manager's status and metrics

Applications
   Gives details about running, failed, and completed applications

The YARN Application Timeline Server focuses on historical data with the following options:

In-depth view of single applications
   Includes durations, task attempts, and resource consumption

Logs
   Access to application and container logs

Visualization
   Visual tools for progress and resource usage

The MapReduce Job History Server is specific to the MapReduce framework. This server provides the following:

Job historical insights
   Details of past MapReduce jobs

Individual tasks
   Task attempts, durations, and logs

Discussion

YARN is a core compute framework installed on the cluster that is leveraged by frameworks like Hive, Tez, and Spark for running workloads. YARN manages resources from the operating system and allows these services to use them in the form of containers. Each container has a CPU and memory allocation. For example, Spark executors run on top of YARN containers. Alternatives to YARN are Mesos or Kubernetes.

The YARN service provides a web UI for browsing through the cluster capacity (cores and memory), jobs (running, pending, etc.), and scheduling information. Here are some of the questions the YARN RM can help you answer:

- What is my cluster's total capacity, allocated capacity (currently in use), and available capacity?

- How many jobs are running in the cluster, what are their current statuses (accepted, running, failed, completed, etc.), and what are their details?

- How are resource queues organized, and what are their current utilization statuses?

To access the YARN RM web UI, enable the web component gateway using the `--enable-component-gateway` option at the time of cluster creation.

The YARN RM web UI has three main menu options:

- Cluster information
- Applications and their detailed information (tasks, counters, etc.)
- Scheduling queues

Let's consider each of these options.

Cluster information path

To navigate to the YARN RM UI, go to the Dataproc cluster home page and click the Web Interfaces tab. Then, click the YARN ResourceManager link, highlighted in Figure 11-13.

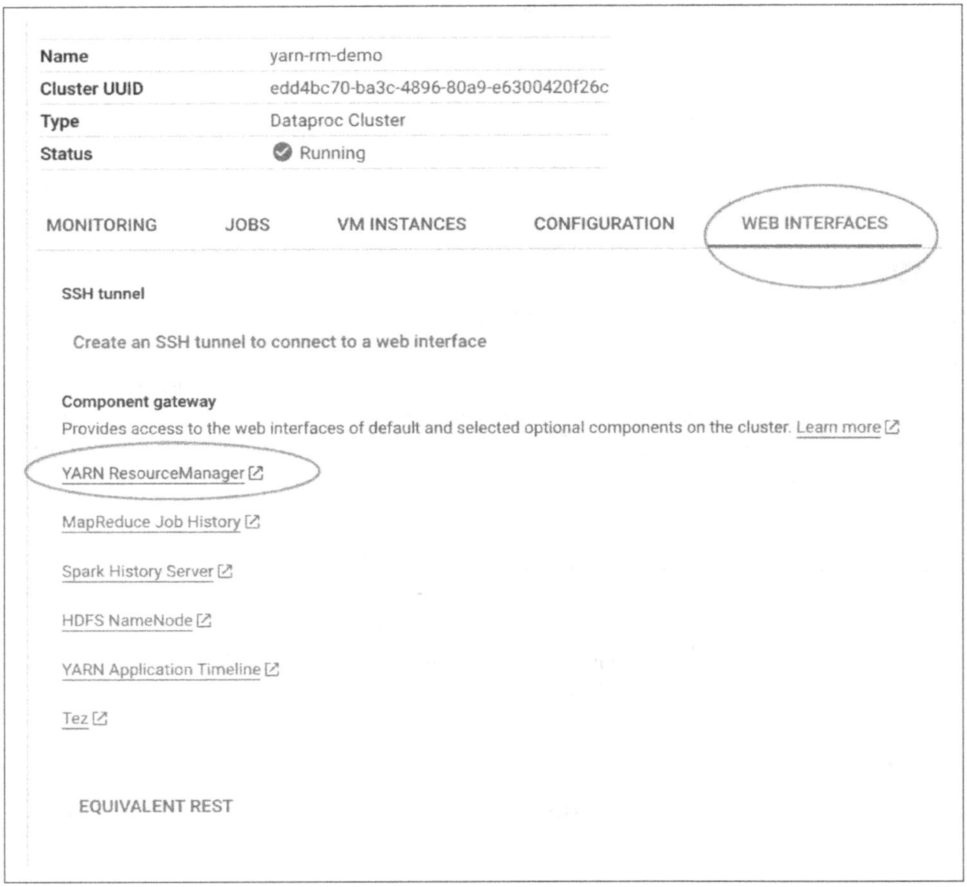

Figure 11-13. Accessing the YARN RM web UI

The home screen of YARN RM gives you information about the cluster's current state and total resources, as shown in Figure 11-14.

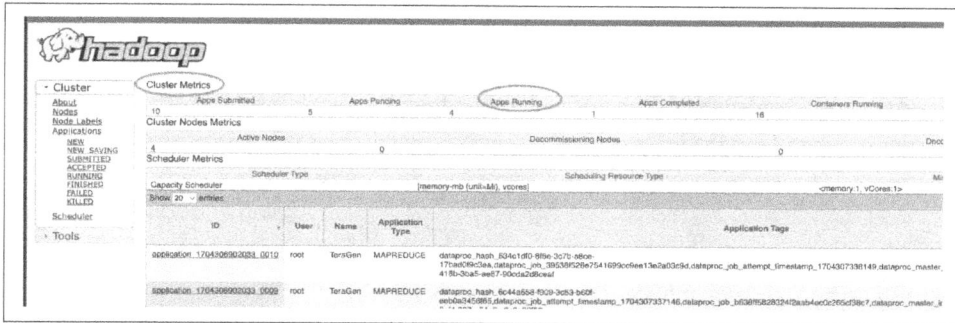

Figure 11-14. Cluster metrics (active and decommission nodes) and application details (pending, running, and completed) in the YARN RM UI

Scrolling to the right on the YARN RM UI reveals details about total cluster resources and the amount currently in use. In Figure 11-15, we can see that the cluster's total resources are 52.91 GB of memory and 16 vCores. The cluster is currently using all of these resources, resulting in 100% usage.

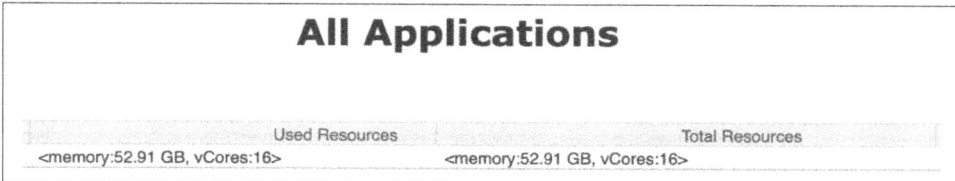

Figure 11-15. The YARN RM UI showing total cluster resources and used resources

Applications information path

Applications running on YARN can include MapReduce jobs, Hive jobs (which run as MapReduce jobs under the hood), Tez jobs, or Spark jobs, as shown in Figure 11-16. For MapReduce and Hive jobs, the YARN RM UI provides complete details. For Spark applications, it offers a link to the Spark history server. Refer to Recipe 11.4 for more information on accessing the Spark history server.

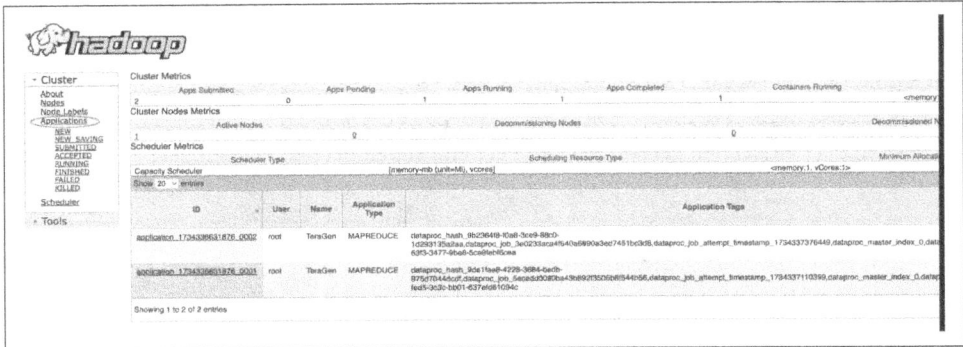

Figure 11-16. Accessing applications in the YARN RM UI

Scheduler home screen

YARN enables grouping of resources using YARN queues. By default, a single queue named "Default" exists, and all applications are submitted to that queue. To achieve multitenancy or job resource isolation, you can create a hierarchy of queues and allocate capacity at each level, as shown in Figure 11-17.

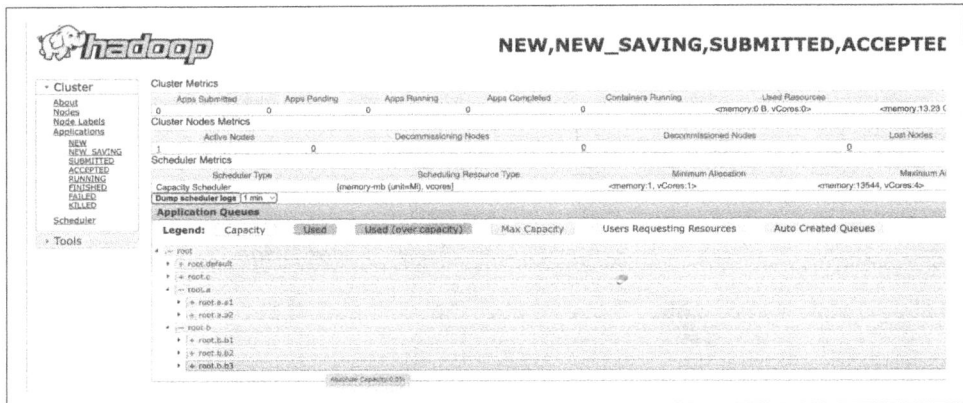

Figure 11-17. Accessing the Scheduler configuration in the YARM RM UI

11.9 Calculating the Cost of a Dataproc Cluster

Problem

You are running a critical Spark job on Dataproc on a GCE 20-worker-node (10 primary and 10 secondary spot instances) cluster of type n2-highmem-16 with 1,000 GB PD SSD and two local SSD disks. This job runs for one hour and is scheduled to run every day. The cluster is ephemeral in nature. You want to forecast the monthly cost of this Dataproc cluster.

Solution

Leverage a pricing calculator to forecast the pricing. The formula for this is:

Dataproc cluster cost = Licensing fee for Google Cloud Dataproc + Compute Engine cost (master, primary worker, secondary worker) + disk usage cost

You would also be charged for any other GCP services used, such as GCS for storing files, network cost (while reading from different regions or the internet), BigQuery cost (if interacting with BigQuery services), and so on.

Table 11-5 shows the pricing breakdown generated by the Google Cloud pricing calculator (*https://oreil.ly/-aTKH*).

Table 11-5. Dataproc Compute Engine pricing breakdown

Component	Quantity	Unit price	Total price	Details
Licensing fee for Google Cloud Dataproc (CPU cost)	10080	0.01	100.8	
n2-highmem-16 core master	480	0.031611	15.17328	1x Dataproc master node
SSD-backed local storage master	30.82191781	0.08	2.465753425	1x Dataproc master node
n2-highmem-16 RAM master	3840	0.004237	16.27008	1x Dataproc master node
n2-highmem-16 core worker	4800	0.031611	151.7328	10x Dataproc worker nodes
SSD-backed local storage workers	308.2191781	0.08	24.65753425	10x Dataproc worker nodes
n2-highmem-16 RAM worker	38400	0.004237	162.7008	10x Dataproc worker nodes
Spot preemptible n2-highmem-16 core secondary workers	4800	0.00818	39.264	10x Spot Dataproc worker nodes
SSD-backed local storage attached to Spot preemptible VMs	308.2191781	0.0634	19.54109589	10x Spot Dataproc worker nodes
Spot preemptible n2-highmem-16 RAM secondary workers	38400	0.001095	42.048	10x Spot Dataproc worker nodes
SSD-backed PD capacity	863.0136986	0.187	161.3835616	
		Total price:	**736.0369052**	

Note: Sustained-use discount is not included. You may need to apply discounts separately for each SKU.

Discussion

You are not charged based on the number of jobs you run on Dataproc on GCE. Dataproc on GCE pricing is based on the size of the Dataproc clusters and the duration of time that they run. The duration of a cluster is the length of time between cluster creation and cluster stopping or deletion. If you configure autoscaling, the machines are charged for the duration they run. When they are deleted, they are no longer billed.

Dataproc is billed by the second, for example:

Licensing fee for Google Cloud Dataproc (CPU cost) = $0.010 × (size of Dataproc clusters) × duration

Licensing fee for Google Cloud Dataproc (CPU cost) = $0.010 × (21 × 16) × 30 = $100.80

The size of your Dataproc clusters is the total number of vCPUs. There are 21 nodes. In this example, the size of the Dataproc clusters is 16 × 21 = 336. The duration is expressed in fractional hours. The job runs for an hour every day, so there are 30 fractional hours in total.

Leveraging secondary spot workers in addition to primary workers in a 50:50 ratio will optimize the price by about 50%. This ratio is great for job-scoped ephemeral clusters. For shared clusters, you have to benchmark thoroughly with all the workloads and come up with the primary-to-secondary Spot ratio.

The pricing also includes costs associated with disk, Compute Engine usage, logging, networking, and writes to GCS.

You can view the actual cost of the cluster from the billing console. You can add custom labels when submitting jobs and track the cost in the billing console by filtering using the labels, as described in Recipe 4.11.

11.10 Optimizing Cost in Dataproc Clusters

Problem

You want to optimize Dataproc clusters for cost savings.

Solution

Here are some key patterns that help when optimizing the cost of Dataproc clusters:

- Choosing the right hardware (machine type, disk)
- Tuning the cluster configuration according to workloads
- Switching to ephemeral clusters or serverless
- Taking advantage of low-cost spot machines
- Assessing cluster oversubscription where applicable

Discussion

Cost optimization and cost monitoring are important aspects when working with cloud environments. Not paying attention to what's happening can quickly drive cloud costs to a higher range. Optimization should be considered throughout your job's lifecycle, starting at the design phase and continuing through the implementation, testing, and support phases. You should always monitor the cost of the cluster and understand which components have a higher cost, whether that is machine type, disks, cloud logging, and so forth.

Choosing the right hardware is the first step toward your cost optimization. For instance, general machine types from the N series are cheaper to compute than optimized machine types from the C series. Assuming the workload will perform the same in both the N machine series and the C machine series, let's consider an example that will make a quick cost comparison.

A cluster with 20 worker nodes runs on average for six hours in 30 days. Here are the total worker machines costs based on type:

> N1-standard-16 costs $0.7599 per hour
>
> C2-standard-16 costs $0.8352 per hour
>
> Total compute cost = per-hour machine price × total worker nodes × hours per day × number of days
>
> N1-standard-16 = $0.7599 × 20 × 6 × 30 = $2,735.64
>
> C2-standard-16 = $0.8352 × 20 × 6 × 30 = $3,006.72

In this example, the C series will cost you about 10% more in machine costs.

The PD SSD disk type costs more than PD standard, PD balanced, or local SSD. Local SSD and PD balanced are also good options for handling workloads requiring a high I/O performance.

Now, let's consider clusters with 20 worker nodes using PD SSD versus PD balanced or local SSD and compare the pricing in Table 11-6. The table lists the 30-day cost of a cluster with 20 worker nodes attached with 100 GB disks running on average for 6 hours per day.

Table 11-6. Disk pricing in the us-central region

Disk type	Price per gigabyte per month
PD SSD	$0.204
PD balanced	$0.120
Local SSD	$0.104

If we look at this mathematically:

Total disk cost = (per-GB disk price per month × disk size in GB (100) × total worker nodes (20) × hours per day (6)) / number of days/hours per month (730)

In this example:

PD SSD cost = ($0.17 × 100 × 20 × 6 × 30) / 730 = $83.83

PD balanced disk cost = ($0.1 × 100 × 20 × 6 × 30) / 730 = $49.31

Local SSD cost = ($0.08 × 100 × 20 × 6 × 30) / 730 = $39.45

Cost-effectiveness isn't solely determined by price; it depends on the price-to-performance ratio. Although PD SSD might look expensive, there are use cases that can benefit from SSD performance and run the overall job faster.

Choosing the right architecture for your Dataproc cluster

Table 11-7 provides a comparative analysis of the costs associated with persistent, ephemeral, and serverless options for executing a Dataproc job.

Table 11-7. Cost analysis of Dataproc cluster options

	Persistent Dataproc cluster	Ephemeral cluster	Serverless
Compute capacity	100 vCPU 400 GB Ram 2 TB disk storage	100 vCPU 400 GB Ram 2 TB disk storage	100 vCPU 400 GB Ram 2 TB disk storage
Cluster configuration	One n1-standard-4 master Seven n1-standard-16 worker machines each with 100 GB PD and one local SSD of 375 GB attached to each node	One n1-standard-4 master Seven n1-standard-16 worker machines each with 100 GB PD and one local SSD of 375 GB attached to each node	100 vCPU 4 GB RAM per CPU 250 GB PD storage
Cluster runtime	24 hours/day (720 hours)	12 hours/day (360 hours)	12 hours/day (360 hours)
Compute cost	$3,967	$1,983	$1,295 (vCPU and RAM DCU's cost)
Storage cost	$268	$134	$739
Dataproc licensing fee	$835	$417	None (included as part of DCU pricing)
Total cost per 30 days	~$5,070	~$2,535	~$2,035

After examining Table 11-7, we can see that, from a cost perspective, transitioning to either an ephemeral or a serverless cluster is preferable to operating persistent clusters. While ephemeral may appear slightly more costly than serverless in the example, it provides customization options, including autoscaling, Spot VMs, and configuration adjustments for optimizing expenses.

> Cost comparison only covers a few specific components. For a full list of components that may be involved, refer to Recipe 11.9.

Tuning the cluster configuration

Tuning the cluster configuration helps set the best default configuration for all jobs running in the cluster. By default, Dataproc automatically detects and configures the default cluster configuration. This includes memory settings for YARN, HDFS, and so on. On top of these, you can further customize your cluster to potentially enhance the performance of your application.

For Spark workloads, the following play a role in cost savings:

- Executor memory
- Executor cores
- Executor overhead memory

For instance, allocating 16 GB RAM for each executor that is only processing a few megabytes of data may result in wasted resources and increased costs.

Oversubscribing the cluster

YARN allocates containers based on the memory available. Let's consider an example where your workloads need containers or executors of one core to 4 GB memory:

- N2-standard-16 costs $0.776944 per hour and provides 16 parallel tasks (64 GB total memory divided by 4 GB).
- N2-highmem-16 costs $1.048112 per hour and provides 32 parallel tasks (128 GB total memory divided by 4 GB).

In a comparative analysis of these options, pricing was augmented by 134% while capacity was increased by 200% (from 16 parallel tasks to 32 parallel tasks). This oversubscription can lead to a 200% increase in CPU utilization from current levels. For jobs that are not CPU intensive, this option may be more suitable.

If you want to increase CPU utilization by only 150%, then choose an average memory per core of 5 GB. You will still get a cost-to-cluster capacity savings of 16% (a 150% capacity increase and a 134% price increase).

Optimizing logging

Optimize your logging costs by controlling which logs are sent to Cloud Logging. While all application logs are sent to Cloud Logging by default for searchability, it's often more cost-effective to store them in GCS. For a deeper dive into log types and optimization strategies, including how to optimize logging costs, refer to Chapter 11 and specifically Recipe 11.6.

## CHAPTER 12

Orchestrating Dataproc Workloads

After developing and testing your big data applications, the next crucial step is orchestration, which ties everything together in the data-to-value lifecycle. In GCP, there are several options for orchestrating Spark and Hadoop jobs. These include Cloud Composer, Vertex AI Pipelines, Cloud Functions, and Dataproc workflows. Which option you choose will depend on your preferences and organizational needs.

In this chapter, you'll get hands-on experience and insights into all of these options:

Cloud Composer
Learn how to configure and orchestrate Dataproc jobs using Python DAGs.

Vertex AI Pipelines
Discover how to leverage Vertex AI for running Dataproc Serverless workflows.

Cloud Functions
Understand how to use Cloud Functions for lightweight, event-driven orchestration.

Dataproc workflows
Understand managing and automating your data-processing tasks.

12.1 Understanding the Prerequisites for Installing Cloud Composer

Problem

Setting up Cloud Composer comes with a list of things you need to have in place beforehand. You need to ensure that you have met all the requirements before beginning the installation.

Solution

Before installing Cloud Composer, a managed Apache Airflow service on Google Cloud, ensure the following prerequisites are met:

- Enable the Composer service in a Google Cloud project
- Determine the appropriate sizes for the scheduler, trigger, web server, and worker
- Select the network configuration
- Set up the necessary service accounts and assign IAM roles

Discussion

In a newly created project, the user has to enable the Cloud Composer service first. To enable this service, navigate to the API Library, search for "Cloud Composer API," and enable it.

Cloud Composer is an orchestration tool built on top of open source Apache Airflow. Its architecture includes the following main components:

Airflow web server
> Provides the web UI

Airflow database
> Stores persistent information about scheduling, connections, job status, and so on, which is hosted in a Cloud SQL instance within a tenant project

Airflow scheduler
> Parses, schedules, and runs jobs

Airflow workers
> Execute the tasks

Cloud Storage bucket
> Stores the orchestration configuration files (DAGs), logs, and custom plug-ins; often called an *environment bucket*

When you create a Cloud Composer environment in a customer project, Cloud Composer will distribute Airflow components between the tenant and the customer projects, as shown in Figure 12-1. A *tenant project* is the Google-managed project that isolates the resources and data of individual customers. In Cloud Composer 3, Airflow database instances and Airflow resources (web server, schedulers, workers) reside within the tenant project. The environment's bucket, logs, and metrics are stored in the *customer project*. Prior versions of Composer (versions 2 and 1) had more individual components distributed across both the tenant and the customer projects.

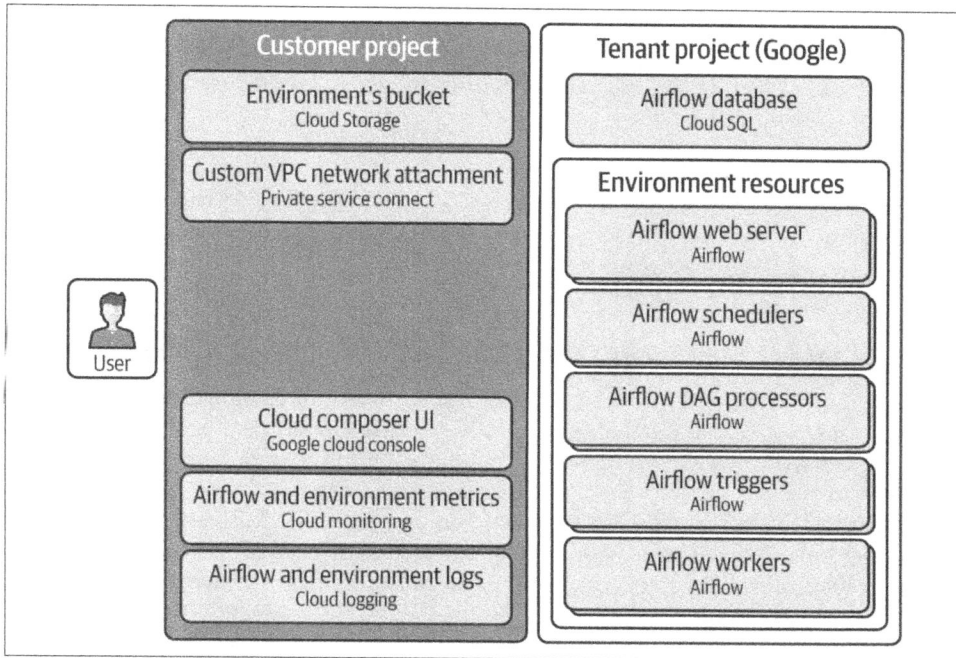

Figure 12-1. Google Cloud Composer components: tenant and customer project separation

Configuring IAM permissions

To successfully create a Cloud Composer environment, specific IAM permissions are required to grant access to necessary resources. Ensure that your user or service account has the following roles assigned:

- Network user role (`roles/compute.networkUser`) on the VPC

- Storage object creator role (`roles/storage.objectCreator`) on the environment's storage bucket

- Composer administrator role (`roles/composer.admin`) or a custom role with the `composer.environments.create` permission

Planning the network and subnet IP ranges

The Composer environment is hosted in GKE, and each service runs as a Pod. You'll find core Airflow components like the Airflow web server, schedulers, and workers all running as individual Pods. The GKE cluster requires subnets with specific IP ranges:

Primary IP range

This range is used for the GKE nodes themselves. Each node within the cluster will be assigned an IP address from this range. It is also used by Kubernetes services.

Secondary IP range (Pod IP range)

This range is dedicated to the Pods running within the GKE cluster. Each Airflow component (web server, scheduler, worker) and other necessary Pods will be assigned an IP address from this secondary range.

What Are the Primary and Secondary IP Ranges of a Subnet?

A VPC network provides isolated network resources within Google Cloud, and subnets are subdivisions of that network. In a VPC, each subnet has a primary IP range that provides IP addresses to nodes and handles core subnet functions. Additionally, subnets can have optional secondary IP ranges used exclusively for assigning alias IP addresses, often leveraged in GKE for Pod and service IP allocation. While the primary range is fixed after subnet creation, secondary ranges offer flexibility for modification, allowing you to adapt your network configuration as needed.

Assessing your scheduling and orchestration needs will help determine the appropriate sizing for your environment's components, ensuring optimal performance and cost efficiency. Understanding the scale and frequency of your workflows will allow you to correctly size the scheduler, trigger, web server, and worker components. Here are the key metrics you can look for while assessing the orchestration requirements:

Workflow volume

How many workflows (DAGs) will you be running concurrently? A high volume of workflows necessitates a robust scheduler and worker pool.

Workflow size and complexity

Consider the complexity and resource demands of your individual workflows. Large, data-intensive tasks require more worker resources.

Workflow frequency

How often will your workflows be triggered? Frequent, short-interval schedules place a higher load on the scheduler and trigger.

Concurrency and parallelism

How many tasks within your workflows need to run concurrently? This will affect the required worker resources.

Peak load

Identify potential peak-load periods. Will certain times of the day or month experience a surge in workflow activity?

Growth projections

Anticipate future growth in your data and orchestration needs. Plan for scalability by considering potential increases in workflow volume and complexity.

With the necessary IAM permissions, network configurations, and capacity planning in place, you're now ready to create your Cloud Composer instance.

12.2 Deploying a Cloud Composer Environment

Problem

You want to deploy a Cloud Composer environment.

Solution

Use the following gcloud command to create a Cloud Composer environment with the minimum required parameters:

```
gcloud composer environments create <ENVIRONMENT_NAME> \
  --location <LOCATION>
```

This basic command establishes a cluster with the standard settings. For advanced configurations and tailored deployments, use the following command:

```
gcloud composer environments create <ENVIRONMENT_NAME> \
  --location <LOCATION> \
  --image-version <IMAGE_VERSION> \
  --machine-type <MACHINE_TYPE> \
  --node-count <NODE_COUNT> \
  --disk-size <DISK_SIZE_GB> \
  --python-version <PYTHON_VERSION> \
  --env-variables <ENV_VARIABLES> \
  --network <NETWORK> \
  --subnetwork <SUBNETWORK> \
  --enable-private-environment \
  --enable-scheduled-snapshot-creation \
  --snapshot-creation-schedule <SNAPSHOT_CREATION_SCHEDULE> \
  --maintenance-window-start <MAINTENANCE_WINDOW_START> \
  --maintenance-window-end <MAINTENANCE_WINDOW_END> \
  --maintenance-window-recurrence <MAINTENANCE_WINDOW_RECURRENCE>
```

Discussion

Before you start creating a Cloud Composer environment, ensure that you have understood the prerequisites in Recipe 12.1 and have all the necessary items ready. Let's look at how to create Cloud Composer from the web UI.

Search for Composer and navigate to the home page of the Cloud Composer service, as shown in Figure 12-2.

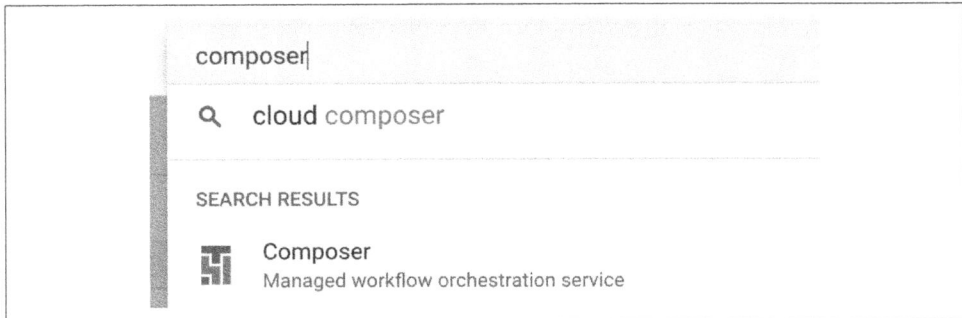

Figure 12-2. Searching for the Composer service in the GCP console

Click the +Create button and select Composer 2, as shown in Figure 12-3.

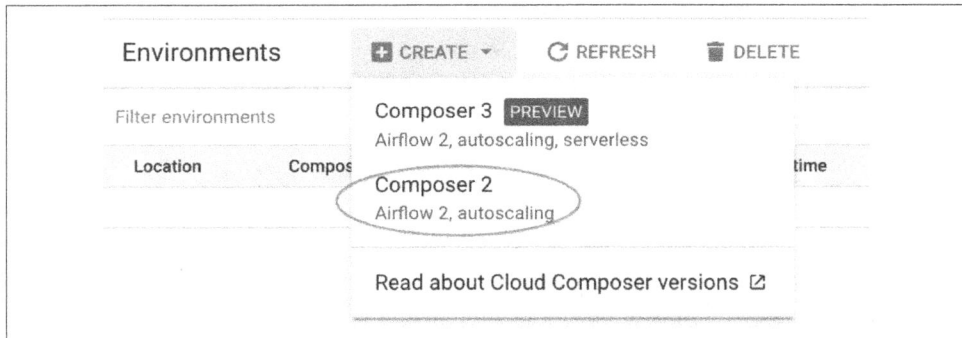

Figure 12-3. Selecting Composer 2 as the version to be created

Common environment details you need to enter or select include the name, location, image version, and service account, as shown in Figure 12-4.

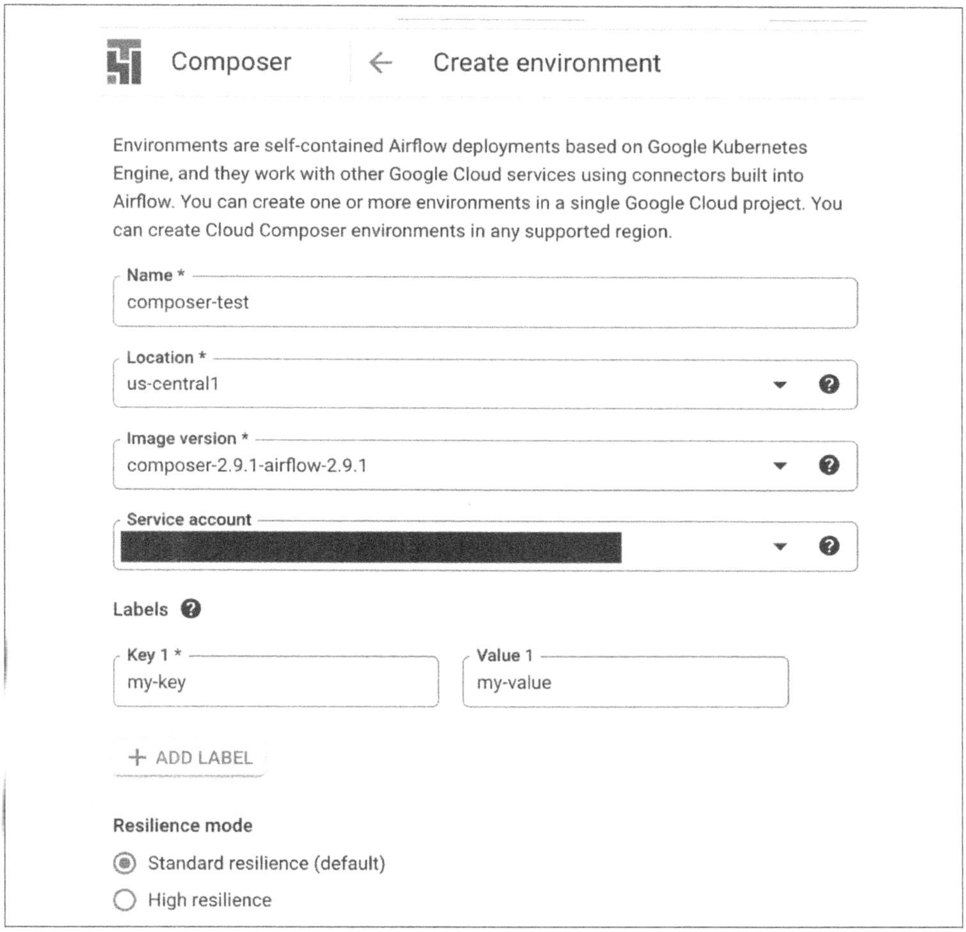

Figure 12-4. Composer create options

Next, choose the size of the environment: small, medium, large, or custom (see Figure 12-5).

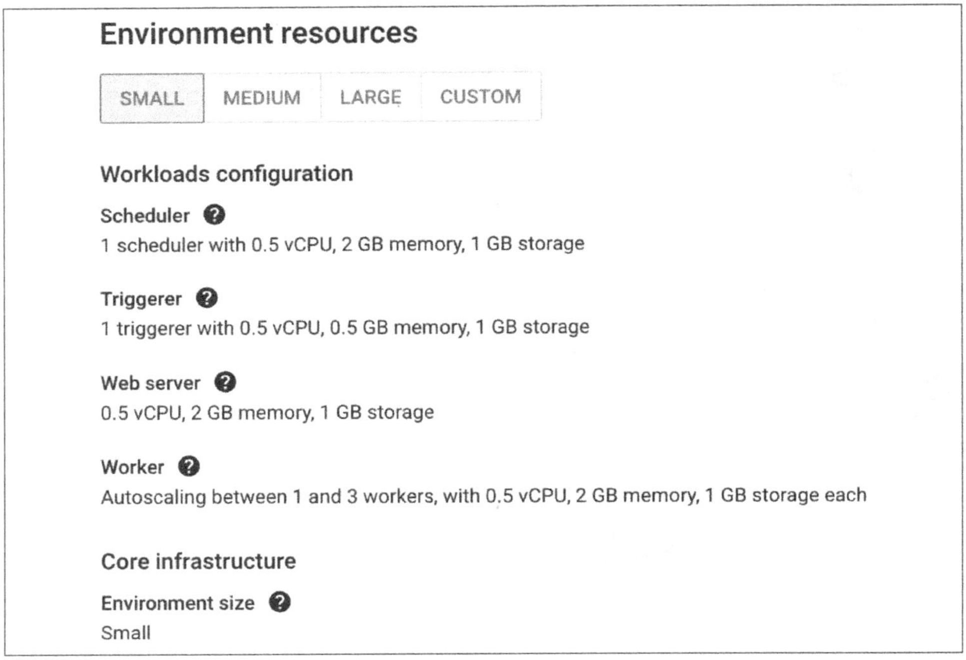

Environment resources

| SMALL | MEDIUM | LARGE | CUSTOM |

Workloads configuration

Scheduler ❓
1 scheduler with 0.5 vCPU, 2 GB memory, 1 GB storage

Triggerer ❓
1 triggerer with 0.5 vCPU, 0.5 GB memory, 1 GB storage

Web server ❓
0.5 vCPU, 2 GB memory, 1 GB storage

Worker ❓
Autoscaling between 1 and 3 workers, with 0.5 vCPU, 2 GB memory, 1 GB storage each

Core infrastructure

Environment size ❓
Small

Figure 12-5. Choosing the environment based on size

What Environment Size Should You Select?

Your environment size controls the capacity of the scheduler, trigger, web server, and worker components of the Cloud Composer. For a small proof of concept, select small. For development or low-level environments, select medium. For production environments, select large. You can also choose custom to set a specific capacity for each component. Selecting the small, medium, or large option affects only the capacity increase but will not impact the resilience required for high-level environments. For higher-level environments like production, enable the "High resilience" option shown in Figure 12-4.

Configure your network where the Composer instance will be created, as shown in Figure 12-6.

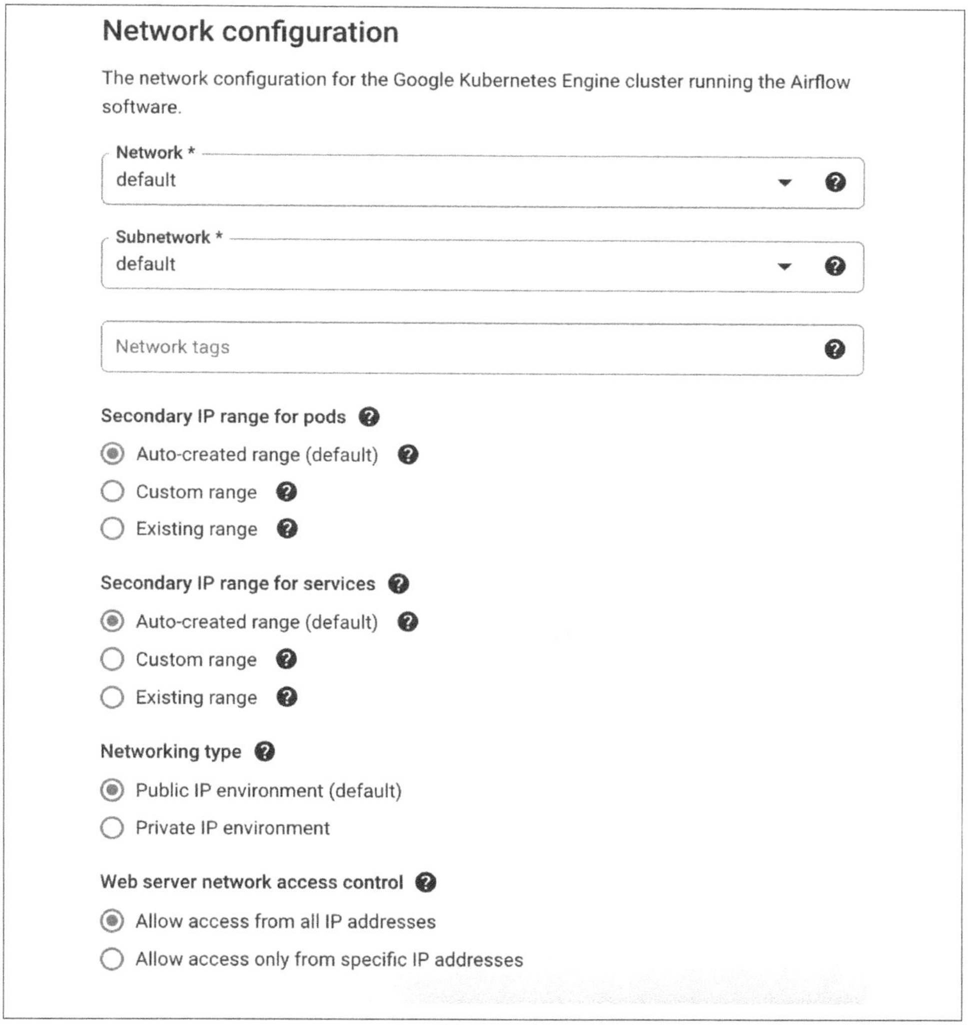

Network configuration

The network configuration for the Google Kubernetes Engine cluster running the Airflow software.

Network *
default   ▼ ❷

Subnetwork *
default   ▼ ❷

Network tags   ❷

Secondary IP range for pods ❷
◉ Auto-created range (default) ❷
○ Custom range ❷
○ Existing range ❷

Secondary IP range for services ❷
◉ Auto-created range (default) ❷
○ Custom range ❷
○ Existing range ❷

Networking type ❷
◉ Public IP environment (default)
○ Private IP environment

Web server network access control ❷
◉ Allow access from all IP addresses
○ Allow access only from specific IP addresses

Figure 12-6. Network configuration for a Composer instance

Optionally, you can add advanced configuration, as shown in Figure 12-7.

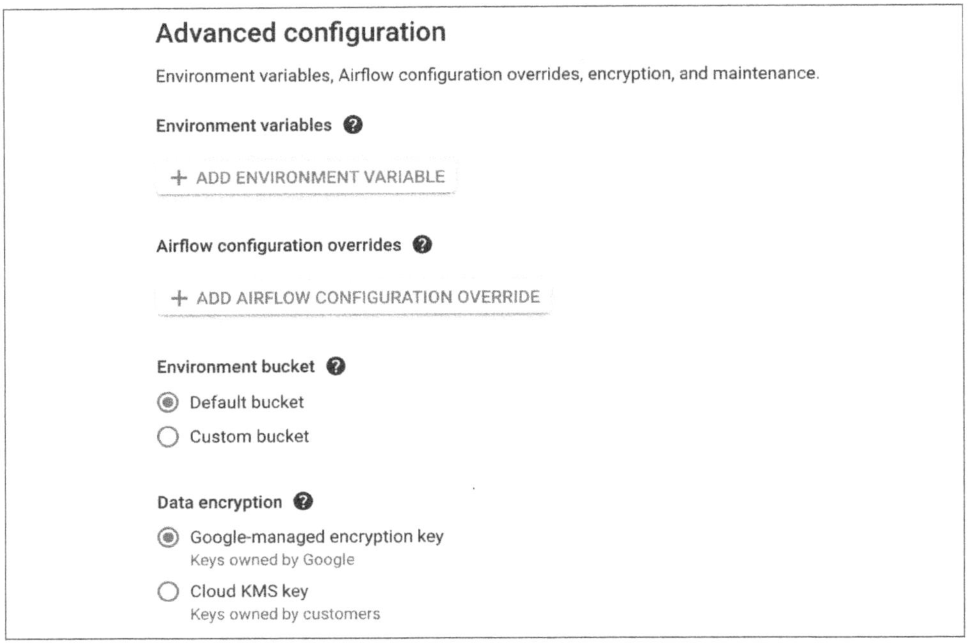

Advanced configuration

Environment variables, Airflow configuration overrides, encryption, and maintenance.

Environment variables

+ ADD ENVIRONMENT VARIABLE

Airflow configuration overrides

+ ADD AIRFLOW CONFIGURATION OVERRIDE

Environment bucket
- Default bucket
- Custom bucket

Data encryption
- Google-managed encryption key
  Keys owned by Google
- Cloud KMS key
  Keys owned by customers

Figure 12-7. Advanced configuration for a Composer instance

Let's look at each of the advanced configurations:

Environment variables
- Purpose: define custom variables for use in Airflow DAGs and other components
- Example: AIRFLOW_VAR=my_custom_value

Airflow configuration overrides
- Purpose: modify default Airflow configuration settings
- Example: core-dags_are_paused_at_creation=True

Environment bucket
- Purpose: choose the storage location for Cloud Composer data
- Example: *gs://my-custom-bucket* (for the custom bucket option)

Data encryption
- Purpose: select the method for encrypting data at rest
- Example: *projects/my-project/locations/my-location/keyRings/my-keyring/crypto-Keys/my-key* (for the Cloud KMS key option)

The options shown in Figure 12-8 control the environment's maintenance, data lineage, and backup procedures.

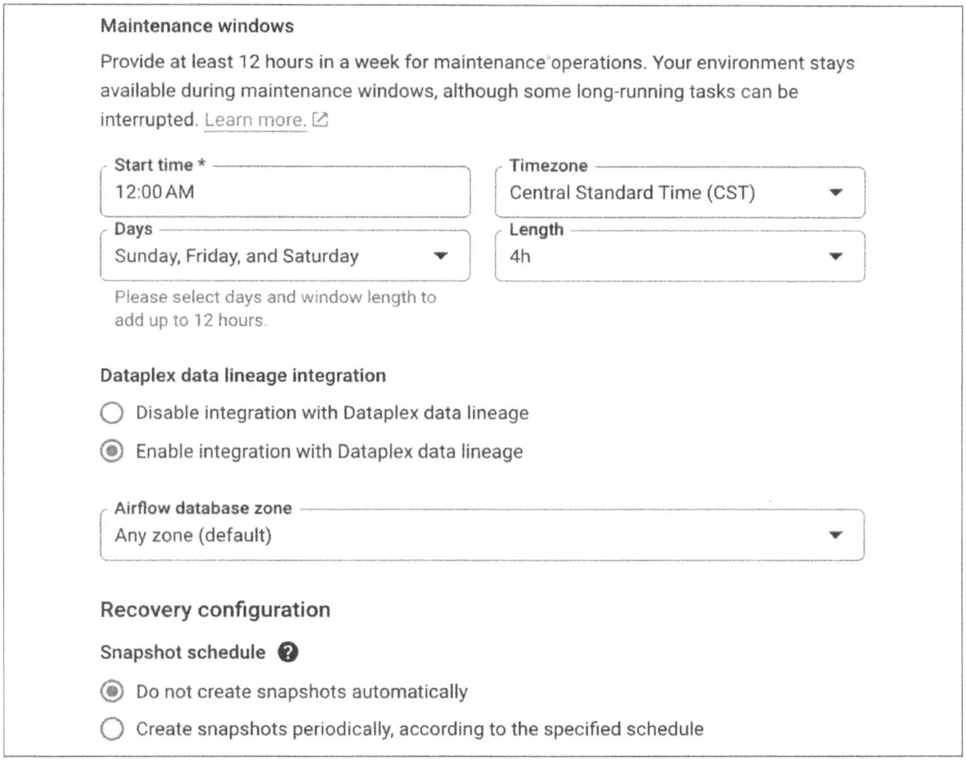

Figure 12-8. Composer instance settings for maintenance, lineage, and recovery

Let's walk through what each of these fields means:

Maintenance windows

Allows you to specify a time period during the week when maintenance operations can be performed on your Cloud Composer environment

Dataplex data lineage integration

Controls whether your Cloud Composer environment integrates with Dataplex data lineage. Here are the possible options:

Disable integration

Data lineage information won't be tracked in Dataplex.

Enable integration

Data lineage will be tracked, providing insights into data transformations within your Airflow workflows.

Airflow database zone
Determines the zone where the Airflow metadata database is located

Recovery configuration
Deals with creating snapshots of your Cloud Composer environment for backup and recovery. Possible options include:

Do not create snapshots automatically
No automatic snapshots will be created.

Create snapshots periodically, according to the specified schedule
Snapshots will be taken based on a schedule you define (not shown in the figure).

Now that these settings have been determined, click Create to create a Composer environment. This could take approximately 45 minutes.

Create a custom network for Composer using the following command:

```
gcloud compute networks create composer-network --subnet-mode custom
```

Create a subnet and attach the primary and secondary ranges:

```
gcloud compute networks subnets create composer-subnet \
  --network composer-network \
  --region <YOUR_REGION> \
  --range 10.0.0.0/24 \
  --secondary-range pods=10.0.1.0/24,services=10.0.2.0/24
```

Then, enable firewall rules:

```
gcloud compute firewall-rules create composer-allow-internet-egress \
  --direction EGRESS --priority 65534 --network composer-network --action \
  ALLOW --rules tcp,udp --destination-ranges 0.0.0.0/0

gcloud compute firewall-rules create composer-allow-google-ingress \
  --direction INGRESS --priority 65534 --network composer-network --action\
  ALLOW --rules all --source-ranges 35.191.0.0/16,130.211.0.0/22
```

Finally, to create the cluster with gcloud, run the following command:

```
gcloud compute networks subnets create composer-subnet \
    --network composer-network \
    --region <YOUR_REGION> \
    --range 10.0.0.0/24 \
    --secondary-range pods=10.0.1.0/24,services=10.0.2.0/24
```

12.3 Scheduling a Job in Composer

Problem

You want to schedule a job in Cloud Composer that will run every day at 8 A.M. Central Time.

Solution

To schedule a job at 8 A.M. every day, you can configure it in the DAG definition as follows:

```
with DAG(
    'use_composer_variable',
    default_args=default_args,
    schedule_interval='0 8 * * *',    ❶
    catchup=False,    ❷
    timezone='America/Chicago'    ❸
) as dag:
```

❶ Defines how often the DAG should run. In this instance, the cron expression sets the job to run every day at 8 A.M.

❷ Prevents the DAG from trying to "catch up" on past dates.

❸ Sets the time zone to Central Time.

Discussion

In the realm of Cloud Composer, a DAG—or *directed acyclic graph*—serves as the blueprint for your workflow orchestration. Essentially, it's a Python script that defines the following:

Tasks
These represent the individual units of work in your workflow, like running a Dataproc job, executing a SQL query, or moving files.

Dependencies
The DAG specifies the order in which tasks should run, ensuring that a task starts only once its prerequisite tasks are complete.

Schedule
You can set how frequently the DAG should run (e.g., hourly, daily, or on a custom schedule).

There are many key components of a composer DAG:

DAG definition
> This sets up the DAG's basic properties, like its `dag_id`, `schedule_interval`, `start_date`, and any default arguments for its tasks.

Operators
> These are the building blocks of tasks. Each operator represents a specific type of action (e.g., DataprocSubmitJobOperator for running a Dataproc job, BashOperator for executing shell commands).

Task instances
> When a DAG runs, each task within it creates a task instance, which is a specific execution of that task at a particular point in time.

Task dependencies
> These define the relationships between tasks, ensuring that they run in the correct order.

Scheduling a job using Composer involves the following steps:

1. Create a DAG. Write a Python script that defines your DAG, including a DataprocSubmitJobOperator task. This operator encapsulates the details of your Dataproc job, such as the cluster to use, the job configuration, and any required parameters.

2. Upload the DAG. Store your DAG file in the designated Cloud Storage bucket associated with your Composer environment.

3. Based on the defined schedule, Composer's scheduler triggers the DAG. Alternatively, you can manually run the Composer.

In this example, we are going to run four jobs:

TeraGen
> This is the initial task and has no dependencies.

TeraSort and WordCount
> These tasks depend on the successful completion of TeraGen. They will begin only after TeraGen finishes.

TeraValidate
> This task depends on both TeraSort and WordCount. It will execute only after both preceding tasks are completed.

A visualization of how this DAG would look is shown in Figure 12-9.

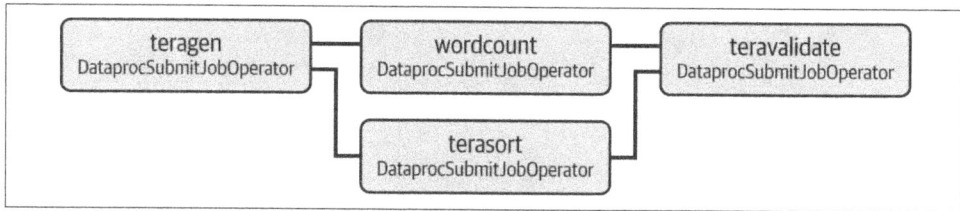

Figure 12-9. Composer DAG dependency diagram

First, create a Python file with Airflow DAG code:

```
from airflow import DAG
from airflow.providers.google.cloud.operators.dataproc import (
    DataprocSubmitJobOperator
)
from airflow.utils.dates import days_ago
from pendulum import timezone

# Default arguments
default_args = {
    'owner': 'airflow',
    'depends_on_past': False,
    'start_date': days_ago(1),
    'email_on_failure': False,
    'email_on_retry': False,
    'retries': 1,
}

# Set timezone to Central Time (CT)
ct_timezone = timezone('America/Chicago')

# Common Dataproc job configuration details
project_id = 'nsadineni'
region = 'us-central1'  # e.g., 'us-central1'
cluster_name = 'composer-examples'

# Define the Dataproc jobs
teragen_job = {
    "reference": {"project_id": project_id},
    "placement": {"cluster_name": cluster_name},
    "hadoop_job": {
        "main_jar_file_uri": (
            "file:///usr/lib/hadoop-mapreduce/hadoop-mapreduce-examples.jar"
        ),
        "args": [
            "teragen",
            "1000000",  # Number of rows to generate
            "/user/hadoop/teragen-output"
        ],
```

```python
        },
}

terasort_job = {
    "reference": {"project_id": project_id},
    "placement": {"cluster_name": cluster_name},
    "hadoop_job": {
        "main_jar_file_uri": (
            "file:///usr/lib/hadoop-mapreduce/hadoop-mapreduce-examples.jar"
        ),
        "args": [
            "terasort",
            "/user/hadoop/teragen-output",
            "/user/hadoop/terasort-output"
        ],
    },
}

wordcount_job = {
    "reference": {"project_id": project_id},
    "placement": {"cluster_name": cluster_name},
    "hadoop_job": {
        "main_jar_file_uri": (
            "file:///usr/lib/hadoop-mapreduce/hadoop-mapreduce-examples.jar"
        ),
        "args": [
            "wordcount",
            "/user/hadoop/teragen-output",
            "/user/hadoop/wordcount-output"
        ],
    },
}

teravalidate_job = {
    "reference": {"project_id": project_id},
    "placement": {"cluster_name": cluster_name},
    "hadoop_job": {
        "main_jar_file_uri": (
            "file:///usr/lib/hadoop-mapreduce/hadoop-mapreduce-examples.jar"
        ),
        "args": [
            "teravalidate",
            "/user/hadoop/terasort-output",
            "/user/hadoop/teravalidate-output"
        ],
    },
}

# Define the DAG
with DAG(
    'hadoop_terasuite_pipeline_dataproc',
    default_args=default_args,
```

```python
    description=(
      'A DAG to demonstrate a composed Hadoop job sequence using Dataproc',
    )
    schedule_interval='0 8 * * *',   # Run every day at 8 AM CT
    start_date=days_ago(1),
    catchup=False,
    tags=['example'],
    #timezone=ct_timezone,   # Set the DAG's timezone to Central Time (CT)
) as dag:

    # Job A: TeraGen
    teragen = DataprocSubmitJobOperator(
        task_id='teragen',
        job=teragen_job,
        region=region,
        project_id=project_id,
    )

    # Job B: TeraSort
    terasort = DataprocSubmitJobOperator(
        task_id='terasort',
        job=terasort_job,
        region=region,
        project_id=project_id,
    )

    # Job C: WordCount
    wordcount = DataprocSubmitJobOperator(
        task_id='wordcount',
        job=wordcount_job,
        region=region,
        project_id=project_id,
    )

    # Job D: TeraValidate
    teravalidate = DataprocSubmitJobOperator(
        task_id='teravalidate',
        job=teravalidate_job,
        region=region,
        project_id=project_id,
    )

    # Define the task dependencies
    teragen >> [terasort, wordcount] >> teravalidate
```

Then, upload the DAG file to the designated GCS bucket. The Composer environment home page will have a link to the Cloud Storage bucket location where it stores all DAG files. Click on this link as shown in Figure 12-10 and upload the DAG file to the GCS bucket.

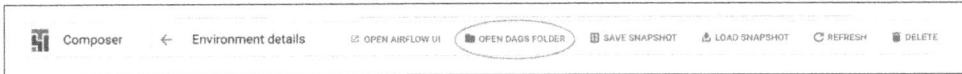

Figure 12-10. Navigating to the DAGs folder from the Composer web UI

Cloud Composer proactively manages your workflow orchestration by regularly scanning its associated GCS bucket for new DAGs. Newly uploaded DAG files are eventually discovered and processed. Once a new DAG is detected during a scan, Composer's Airflow scheduler parses the file and makes it available for execution according to its specified schedule.

12.4 Parameterizing Variables

Problem

You have different environments (dev, test, prod) that might require different configurations, such as source table and target table names. How do you set variables in Cloud Composer?

Solution

Navigate to Variables in the Airflow UI in the existing Cloud Composer environment, as shown in Figure 12-11.

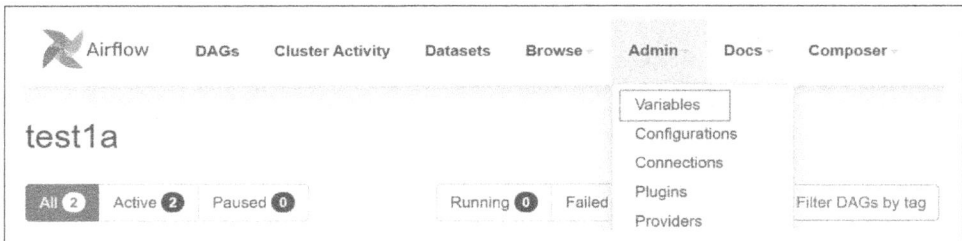

Figure 12-11. Airflow UI Admin tab with Variables highlighted

Create the variable by clicking on the +. Then, add the key and values, as shown in Figure 12-12.

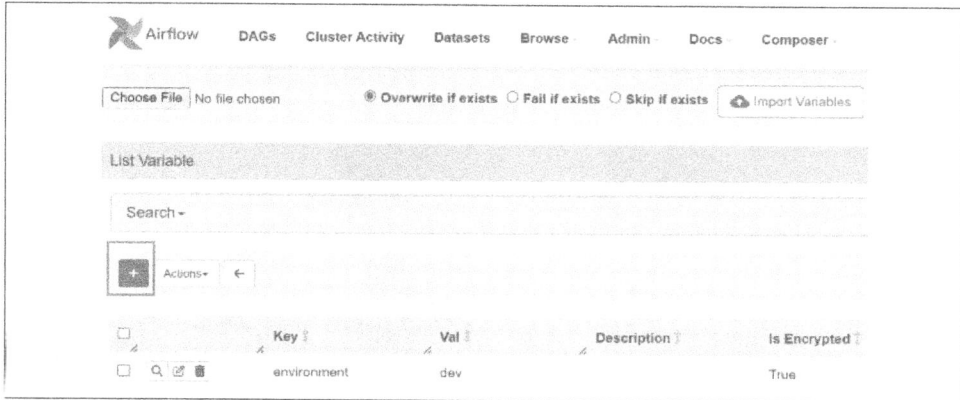

Figure 12-12. Adding a new variable

Discussion

It's not a good practice to hardcode specific configurations in the DAG. Instead, you might parameterize variables in various scenarios like different settings across dev, test, and prod to access credentials or API keys securely.

If you don't have a Cloud Composer environment configured, refer to Recipe 12.2 to configure one. For bulk import of variables, have all the variables in a JSON file and import them, as shown in Figure 12-13. Here is the sample JSON format:

```
{"source_table": "organization_raw",
 "target_table": "organization_gold"}
```

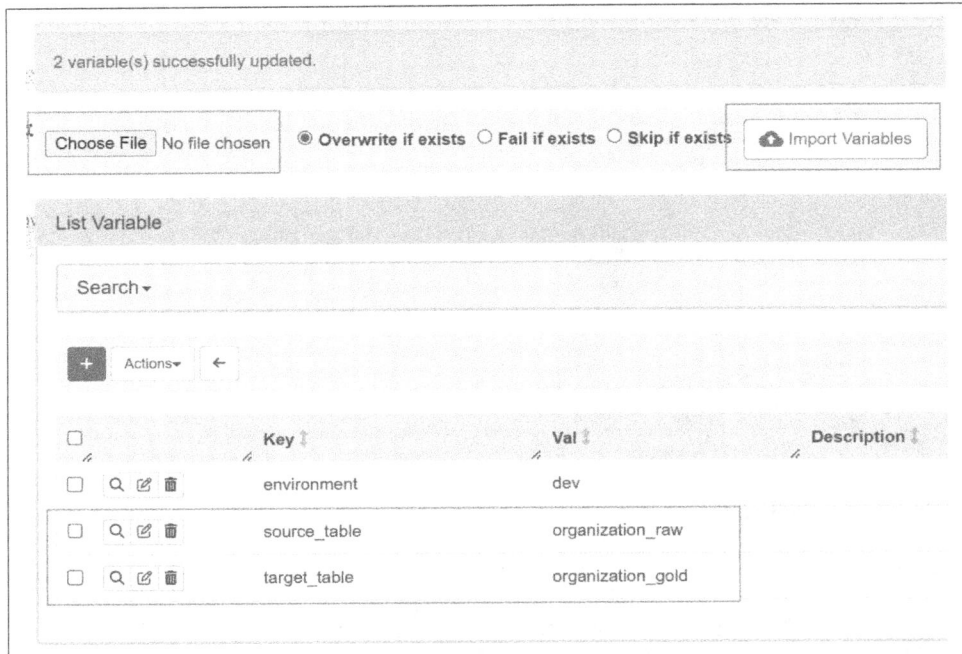

Figure 12-13. Two variables added to Airflow UI

In your DAG, you can access the variables with `Variable.get('environment')`, as shown in the following code snippet:

```
from airflow import DAG
from airflow.operators.python_operator import PythonOperator
from airflow.models import Variable
from datetime import datetime

default_args = {
    'owner': 'airflow',
    'start_date': datetime(2023, 1, 1),
    'retries': 1
}

with DAG('use_composer_variable',
        default_args=default_args,
        schedule_interval='@daily',
        catchup=False) as dag:

    def print_variable():
        sample_value = Variable.get('environment')
        print(f"The value of 'environment' is: {sample_value}")

    print_var_task = PythonOperator(
        task_id='print_composer_variable',
```

```
        python_callable=print_variable
    )

    print_var_task
```

Add the sample DAG to your Cloud Composer environment's GCS bucket.

> Jinja templating lets you dynamically insert variables at runtime. While Airflow variables are useful for small-scale parameterization, overusing them can affect performance and maintainability in larger environments. Instead, consider using Jinja templates in your DAGs for dynamic configurations. Jinja allows you to reference variables directly within task parameters, reducing runtime overhead and improving readability in higher-level environments like production.

Trigger the DAG and run it as shown in Figure 12-14. You can then check the details of the run to validate if the variables are configured correctly, as shown in Figure 12-15.

Figure 12-14. The triggered DAG and its status shown on the home page

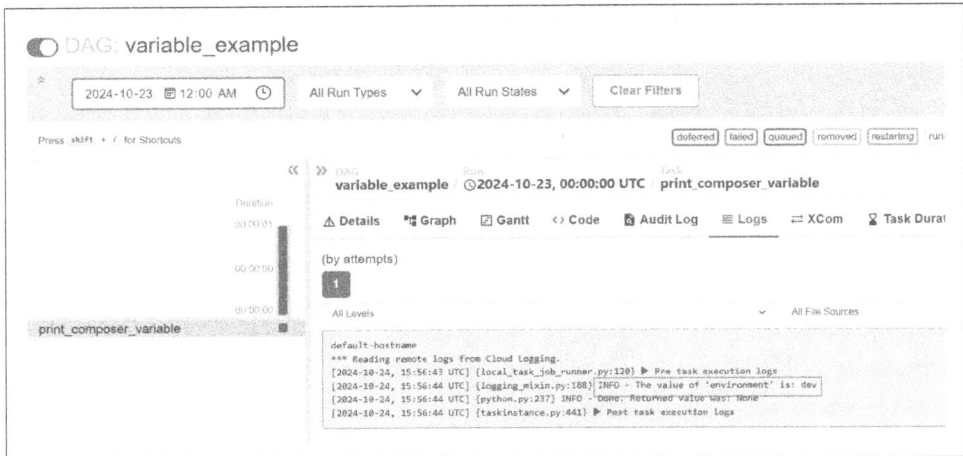

Figure 12-15. DAG logs showing the usage of configured variables

12.5 Scaling Up a Cloud Composer Environment

Problem

Your Cloud Composer workloads have increased from 250 DAGs to 700 DAGs, and you want to scale up the existing Cloud Composer environment.

Solution

Cloud Composer versions 2 and 3 support autoscaling. To scale up, tweak the environment size using the following gcloud command:

```
gcloud composer environments update <environment_name_here> \
  --location <region_here> --environment-size large
```

Discussion

Optimizing a Cloud Composer environment requires a balanced approach between infrastructure changes (resizing workers) and DAG-level optimizations (adjusting parsing intervals). Carefully monitoring resource utilization and scheduler performance will help ensure a stable and efficient environment.

Infrastructure

Before scaling up, it's crucial to assess the health status of your Composer environment. If the status is not green, navigate to the Monitoring tab and click Workers to check CPU and memory utilization, as shown in Figure 12-16. If either metric is consistently high, resizing the Composer environment may be necessary.

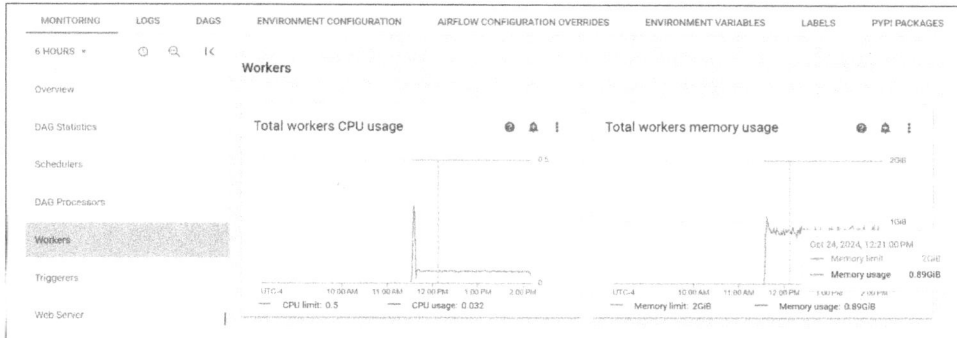

Figure 12-16. Monitoring instance metrics like CPU and memory usage

Refer to Table 12-1 for the minimum and maximum workers for each environment size. If the current environment is small, you can scale to medium and monitor the health and compute utilization.

Table 12-1. Resource breakdown by worker type and autoscaling limits

Worker type	Autoscaling range	vCPU	Memory	Storage
Small	1–3 workers	0.5	2 GB	10 GB
Medium	2–6 workers	2	7.5 GB	20 GB
Large	3–12 workers	4	15 GB	50 GB

You can also choose a custom size for workers, as shown in Figure 12-17.

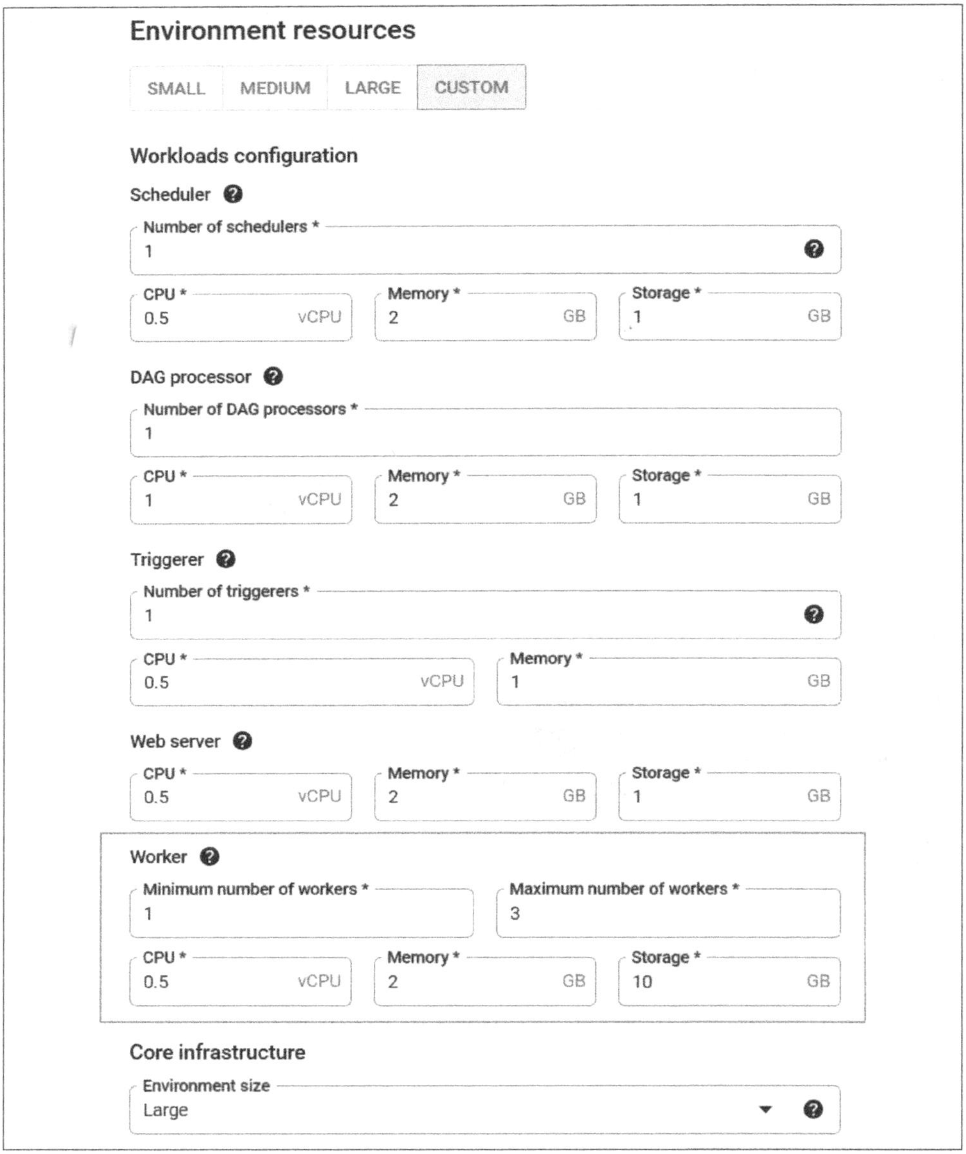

Environment resources

| SMALL | MEDIUM | LARGE | CUSTOM |

Workloads configuration

Scheduler ❓

Number of schedulers *
1 ❓

| CPU * | Memory * | Storage * |
| 0.5   vCPU | 2   GB | 1   GB |

DAG processor ❓

Number of DAG processors *
1

| CPU * | Memory * | Storage * |
| 1   vCPU | 2   GB | 1   GB |

Triggerer ❓

Number of triggerers *
1 ❓

| CPU * | Memory * |
| 0.5   vCPU | 1   GB |

Web server ❓

| CPU * | Memory * | Storage * |
| 0.5   vCPU | 2   GB | 1   GB |

Worker ❓

| Minimum number of workers * | Maximum number of workers * |
| 1 | 3 |

| CPU * | Memory * | Storage * |
| 0.5   vCPU | 2   GB | 10   GB |

Core infrastructure

Environment size
Large ▼ ❓

Figure 12-17. Creating a Composer instance with a custom configuration

DAG-level optimizations

If you have a lot of DAGs, you can increase or decrease the DAG file parsing interval (`min_file_process_interval`) based on how frequently the DAGs will get modified. If you have a large number of infrequently modified DAGs, increase the interval. By default, `min_file_process_interval` is 30 seconds. Increasing this will reduce the scheduler load to continuously parse and render in the UI. If you are in development making frequent changes to the DAG folder, reducing the load will optimize productivity. Navigate to "Airflow configuration overrides" to add the value for `min_file_process_interval`, as shown in Figure 12-18.

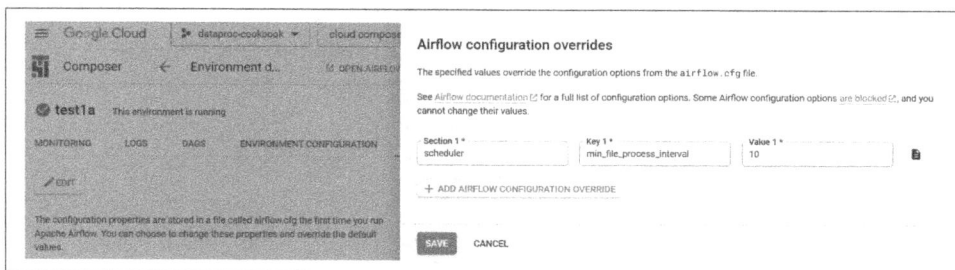

Figure 12-18. Configuring the minimum file process interval in overrides

12.6 Running Spark Jobs Using Vertex AI Machine Learning Pipelines

Problem

The data science team within your organization currently uses Vertex AI Pipelines for orchestrating model training workflows. They want to use the same platform Vertex AI Pipelines to orchestrate a PySpark data-processing job on Dataproc Serverless.

Solution

Navigate to Vertex AI Pipelines and click Create Run. In the "Run source" section, select "Dataproc: PySpark Batch" from the Template Gallery drop-down menu, as shown in Figure 12-19.

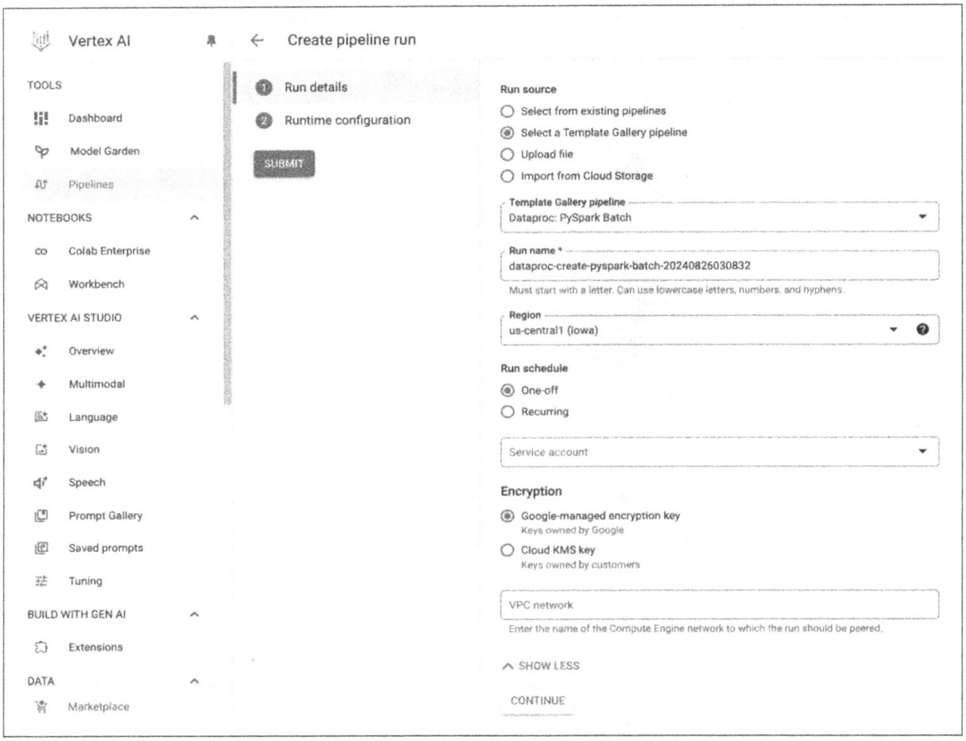

Figure 12-19. The Dataproc PySpark batch template in Vertex AI

Discussion

Vertex AI Pipelines are a robust way to orchestrate independent Spark jobs on Dataproc Serverless. For running other types of Spark jobs, such as Spark Scala, Spark SQL, or SparkR, you can choose the appropriate templates from the drop-down menu, as shown in Figure 12-20.

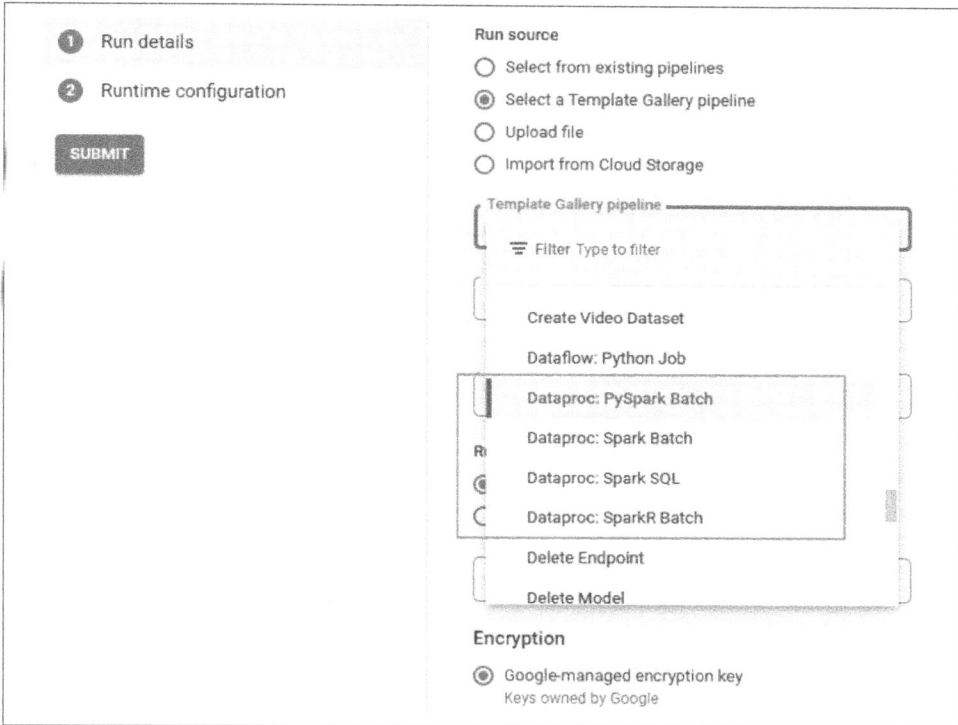

Figure 12-20. Predefined pipeline templates in Vertex AI

Navigate to the runtime configuration and upload the PySpark code to be submitted, as shown in Figure 12-21. Set the output directory and pass any necessary arguments for the Spark job.

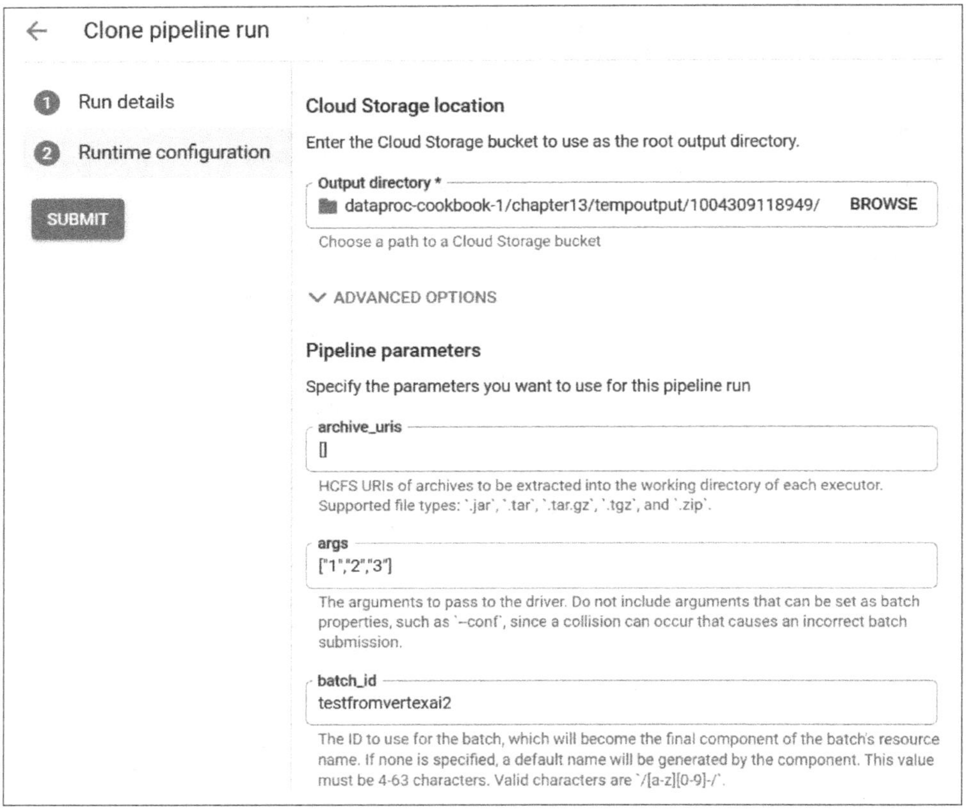

Figure 12-21. Runtime configuration output directory and arguments setup

To configure Spark properties, add them under `runtime_config_properties`, as shown in Figure 12-22.

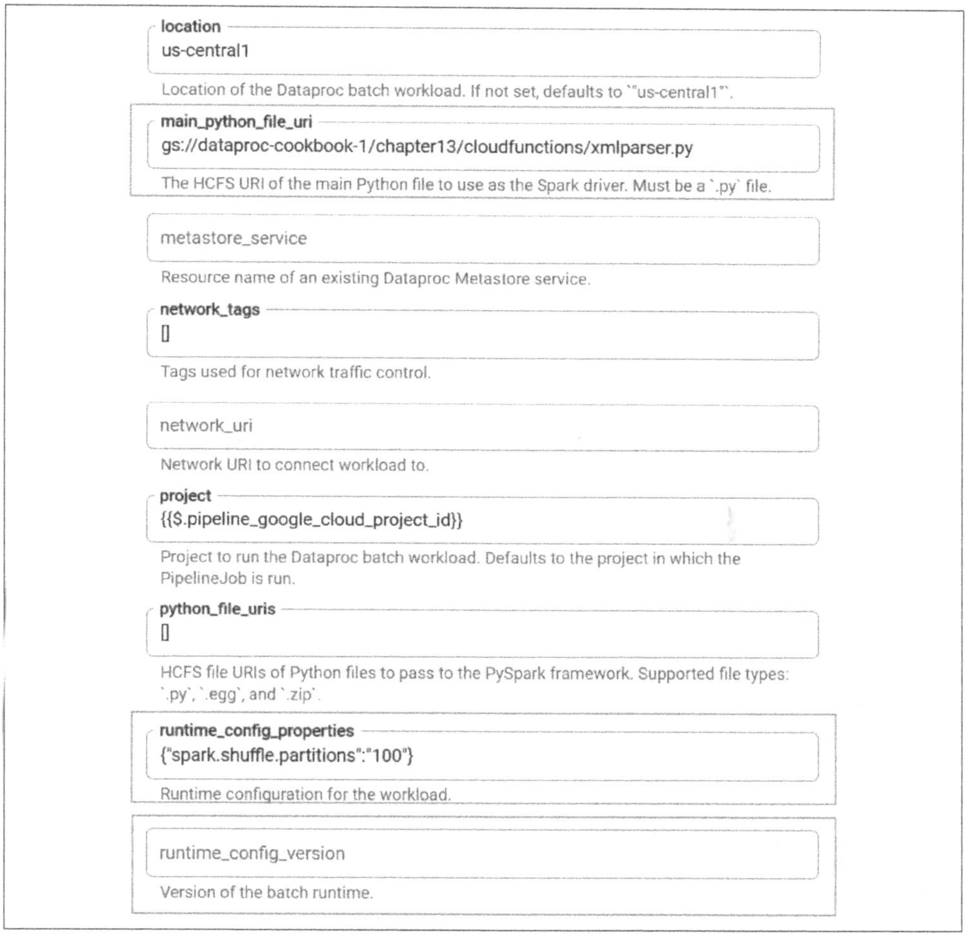

location
us-central1

Location of the Dataproc batch workload. If not set, defaults to `"us-central1"`.

main_python_file_uri
gs://dataproc-cookbook-1/chapter13/cloudfunctions/xmlparser.py

The HCFS URI of the main Python file to use as the Spark driver. Must be a `.py` file.

metastore_service

Resource name of an existing Dataproc Metastore service.

network_tags
[]

Tags used for network traffic control.

network_uri

Network URI to connect workload to.

project
{{$.pipeline_google_cloud_project_id}}

Project to run the Dataproc batch workload. Defaults to the project in which the PipelineJob is run.

python_file_uris
[]

HCFS file URIs of Python files to pass to the PySpark framework. Supported file types: `.py`, `.egg`, and `.zip`.

runtime_config_properties
{"spark.shuffle.partitions":"100"}

Runtime configuration for the workload.

runtime_config_version

Version of the batch runtime.

Figure 12-22. Configuring the PySpark file path and Spark properties in runtime configuration

It is mandatory to specify the `runtime_config_version`. Check for the available versions in the Dataproc batch, as shown in Figure 12-23, and select the appropriate runtime for the job.

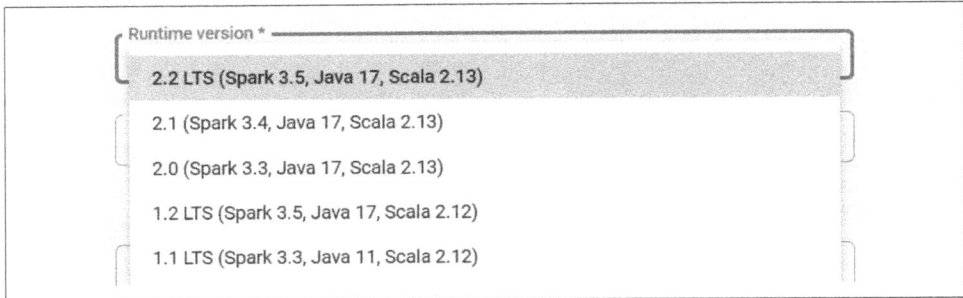

Figure 12-23. Dataproc batch runtime versions

This process applies similarly to Scala Spark and SparkR jobs. You can also specify details like the Hive Metastore and Spark history server configuration before submitting the template.

12.7 Scheduling a Dataproc Job in Event Driven Using a Cloud Function

Problem

You have a PySpark job that you want to run on Dataproc Serverless. The goal is to trigger this job automatically when a new file is added to a specific landing bucket in GCS.

Solution

To achieve this, you'll use a Cloud Function that is triggered by new file events. First, open Cloud shell and navigate to a directory containing your *main.py* and *requirements.txt* files. Then, deploy the Cloud Function, configuring it to trigger on new file events in the landing bucket by running the following gcloud command:

```
gcloud functions deploy <function-name-here> \
  --gen2 \
  --runtime=python312 \
  --region=<region-here> \
  --source=<source-directory-here> \
  --entry-point=<entry-function-name> \
  --trigger-event-filters="type=google.cloud.storage.object.v1.finalized" \
  --trigger-event-filters="bucket=<landing-bucket-here>" \
  --trigger-location=<trigger-location-here>
```

Discussion

You can view the Function at *https://console.cloud.google.com/functions/details/ <region-here>/<function-name-here>?project=<project-id-here>* when the deployment is done.

To test the Function, follow these steps:

1. Upload a file to the landing GCS bucket to trigger the Function.

2. In the console, navigate to the logs section to verify if the Function was triggered or to troubleshoot any issues, as shown in Figure 12-24.

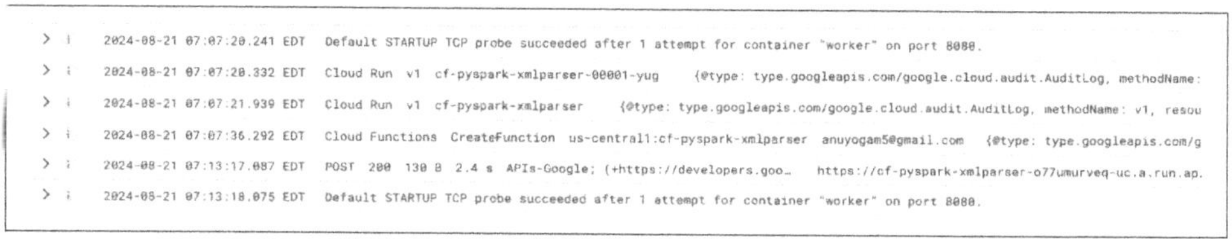

Figure 12-24. Successful POST request in Cloud Function logs

Let's consider an example scenario. In this example, the Cloud Function triggers a PySpark job that reads an XML file and writes it in Parquet format. Here's a sample *main.py* script:

```
def run_dataproc_job(request, context):
    # Initialize variables
    project_id = 'dataproc-cookbook-425300'
    region = 'us-central1'

    # Generate a unique batch ID using the current timestamp
    batch_id = f'xml-parse-job-{datetime.now().strftime("%Y%m%d%H%M%S")}'

    main_python_file_uri = (
        'gs://dataproc-cookbook-1/chapter13/cloudfunctions/xmlparser.py'
    )
    input_file_uri = 'gs://dataproc-cookbook-1/chapter13/cloudfunctions/menu.xml'
    output_dir_uri = 'gs://dataproc-cookbook-1/chapter13/cloudfunctions/output'
    rowTag= "food"

    # Create the Dataproc batch job request
    dataproc = build('dataproc', 'v1', cache_discovery=False)
    parent = f'projects/{project_id}/locations/{region}'

    batch = {
        'pysparkBatch': {
            'mainPythonFileUri': main_python_file_uri,
            'args': [input_file_uri, output_dir_uri, rowTag]
        },
        'environmentConfig': {
```

```
        'executionConfig': {
            'serviceAccount': (
                '1004309118949-compute@developer.gserviceaccount.com'
            )
        }
    }
}

# Submit the job to Dataproc Serverless
request = dataproc.projects().locations().batches().create(
    parent=parent, batchId=batch_id, body=batch)
response = request.execute()

return f"Submitted Dataproc job with ID: {response['name']}"
```

Here is the gcloud command to submit the sample PySpark code to Dataproc
Serverless:

```
gcloud functions deploy cf-pyspark-xmlparser \
    --gen2 \
    --runtime=python312 \
    --region=us-central1 \
    --source=cloudfunctions/code/ \
    --entry-point=run_dataproc_job \
    --trigger-event-filters="type=google.cloud.storage.object.v1.finalized" \
    --trigger-event-filters="bucket=dataproc-cookbook-landing" \
    --trigger-location=us
```

After deployment, navigate to the Dataproc Batches console to verify if the Spark job
was successfully submitted, as shown in Figure 12-25.

Figure 12-25. Status of the Spark job on the Dataproc Batches console

To update the source code, modify it and redeploy using the gcloud command. This
will deploy a new revision.

12.8 Using Dataproc Workflow Templates

Problem

You have a long-running Dataproc cluster and want to orchestrate Spark jobs on this existing cluster using workflows.

Solution

Create a workflow template and specify the label details of the existing Dataproc cluster in the appropriate cluster pool. This will allow you to run Spark jobs on the existing cluster, as shown in Figure 12-26.

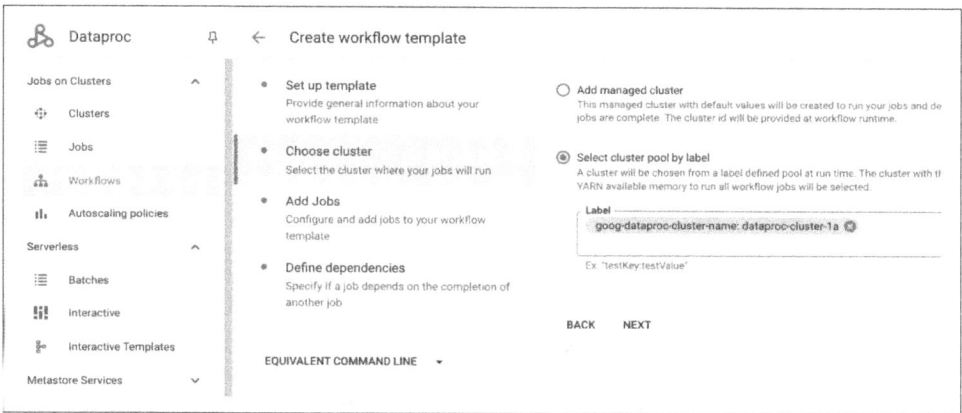

Figure 12-26. Specifying the label of the Dataproc cluster to run the job(s)

Discussion

In the previous recipes, we explored different ways to orchestrate big data jobs. Dataproc workflows offer a simple way to manage and orchestrate one or more Hadoop or Spark jobs on an existing or ephemeral Dataproc cluster. They are particularly useful when working exclusively with big data workloads such as Hive, Spark, and MapReduce. You can also use Dataproc workflows to define dependencies between jobs.

However, in production scenarios, you often have end-to-end data pipelines where a Spark job is triggered based on specific conditions, such as file arrivals or scheduled events. The job runs on Dataproc, and upon completion, a series of analytical queries are executed in BigQuery to generate reports. In such cases, Cloud Composer is an excellent choice for orchestration.

To create a new workflow template, navigate to the Workflows section of the console and click "Create workflow template." Assign a template ID, as shown in Figure 12-27.

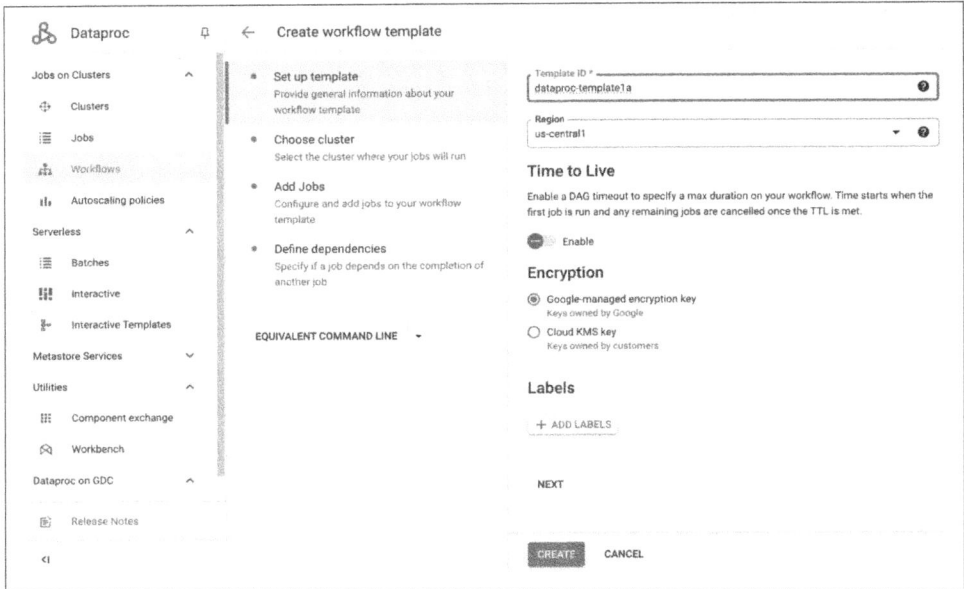

Figure 12-27. Assigning a unique template ID to a workflow template

If needed, you can configure a managed ephemeral cluster for your specific jobs by clicking "Configure managed cluster," as shown in Figure 12-28.

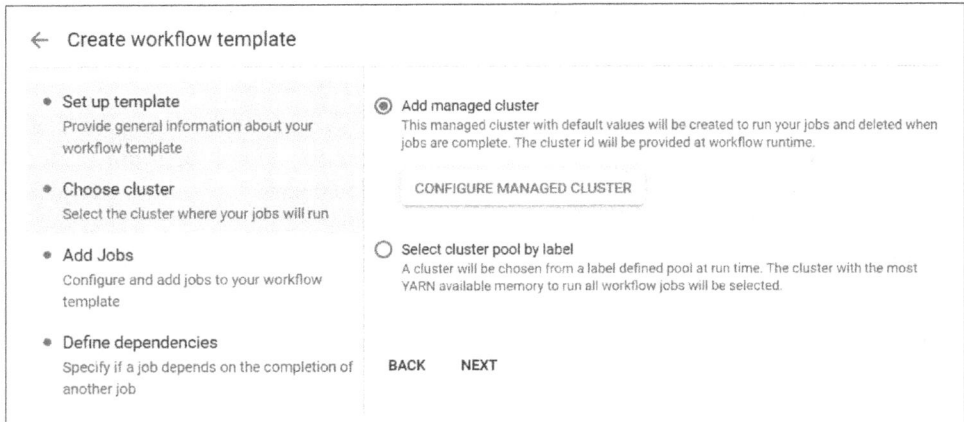

Figure 12-28. Selecting "Add managed cluster" to run on an ephemeral cluster

Add one or more jobs to the workflow. You can select job types like Hadoop, Spark, Hive, or Pig and define job dependencies, as shown in Figure 12-29.

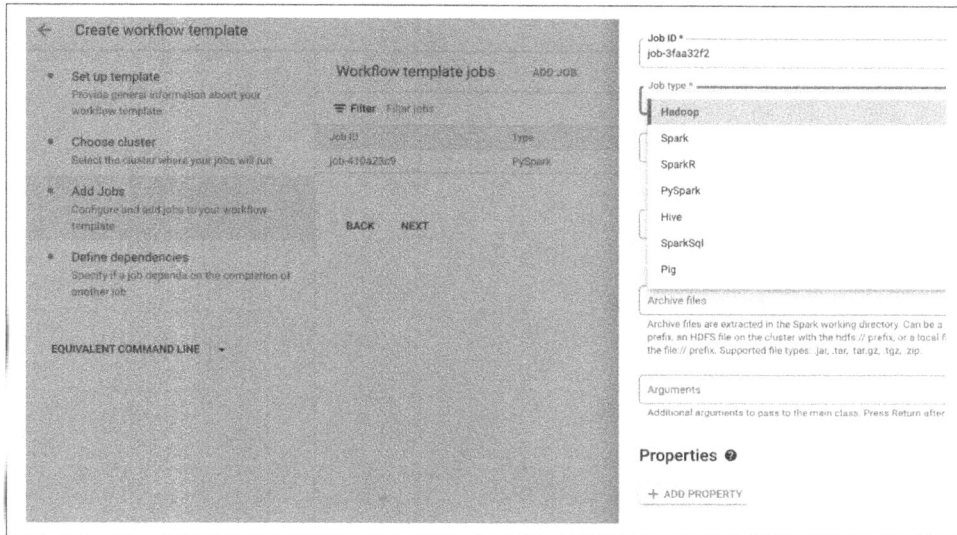

Figure 12-29. Selecting the job type to run

Once the template is successfully created, it will appear in the console, as shown in Figure 12-30.

Template ID	Region	Creation time	Cluster type	Total jobs	Action
dataproc-template1a	us-central1	Oct 20, 2024, 2:57:13 PM	Auto managed cluster	1	RUN

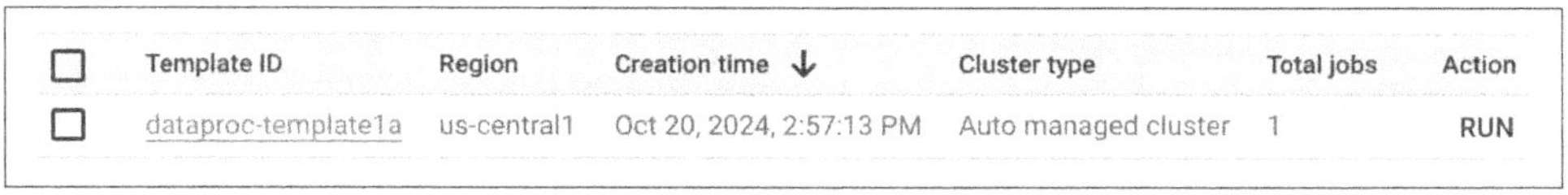

Figure 12-30. The created template in the console

To manually trigger the workflow, click Run in the template.

To automate the execution of the workflow, use Cloud Scheduler or Cloud Composer. For periodic execution, you can use the DataprocInstantiateWorkflowTemplateOperator to instantiate the workflow template, as shown in Figure 12-31.

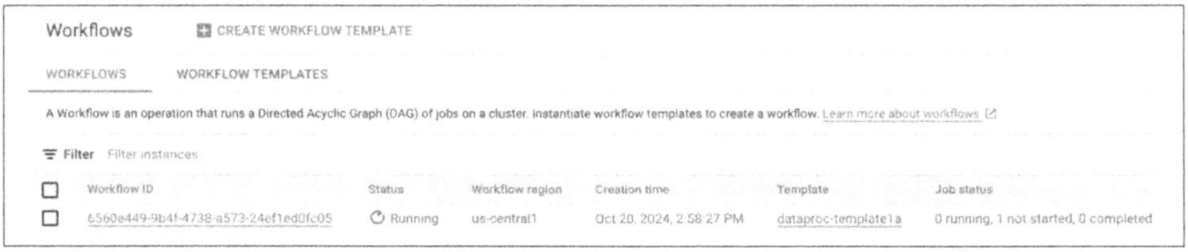

Figure 12-31. Progress of the running template

## CHAPTER 13

Using Spark Notebooks on Dataproc

Interactive development tools like notebooks help data engineers easily collaborate as well as explore and analyze data, thereby improving developer productivity. Choosing the right notebook environment for developing and running Spark applications on GCP is essential to optimizing workflows and leveraging GCP's capabilities.

This chapter explores the various notebook environments available on GCP and provides guidance on selecting the most suitable one for your Spark workloads. We will demonstrate how to set up a Jupyter Notebook on Dataproc on GCE clusters and run different Spark workloads, such as Spark Scala and PySpark, using various notebook kernels.

We will also cover advanced topics, including managing libraries, configuring Spark properties, and working with Dataproc-enabled Vertex AI Workbench instances. By the end of this chapter, you'll have a comprehensive understanding of how to efficiently configure, execute, and manage Spark jobs using notebooks on GCP's Dataproc service.

13.1 Deciding Which Notebook Environments to Choose

Problem

There are various notebook environments in GCP, including Jupyter, JupyterLab, Apache Zeppelin, and Colab. Which one should you choose for running Spark applications?

Solution

Use Dataproc-enabled Vertex AI Workbench (Recipe 13.5) for developing and running Spark applications on GCP. It provides a managed JupyterLab Notebook environment to develop and run Spark applications interactively.

Discussion

Dataproc-enabled Vertex AI Workbench instances have JupyterLab versions 3 and 4 preinstalled and are configured with GPU-enabled machine learning frameworks. They also include a built-in scheduler, making it easy to orchestrate and schedule big data jobs. You can run notebooks on either serverless Spark or an existing Dataproc cluster. This makes Dataproc-enabled Vertex AI Workbench an excellent choice for developing and running Spark applications in notebooks.

Let's walk through the differences between the various notebook environments on GCP:

Jupyter and JupyterLab
These environments make it easy to develop and run Spark applications on Dataproc on GCE with Jupyter enabled. The Project Jupyter community comprises several projects, including Jupyter Notebook and JupyterLab. Jupyter Notebook is a web-based, interactive notebook environment for data analytics, collaboration, and easy distribution. You can write code, narrative texts, visualizations, and widgets inside the notebook. JupyterLab is the latest web-based, interactive notebook development environment, with more functionalities like drag-and-drop cells to rearrange your notebook in addition to what classic Jupyter Notebook offers.

Zeppelin
Supports big data processing with Spark natively but is not supported in Dataproc Workbench and requires configuration on a Dataproc cluster using GCE.

Colab
Great for collaboration and quick prototyping with GCP integration but is not currently supported in Dataproc Workbench or on Dataproc with GCE.

Refer to Table 13-1 for a more detailed comparison of the notebook environments on GCP.

Table 13-1. *High-level comparison of notebook environments on GCP*

Capabilities	Dataproc Jupyter	Dataproc JupyterLab	Zeppelin	Colab
Ease of use	User-friendly interface; minimal setup required	Enhanced interface with additional features like moving cells	Steep learning curve, as it is not supported in Workbench; requires configuration with Dataproc on GCE	User-friendly and integrates with GCP but not available on Workbench or Dataproc on GCE
Spark integration	Good integration with Spark	Good integration with Spark	Native support for Spark; designed for big data	Limited Spark support
Collaboration features	Basic collaboration through notebook sharing	Advanced collaboration tools, real-time editing, and version control	Limited collaboration features	Excellent collaboration tools, real-time editing, and easy sharing
Suitability for GCP	Easy to set up using Dataproc or Dataproc Workbench	Easy to set up using Dataproc or Dataproc Workbench	Easy to set up using Dataproc	Not supported in Dataproc or Dataproc Workbench
Concurrency	Does not support multiple users editing the notebook concurrently	Does not support multiple users editing the notebook concurrently	Does not support multiple users editing the notebook concurrently	Supports multiple users working on the notebook simultaneously

13.2 Configuring Notebooks on a Dataproc Cluster

Problem

You want to set up a Jupyter Notebook on a Dataproc cluster to develop Spark applications.

Solution

Create a Dataproc cluster on GCE with the component gateway and Jupyter Notebook components enabled. Here is an example gcloud command to configure a notebook on a long-running Dataproc cluster:

```
gcloud dataproc clusters create {dataproc-notebook-cluster-name} \
    --enable-component-gateway \
    --region {region} \
    --no-address \
    --master-machine-type {master-machine-type} \
    --master-boot-disk-type pd-balanced \
    --master-boot-disk-size 500 \
    --num-workers 2 \
    --worker-machine-type {worker-machine-type} \
    --worker-boot-disk-type pd-balanced \
    --worker-boot-disk-size 500 \
    --image-version 2.2-debian12 \
    --optional-components JUPYTER \
    --project {project-name}
```

This is the sample gcloud command to create a cluster with the necessary components. Remember to replace placeholders like *{dataproc-notebook-cluster-name}* and *{region}* with your specific details.

Discussion

After creating the Dataproc cluster with Jupyter Notebook enabled, you can access Jupyter or JupyterLab through the Web Interfaces section of the console, as shown in Figure 13-1.

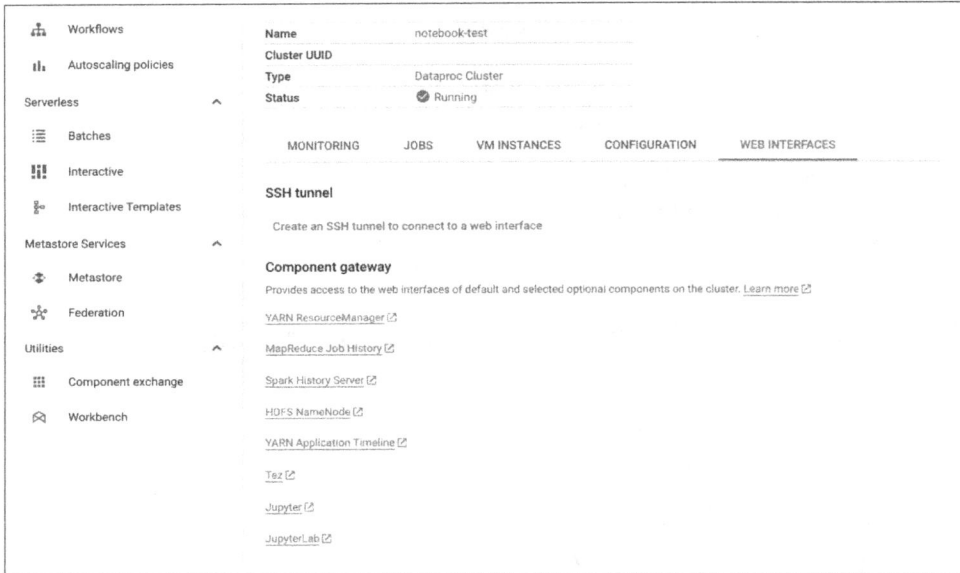

Figure 13-1. The Web Interfaces section in Dataproc displaying component gateways, including JupyterLab

If you want to run a Spark Scala notebook, choose Apache Toree. Apache Toree is an incubation project that provides a Scala kernel for Jupyter Notebook in Dataproc. To run a PySpark application, select PySpark under the Notebooks section, as shown in Figure 13-2.

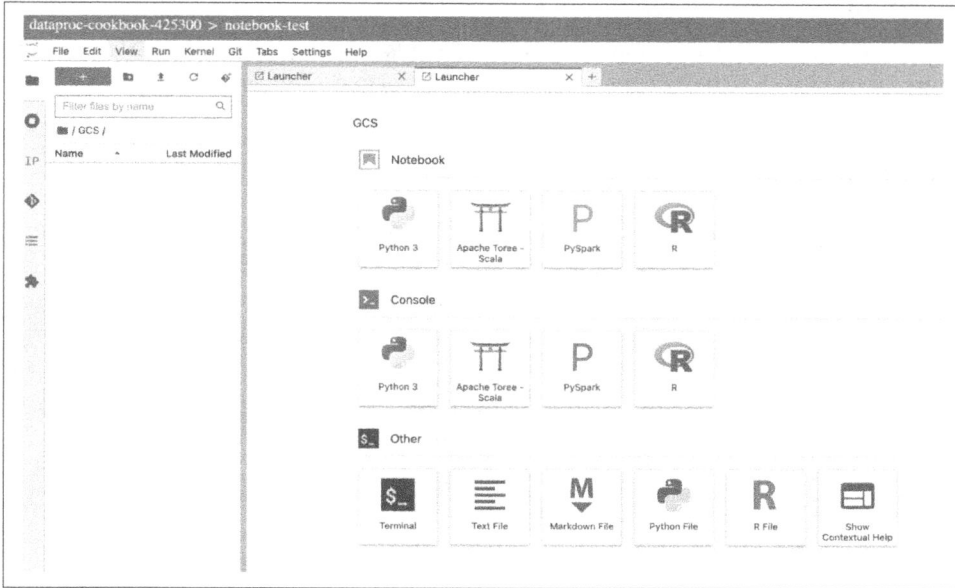

Figure 13-2. JupyterLab launcher with different kernels

Even if you delete the Dataproc cluster, your notebooks will be persisted in GCS at *gs://cluster-staging-bucket/notebooks/jupyter*.

13.3 Running Spark Scala and PySpark Notebooks on Dataproc

Problem

You have Spark Scala or PySpark code that reads JSON files from GCS, performs aggregations, and writes the results in Parquet format. You want to run this code using Dataproc Jupyter Notebooks.

Solution

To run the Spark Scala code, select the Apache Toree kernel in the Notebooks section, as shown in Figure 13-3. For PySpark code, choose the PySpark kernel next to Apache Toree - Scala, also shown in Figure 13-3.

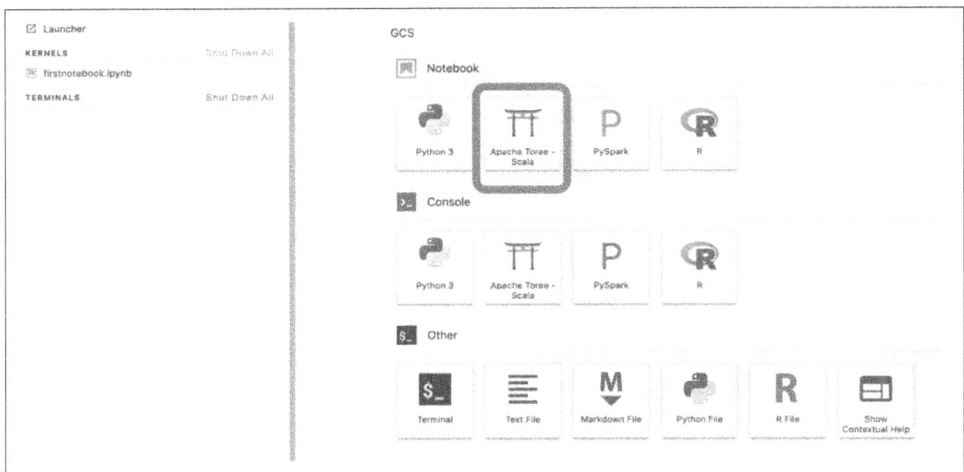

Figure 13-3. Apache Toree and PySpark kernels in the JupyterLab launcher

Discussion

When you click either the Apache Toree or the PySpark kernel, that will open a file named *Untitled*.ipynb*. You can rename this file by selecting Rename Notebook from the File drop-down menu, as shown in Figure 13-4.

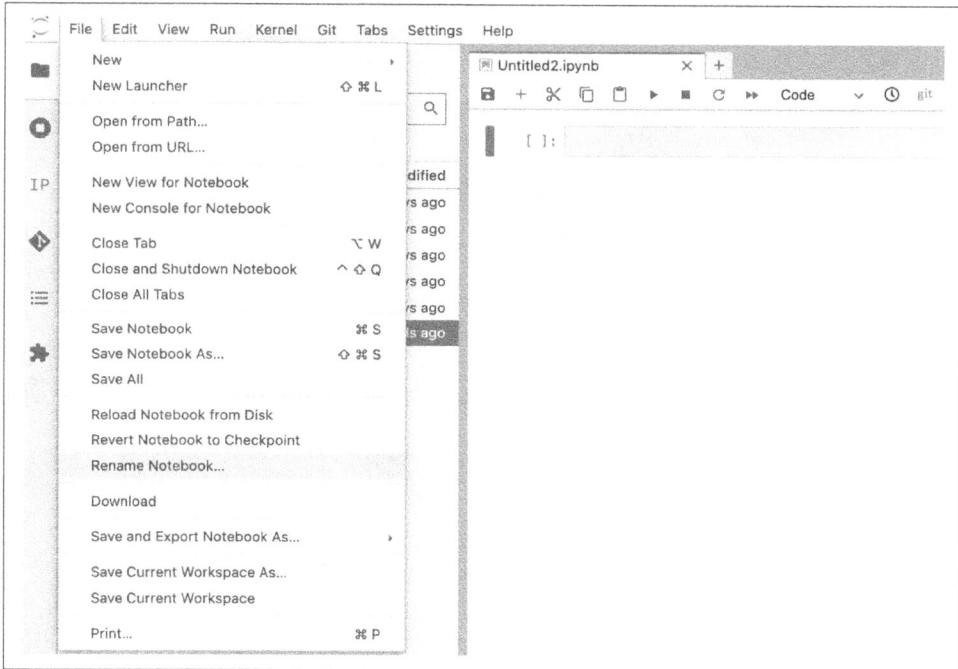

Figure 13-4. Renaming the notebook

Running the first Spark Scala application

Follow these steps to run your first Scala application:

1. Select the Apache Toree - Scala kernel in the Notebooks section (see Figure 13-3).
2. Add the appropriate cells (Code, Markdown, Raw), as shown in Figure 13-5. For Scala code, select Code, write your code, and run the cell, as shown in Figure 13-6.

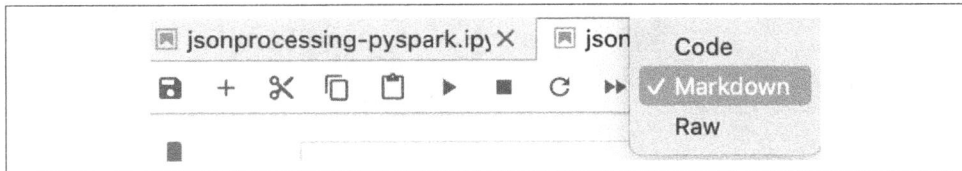

Figure 13-5. Adding cells for Code, Markdown, or Raw

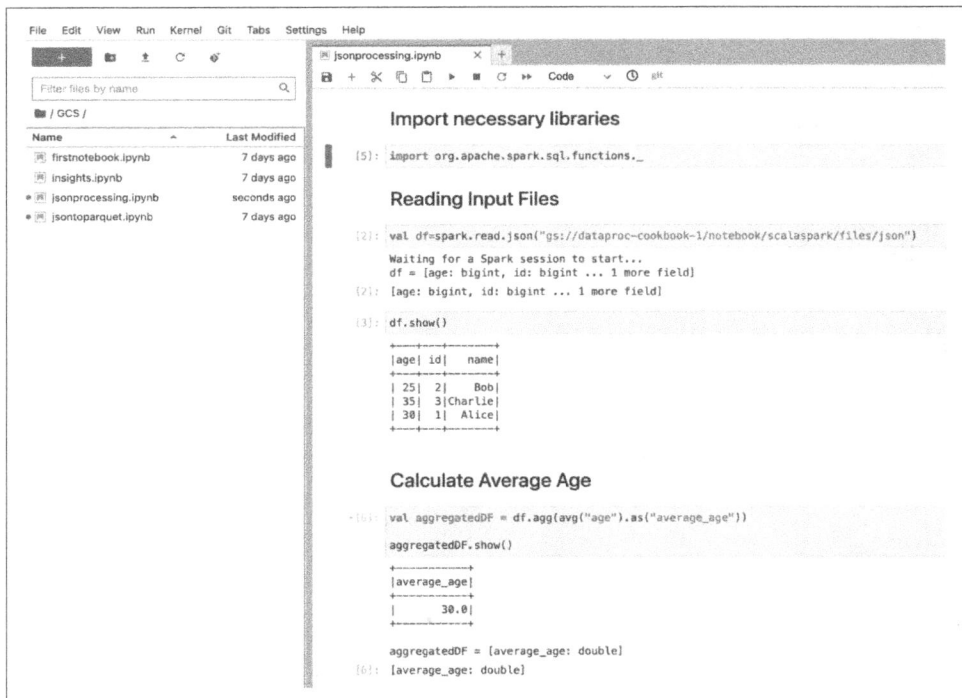

Figure 13-6. Sample Scala Spark notebook

Running the first PySpark application

Follow these steps to run your first PySpark application:

1. To create a new PySpark notebook, click the new launcher (+), as shown in Figure 13-7.

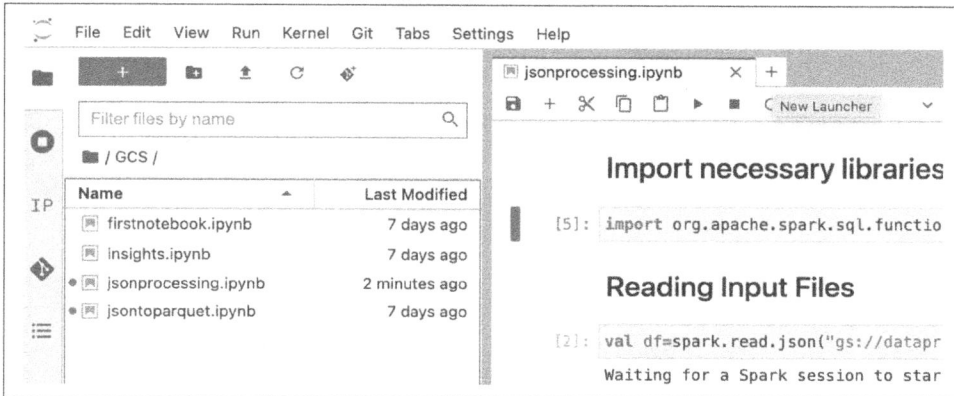

Figure 13-7. Clicking the new launcher to create a new notebook

2. In the Notebook launcher, select the PySpark kernel. Write your PySpark code and run the notebook.

> When you select the Apache Toree kernel (Scala, PySpark, or SparkR), a Spark session will be preconfigured. When you run your code, the Spark application will be submitted in client mode, with tasks executed on the worker nodes. However, when you select the Python kernel, you cannot run a Spark application. You can write Python code, but it will only run on the driver.

After running the Spark application, you can monitor completed applications in the Spark history server, as shown in Figure 13-8.

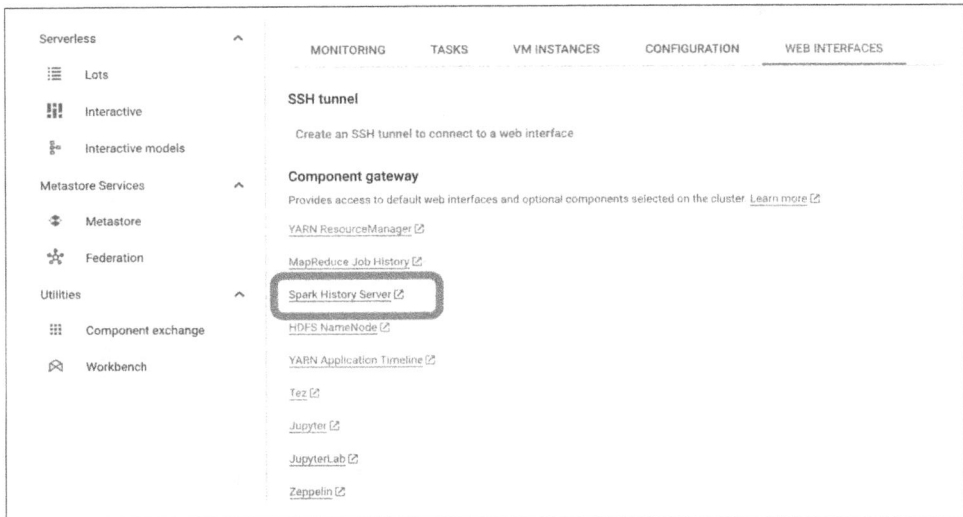

Figure 13-8. Accessing the Spark history server in the Dataproc console

13.4 Managing Libraries and Configs

Problem

You want to install the Delta library and use it to read Delta tables stored in GCS.

Solution

In the Dataproc console, either create a new interactive template or edit an existing one. Under the Properties section, add the custom Delta library by specifying the Maven package in `spark.jars.packages` (e.g., `io.delta:delta-spark_2.13:3.2.1`), as shown in Figure 13-9. If you have custom libraries stored in a GCS bucket, you can use `spark.jars` to include them.

Properties ❷

```
Key 1 *
spark.jars.packages

Value 1
```

`+ ADD PROPERTY`

Labels

`+ ADD LABEL`

SUBMIT   CANCEL

Figure 13-9. Installing Maven packages: pass the package in `spark.jars.packages` when creating the template

For the purposes of this recipe, we'll focus on installing the Delta library. However, this same approach can be used for other Maven packages or custom libraries.

Discussion

Following the steps in this recipe will create a serverless interactive template and session. In the Vertex AI console, you can either create a new notebook from scratch to run the Delta table reading code or select an existing notebook template configured to read Delta tables, as shown in Figure 13-10.

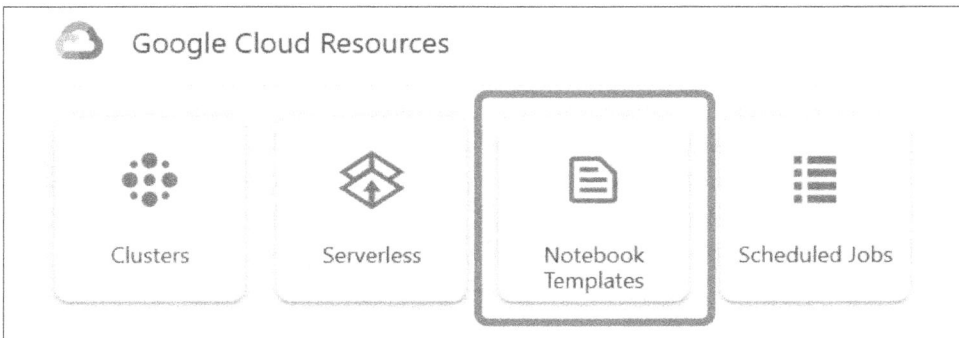

☁ Google Cloud Resources

Clusters   Serverless   Notebook Templates   Scheduled Jobs

Figure 13-10. Accessing existing notebook templates in the launcher

When you open the notebook, shown in Figure 13-11, it will contain the code and steps needed to read Delta files from GCS and display the results.

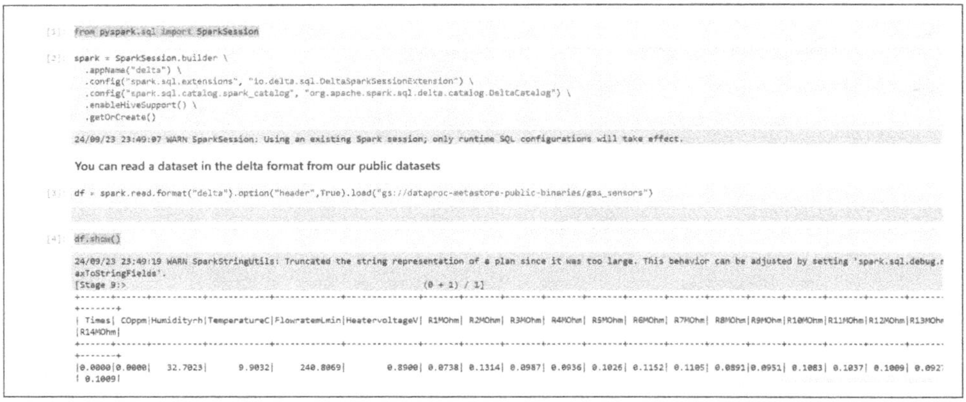

Figure 13-11. Existing Delta read template

If you are running the code in a newly created notebook, use the following PySpark script:

```python
from pyspark.sql import SparkSession

spark = SparkSession.builder \
  .appName("delta") \
  .config("spark.sql.extensions", "io.delta.sql.DeltaSparkSessionExtension") \
  .config(
      "spark.sql.catalog.spark_catalog",
      "org.apache.spark.sql.delta.catalog.DeltaCatalog"
  ) \
  .enableHiveSupport() \
  .getOrCreate()


df = (
  spark.read
  .format("delta")
  .option("header", True)
  .load("gs://dataproc-metastore-public-binaries/gas_sensors")
)


df.show()
```

Before running your notebook, make sure to select the appropriate kernel and choose the interactive template you created earlier, as shown in Figure 13-12. This ensures that your environment is correctly configured for Delta operations and starts the interactive session.

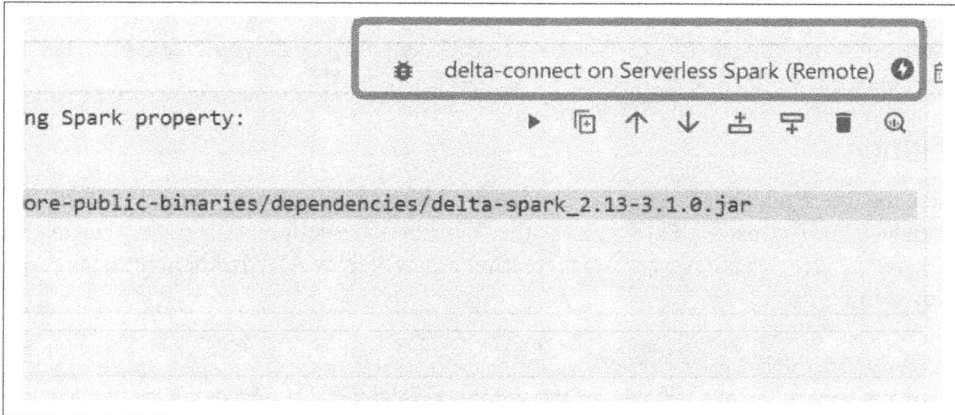

Figure 13-12. Choosing a kernel to run the notebook

13.5 Creating Dataproc-Enabled Vertex AI Workbench Instances

Problem

You want to run Spark jobs on Vertex AI Workbench. How do you create a Vertex AI Workbench instance?

Solution

First, enable the Vertex AI API. Navigate to Vertex AI by searching for it in the Google Cloud console. Then, go to the Workbench section within the Vertex AI console. Click +Create New to start creating a new Vertex AI Workbench instance, as shown in Figure 13-13.

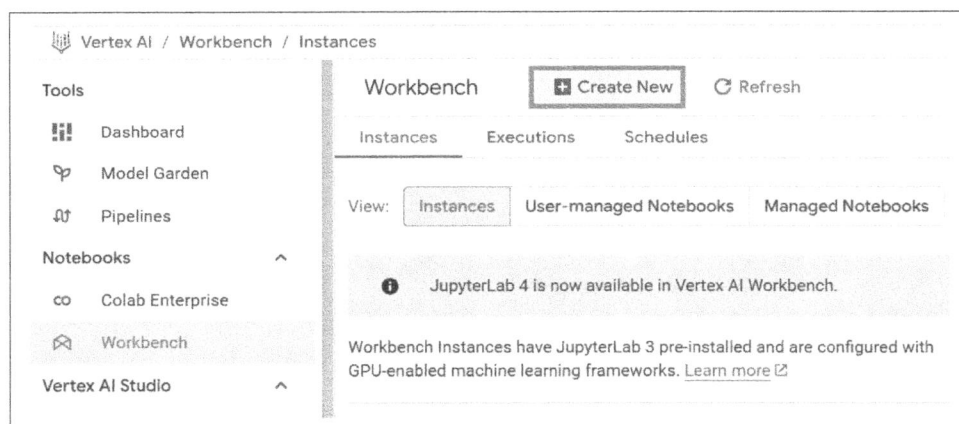

Figure 13-13. Creating a new Vertex AI Workbench instance

Next, enable Dataproc Serverless Interactive Sessions, as shown in Figure 13-14, to run Spark notebooks on Dataproc Spark Serverless and proceed to create the instance.

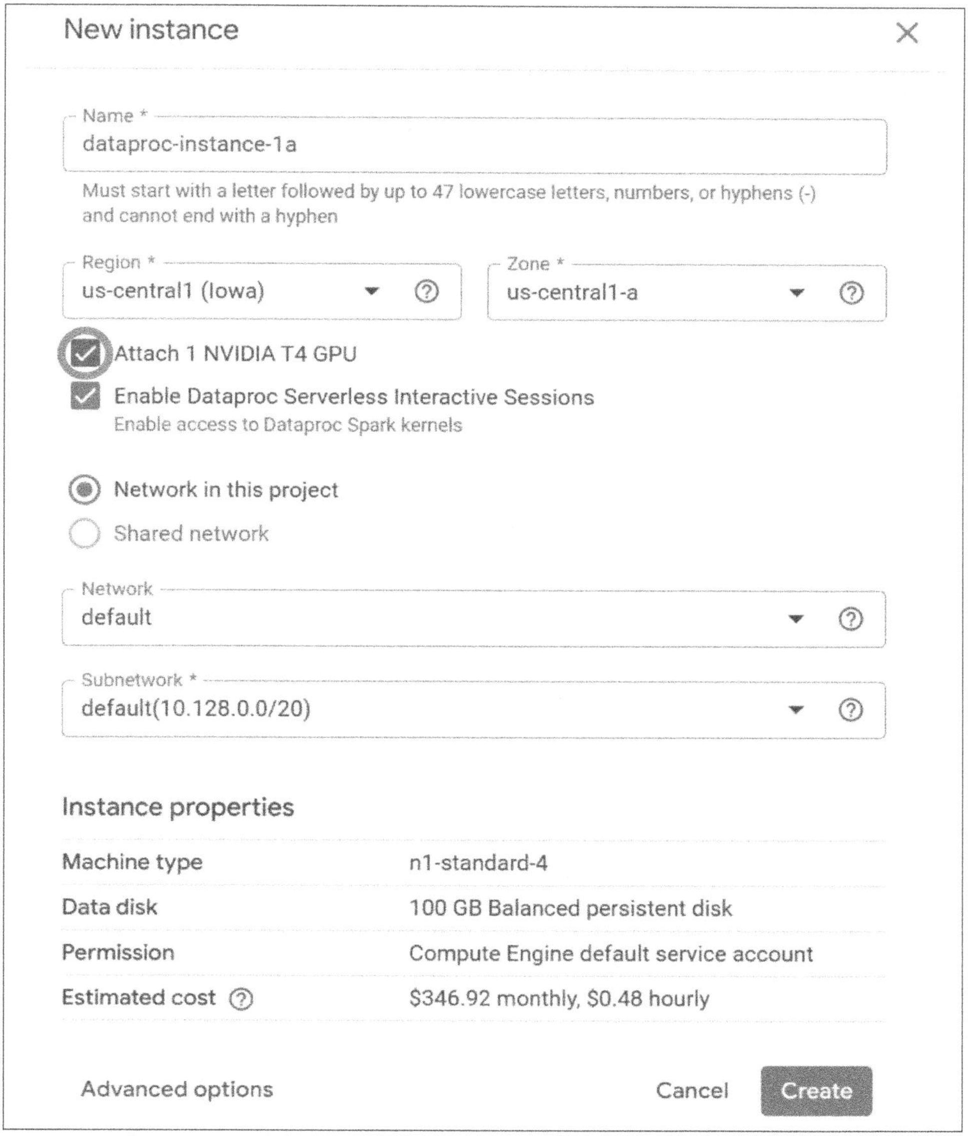

Figure 13-14. Enabling Dataproc Serverless Interactive Sessions

Discussion

During instance creation, you'll be prompted to select a JupyterLab version. Choose JupyterLab 3.x, as shown in Figure 13-15.

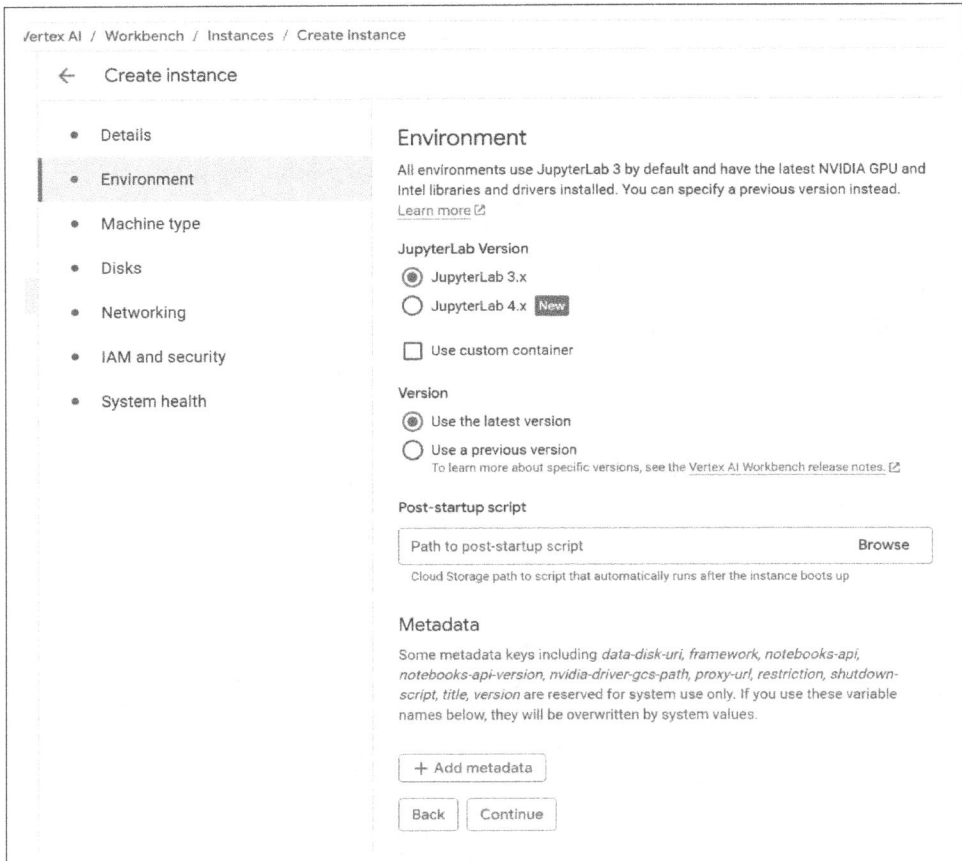

Figure 13-15. Selecting the JupyterLab version for Vertex AI Workbench

Then, select machine types based on your workload requirements, as shown in Figure 13-16.

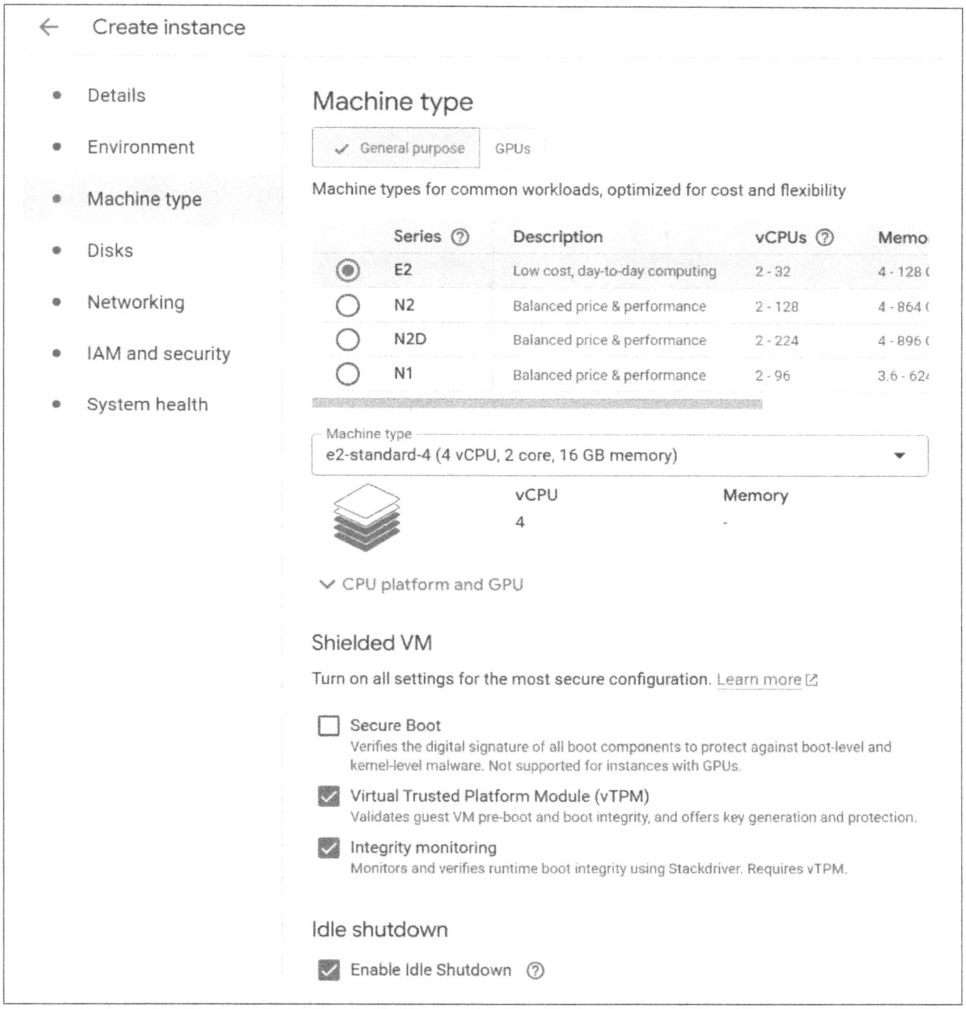

Figure 13-16. *Vertex AI Workbench compute configurations*

Once the instance is successfully created and configured, a green checkmark will appear, indicating that everything is set up correctly, as shown in Figure 13-17. The instance will come preinstalled with JupyterLab 3, providing immediate access to notebook functionality.

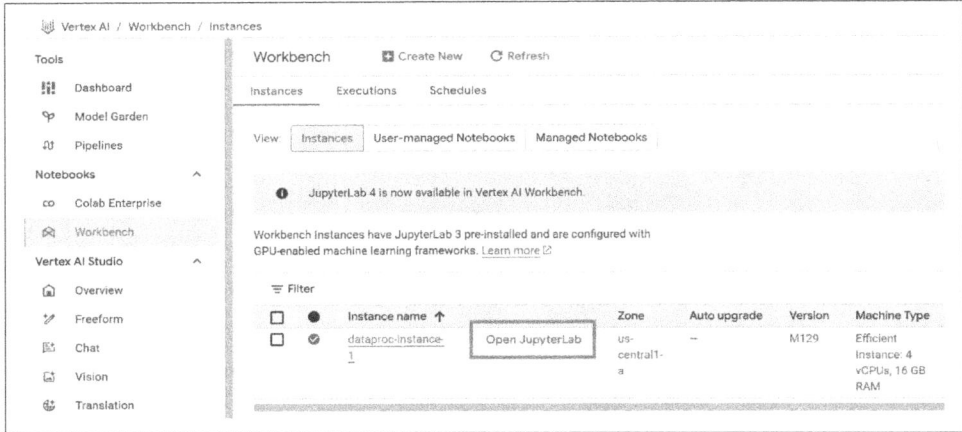

Figure 13-17. Vertex AI Workbench instance creation status

Running Spark notebooks

To launch the JupyterLab environment, click Open JupyterLab (also shown in Figure 13-17). You can run Spark notebooks on existing or new Dataproc clusters or use Dataproc Serverless Spark, as shown in Figure 13-18. Alternatively, you can create and run notebooks using serverless runtime templates by selecting New Runtime Template.

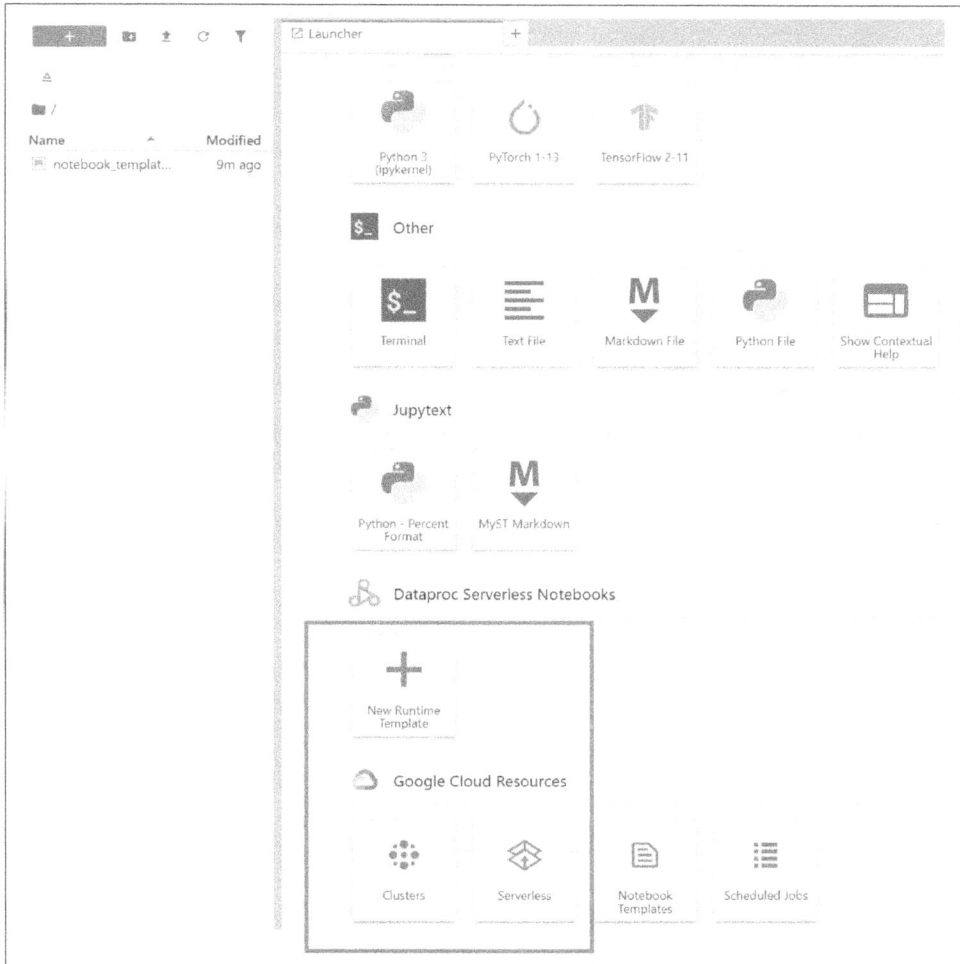

Figure 13-18. Options for running notebooks on existing clusters or Serverless runtime

Integrating with Git

To integrate Git with your notebooks, open the JupyterLab console. In the left panel, select the Git tab. From there, you can either initialize a new Git repository or clone an existing one, as shown in Figure 13-19.

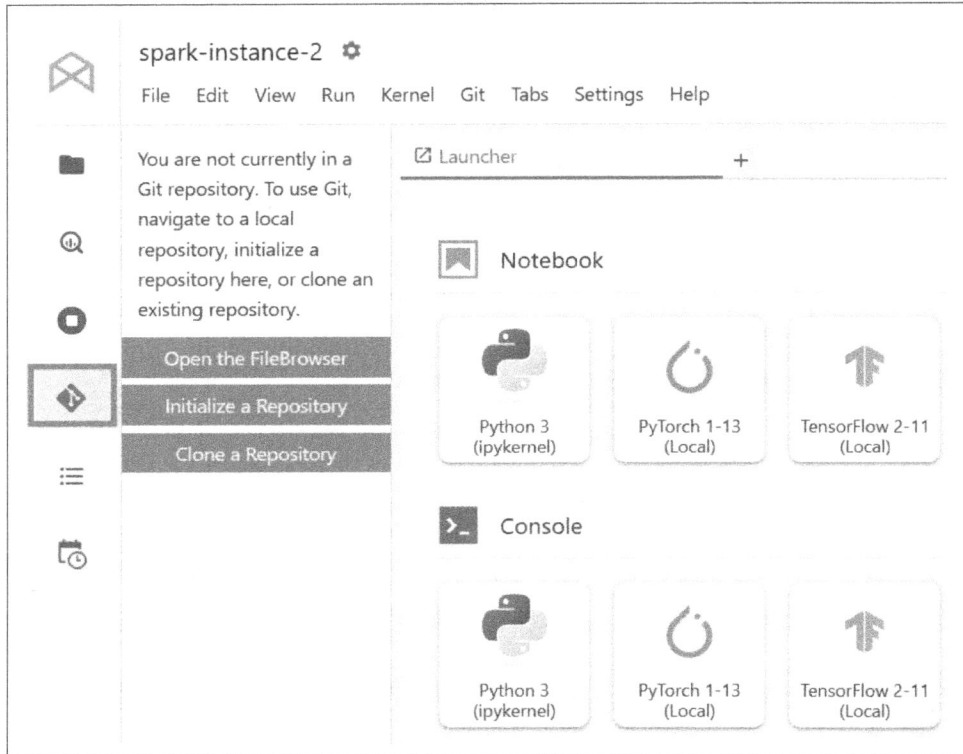

Figure 13-19. Configuring a Git repository

Mounting a GCS bucket

To mount a GCS bucket to the local filesystem, open a terminal and run the following commands using the gcsfuse tool:

```
(base) jupyter@dataproc-notebook-1:~$ MY_BUCKET="dataproc-cookbook-1"
(base) jupyter@dataproc-notebook-1:~$ mkdir -p gcs
(base) jupyter@dataproc-notebook-1:~$ gcsfuse --implicit-dirs $MY_BUCKET \
    "/home/jupyter/gcs"
```

The output is shown in Figure 13-20.

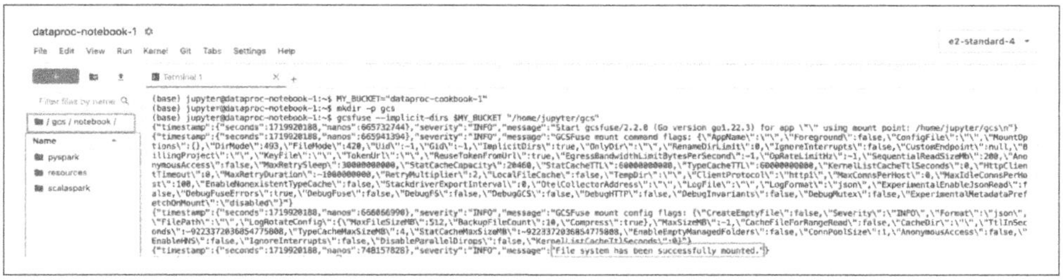

dataproc-notebook-1 ☼   e2-standard-4 ▾
File   Edit   View   Run   Kernel   Git   Tabs   Settings   Help

[terminal/log output]

(base) jupyter@dataproc-notebook-1:~$ MY_BUCKET="dataproc-cookbook-1"
(base) jupyter@dataproc-notebook-1:~$ mkdir -p gcs
(base) jupyter@dataproc-notebook-1:~$ gcsfuse --implicit-dirs $MY_BUCKET "/home/jupyter/gcs"
{"timestamp":{"seconds":1719920188,"nanos":665732744},"severity":"INFO","message":"Start gcsfuse/2.2.0 (Go version go1.22.3) for app \"\" using mount point: /home/jupyter/gcs\n"}
{"timestamp":{"seconds":1719920188,"nanos":665941394},"severity":"INFO","message":"GCSFuse mount command flags: {\"AppName\":\"\",\"Foreground\":false,\"ConfigFile\":\"\",\"MountOptions\":{},\"DirMode\":493,\"FileMode\":420,\"Uid\":-1,\"Gid\":-1,\"ImplicitDirs\":true,\"OnlyDir\":\"\",\"RenameDirLimit\":0,\"IgnoreInterrupts\":false,\"CustomEndpoint\":null,\"BillingProject\":\"\",\"KeyFile\":\"\",\"TokenUrl\":\"\",\"ReuseTokenFromUrl\":true,\"EgressBandwidthLimitBytesPerSecond\":-1,\"OpRateLimitHz\":-1,\"SequentialReadSizeMb\":200,\"AnonymousAccess\":false,\"MaxRetrySleep\":30000000000,\"StatCacheCapacity\":20480,\"StatCacheTTL\":60000000000,\"TypeCacheTTL\":60000000000,\"KernelListCacheTtlSeconds\":0,\"HttpClientTimeout\":0,\"MaxRetryDuration\":-1000000000,\"RetryMultiplier\":2,\"LocalFileCache\":false,\"TempDir\":\"\",\"ClientProtocol\":\"http1\",\"MaxConnsPerHost\":0,\"MaxIdleConnsPerHost\":100,\"EnableNonexistentTypeCache\":false,\"StackdriverExportInterval\":0,\"OtelCollectorAddress\":\"\",\"LogFile\":\"\",\"LogFormat\":\"json\",\"ExperimentalEnableJsonRead\":false,\"DebugFuseErrors\":true,\"DebugFuse\":false,\"DebugFS\":false,\"DebugGCS\":false,\"DebugHTTP\":false,\"DebugInvariants\":false,\"DebugMutex\":false,\"ExperimentalMetadataPrefetchOnMount\":\"disabled\"}"}
{"timestamp":{"seconds":1719920188,"nanos":666066990},"severity":"INFO","message":"GCSFuse mount config flags: {\"CreateEmptyFile\":false,\"Severity\":\"INFO\",\"Format\":\"json\",\"FilePath\":\"\",\"LogRotateConfig\":{\"MaxFileSizeMB\":512,\"BackupFileCount\":10,\"Compress\":true},\"MaxSizeMB\":-1,\"CacheFileForRangeRead\":false,\"CacheDir\":\"\",\"TtlInSeconds\":-9223372036854775808,\"TypeCacheMaxSizeMB\":4,\"StatCacheMaxSizeMB\":-9223372036854775808,\"EnableEmptyManagedFolders\":false,\"ConnPoolSize\":1,\"AnonymousAccess\":false,\"EnableHNS\":false,\"IgnoreInterrupts\":false,\"DisableParallelDirops\":false,\"KernelListCacheTtlSeconds\":0}}
{"timestamp":{"seconds":1719920188,"nanos":748157820},"severity":"INFO","message":"File system has been successfully mounted."}

Figure 13-20. Mounting a GCS bucket to the local filesystem

When you run this code, replace `dataproc-cookbook-1` with the name of your bucket. The gcsfuse command mounts your GCS bucket to the local file system at */home/jupyter/gcs*, allowing you to interact with it as though it were a local directory.

13.6 Executing Notebooks Using Spark Serverless Sessions

Problem

You have developed a Spark application in a JupyterLab notebook on a Dataproc-enabled Vertex AI Workbench and need to run the notebook using a runtime template.

Solution

In the JupyterLab console, you can create runtime templates for running notebooks, as shown in Figure 13-21.

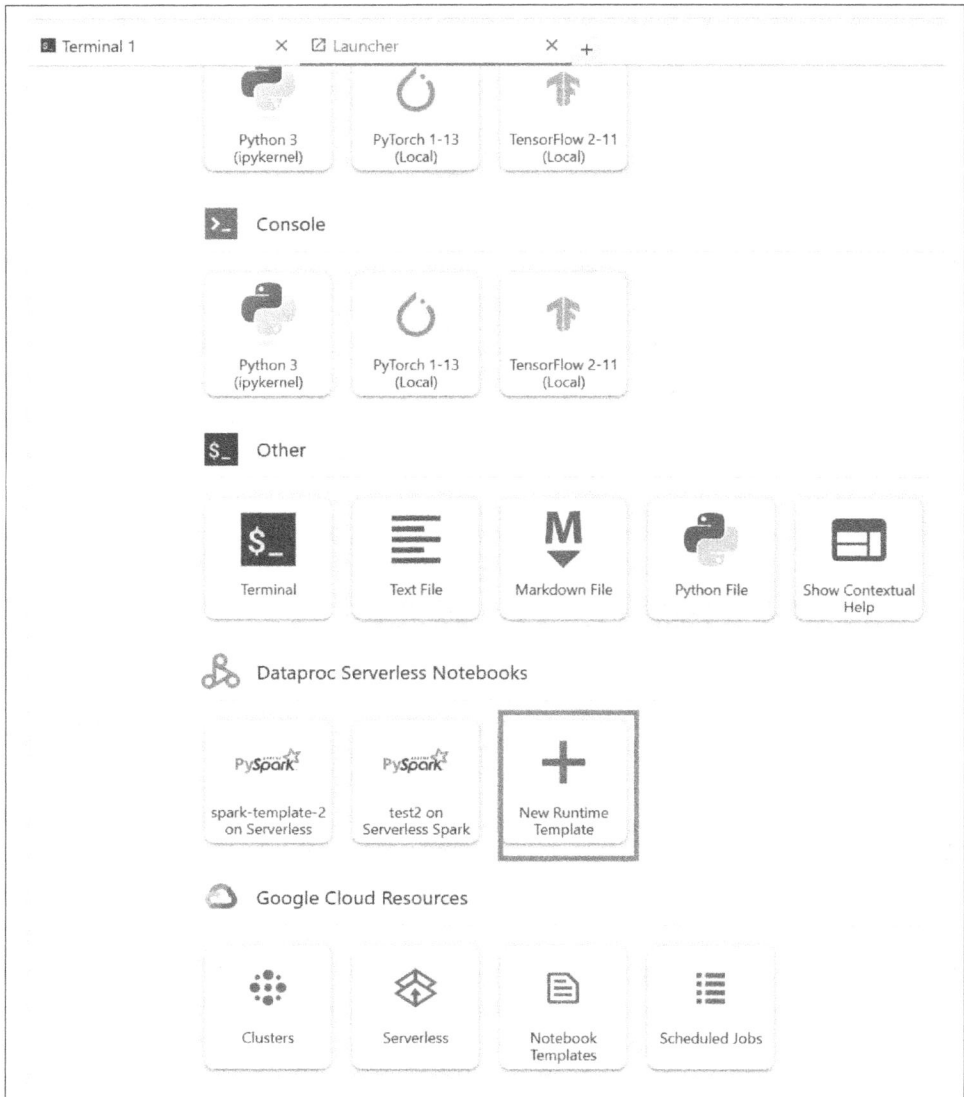

Figure 13-21. Creating a serverless runtime template

Choose an appropriate name and version for the Spark notebook runtime, as shown in Figure 13-22.

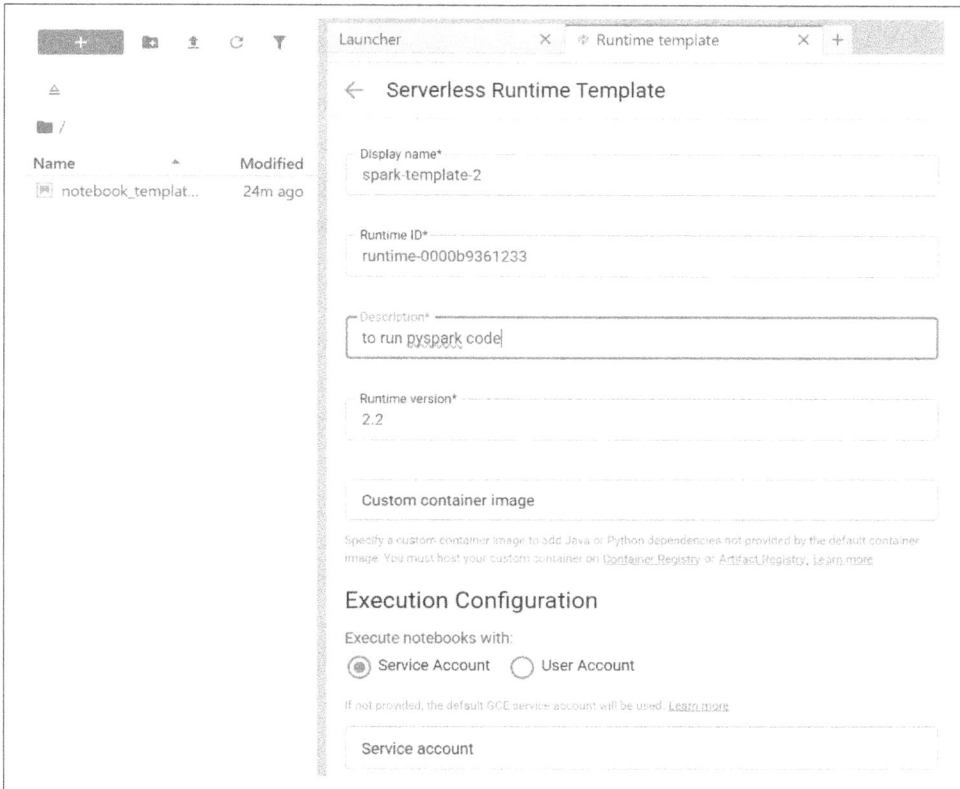

Figure 13-22. Adding a name for the serverless runtime template

Discussion

Once you've created the template, a notification will appear, as shown in Figure 13-23. Then, the template will be available in the JupyterLab console under the Dataproc Serverless Notebooks section, as shown in Figure 13-24.

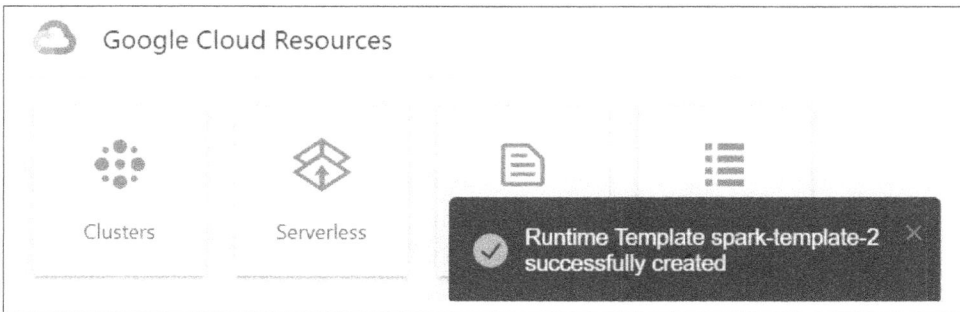

Figure 13-23. Notification that a serverless runtime template was created

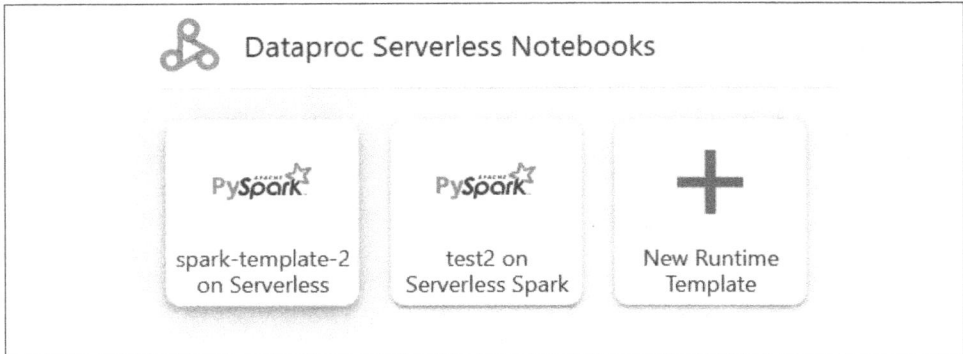

Figure 13-24. Serverless runtime template available in the JupyterLab console after successful creation

Another place to validate that the runtime template was created successfully is in the Dataproc console, where it will appear under Interactive Templates, as shown in Figure 13-25.

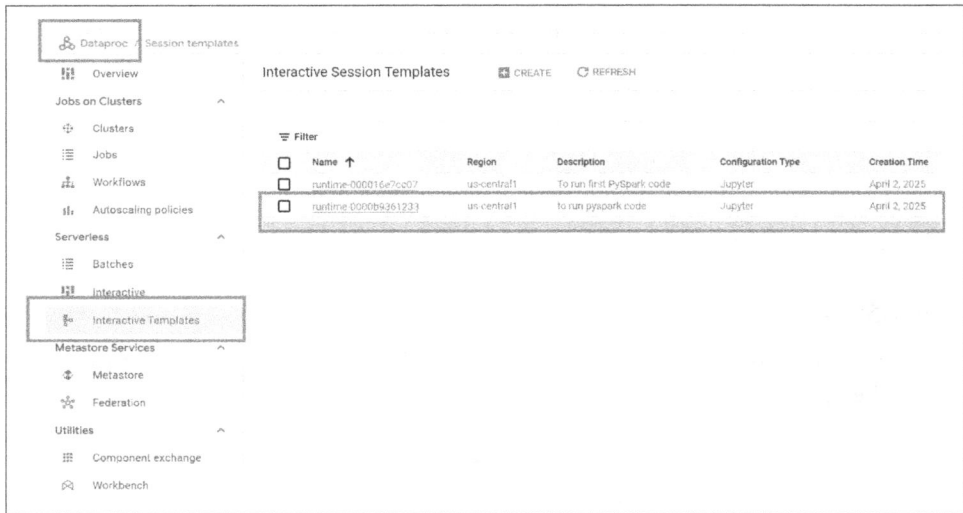

Figure 13-25. Serverless runtime template in the Dataproc Interactive Templates

To create a new notebook, open the File drop-down menu and select New and Notebook, as shown in Figure 13-26.

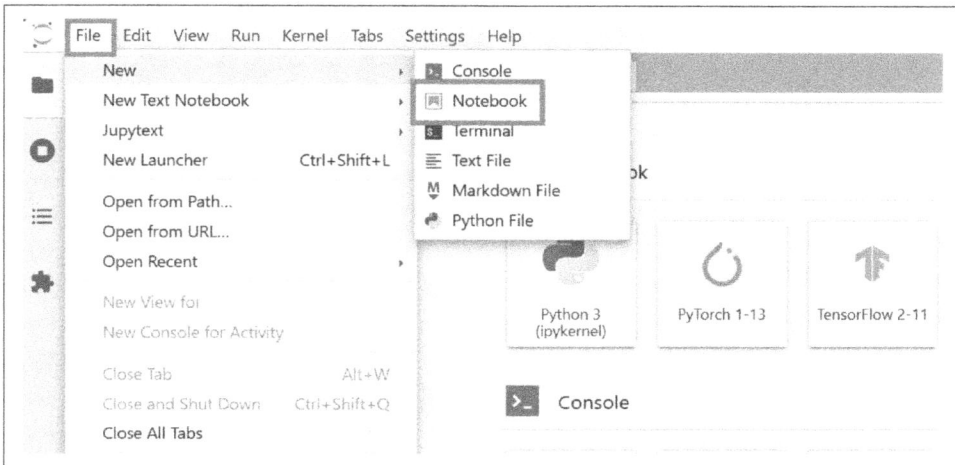

Figure 13-26. Creating a new notebook

Enter the Spark code in the cells. To query a BigQuery table from the notebook, use the #@bigquery magic command, as shown in Figure 13-27.

```python
from pyspark.sql.types import StructType, StructField, StringType, IntegerType

# Define schema
schema = StructType([
    StructField("ID", IntegerType(), True),
    StructField("Name", StringType(), True),
    StructField("Age", IntegerType(), True)
])

# Create DataFrame with 3 rows
data = [
    (1, "Alice", 25),
    (2, "Bob", 30),
    (3, "Charlie", 35)
]
df = spark.createDataFrame(data, schema=schema)

# Show the DataFrame
df.show()
```

```
+---+-------+---+
| ID|   Name|Age|
+---+-------+---+
|  1|  Alice| 25|
|  2|    Bob| 30|
|  3|Charlie| 35|
+---+-------+---+
```

Querying BigQuery Table

▶ Submit Query   This query will process 28.3 KB when run.

```
#@bigquery
SELECT * FROM `bigquery-public-data.ml_datasets.penguins`
```

Figure 13-27. Using the magic command to query a BigQuery table

To run the notebook using a serverless runtime template, select the template from the Kernel drop-down menu, as shown in Figure 13-28. This selection will start an interactive session based on the template's configurations.

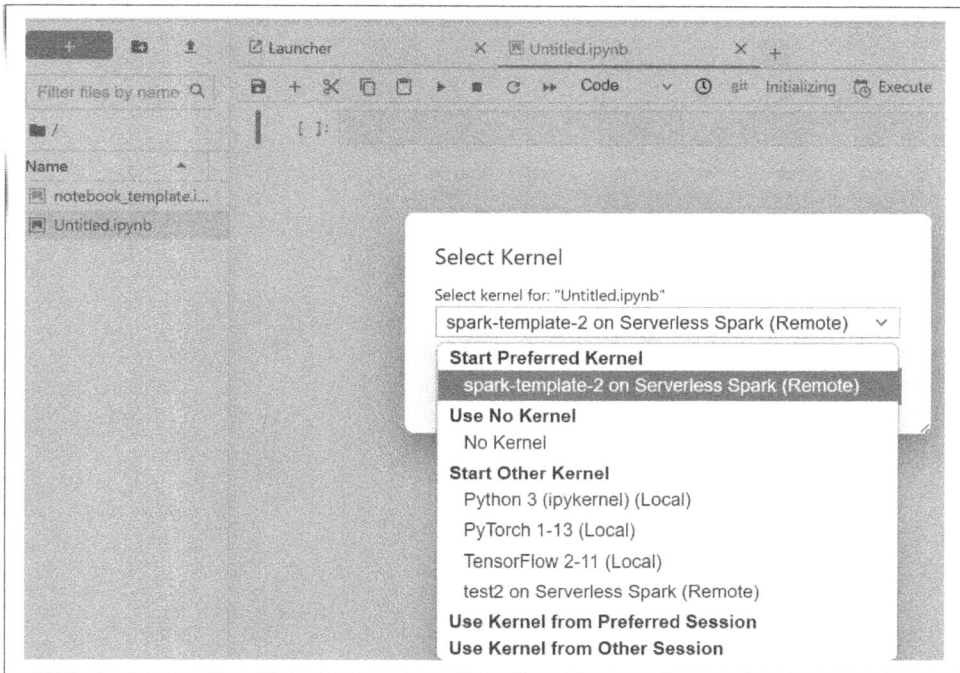

Figure 13-28. Selecting the custom template from the Kernel drop-down menu

When the interactive session starts, by default it will assign the following driver and executor configurations if you haven't customized them:

```
spark.executor.instances = 2
spark.driver.cores = 4
spark.driver.memory = 12200m
spark.executor.cores = 4
spark.executor.memory = 12200m
```

You can adjust these settings based on the requirements of your Spark job.

> If you want to orchestrate or schedule your notebook execution, you can use the built-in scheduler. This option is available at the top-right corner of the notebook interface, shown in Figure 13-29.

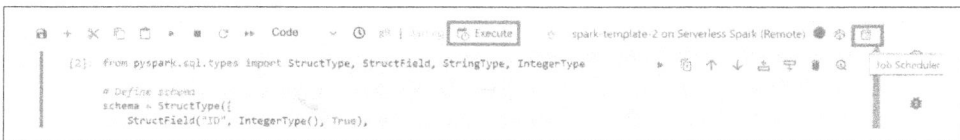

Figure 13-29. Scheduling the notebook to run on the serverless runtime template

## CHAPTER 14

Migrating from On-Premises and Public Cloud Services to GCP

Migrating big data applications from on premises or alternate public clouds is a complex task that requires strategic decision making, migration planning and technical expertise. This chapter provides guidance for the migration. Whether you are moving from an on-prem Cloudera setup, Amazon Elastic MapReduce (EMR), or Azure HDInsight, the objective remains consistent: migrating with minimal disruptions while leveraging GCP's managed services, such as Dataproc, for enhanced performance and efficiency.

In this chapter, you will learn:

- How to plan and execute a successful migration strategy tailored to your organization's big data ecosystem

- Techniques for assessing your current infrastructure using tools like Google's Hadoop Discovery Tool

- Strategies for migrating data and metadata while addressing compatibility and performance challenges

- Best practices for workload optimization postmigration, including choosing appropriate cluster configurations and leveraging modern features like Dataproc Serverless

- Critical architectural and security decisions to ensure reliability, scalability, and compliance

Let's dive into the first critical step: planning your migration.

14.1 Planning Migration

Problem

You need to migrate big data applications (Hive, Spark, MapReduce) from an on-prem Cloudera environment to a managed Hadoop platform on GCP while keeping the current applications running on GCP with minimal code changes.

Solution

A successful data analytics migration to Google Cloud necessitates thorough planning, starting with a deep dive into your current on-prem environment, including data pipelines, code dependencies, and business needs. Define your target state, considering whether a simple "lift and shift" or a full modernization is required. For Hadoop and Spark workloads, Dataproc offers a seamless lift-and-shift path, while BigQuery enables a modernization approach, leveraging serverless architecture and SQL for enhanced scalability and performance.

Key considerations during this planning phase include:

Assessment first
> Prioritize a comprehensive assessment of your existing data analytics landscape to identify dependencies and complexities.

Strategic choice
> Carefully select a migration strategy (lift and shift vs. modernization) based on your long-term goals and resource constraints.

Dataproc versus BigQuery
> Understand the distinct advantages of Dataproc for Hadoop/Spark compatibility and BigQuery for serverless, scalable analytics.

> This recipe will provide an overview of what you will need to plan your migration. Migration strategies and other key components will be covered in the following recipes.

Discussion

When migrating applications, several factors need to be considered, such as the motivation behind the migration. This could include the expiration of the current platform's license, cost reductions, or a shift in the business's strategic direction. Based on these factors, the migration approach can be either lift and shift or lift and replatform:

Lift and shift

> This involves moving workloads to GCP with minimal code changes—for example, migrating Cloudera Spark jobs to Dataproc and running them as is with few modifications.

Lift and replatform

> This approach embraces modernization, such as migrating Hive workloads to BigQuery for better cost optimization and alignment with business strategy.

For any migrations, start with discovery and assessment. The Hadoop Discovery Tool was developed by Google Cloud Professional Services. It is a Python-based tool that uses the Cloudera API, YARN API, filesystem config files, and OS-based native CLI requests to analyze the on-premises Hadoop environment. This tool generates a report with details of the cluster configuration (number of workers, cores, memory, OS, and health status), which types of big data services are running in the cluster and their details, network interface and traffic details, HDFS data distribution metrics, Cloudera Hadoop metrics, YARN metrics, and Spark metrics. You should have admin privileges to run this tool. Follow the installation steps (*https://oreil.ly/cHvrT*) to install the Hadoop Discovery Tool.

Based on your discovery and assessment, plan for the data and metadata migration. We'll cover this process in greater detail throughout this chapter. Specifically, you can find data migration strategies in Recipe 14.2, and the metadata migration strategy is covered in Recipe 14.5.

Once you have planned your data and metadata migration, you'll need to plan the types of workloads to migrate (Hive, Spark, MapReduce, HBase). Here are a few things to consider when planning a migration.

Compute and cluster planning

Begin by determining whether ephemeral or long-running compute instances best suit your needs. Benchmark complex custom big data applications to establish the appropriate compute size and machine type. Consult the machine type throughput documentation to aid in this decision.

Given the cloud-hosted environment, you should evaluate whether to maintain one big cluster like on premises or build separate smaller clusters in GCP for each team or type of work (such as Spark or Hive), which makes it easier to track costs and keep things secure and lets each team work without slowing one another down.

Hive migration

Gather details about Hive versions, table counts, table types, and partition information. Based on these details and organizational goals, decide whether to lift and replatform or modernize. For lift and replatform, Dataproc is the recommended target service. For modernization, consider BigLake or BigQuery. If you are using Hive ACID tables, compact them on premises before migration.

Spark migration

Document the Spark versions and the number of applications. Dataproc is generally the target service; select the appropriate image based on the Spark version. If you are using Spark 2.x, consider upgrading to Spark 3.x to leverage the latest enhancements. Identify and plan the migration of libraries and dependencies. Determine if jobs are batch or streaming. For batch jobs, consider ephemeral clusters to minimize costs and customize configurations for optimal performance if a shared long-running cluster is not required. For batch jobs using Spark version 3.x, you can also leverage Dataproc Serverless.

Additional migration considerations

This section covers some additional topics you should consider when planning your migration.

Architectural decisions. Decide on the architecture choices that make the most sense for your migration. For example, if you choose a medallion architecture, you can have bronze (raw), silver (enriched), and gold (curated) zones in GCS:

Bronze layer
Data lands into this zone from different source systems in raw format. In this zone, it is generally "append only" and immutable.

Silver layer
Data is cleaned, standardized, and stored in GCS.

Gold layer
Data is optimized for analytics and stored in GCS. It can have:

Denormalized tables
Optimized for fast querying in BigQuery, these tables combine fact and dimension data to reduce the need for joins.

Fact and dimension tables
A more normalized approach, useful for maintaining data integrity and flexibility while still being optimized for analytics.

The choice depends on your use case. Denormalized tables are best for performance and ease of use in reporting. Fact and dimension tables provide flexibility and are useful for more complex analytical models.

You'll want to decide on the standardized formats like Delta, Iceberg, or Hudi in the silver and gold zones. Once you make a selection, you can check if the services you interact with have support for these formats. If not, install the latest libraries.

IAM policy redesign. Assess current on-premises IAM strategies and implement GCP IAM policies based on the principle of least privilege for enhanced security.

Reliability and region selection. Specify the desired zone and region for Dataproc clusters. Align the region selection with customer location or data-residency requirements (e.g., us-central1 or us-east1 for US-based data).

Cost optimization. Implement Cloud Storage lifecycle policies to automatically transition data between storage tiers based on access frequency. Use preemptible VMs or committed-use discounts for Dataproc clusters to reduce costs. Implement Google Cloud Monitoring and Logging to track resource utilization and identify cost-optimization opportunities.

Testing and validation. Conduct parallel runs of workloads in both on-premises and GCP environments to validate data accuracy and performance. Execute thorough performance testing to ensure that migrated workloads meet performance requirements. Implement data-validation checks to ensure data integrity during and after migration.

Automation. Use tools like Terraform to automate the provisioning and management of GCP resources. Implement continuous integration and continuous delivery (CI/CD) pipelines to automate the deployment and testing of applications and workloads. Leverage Google Cloud migration tools such as Storage Transfer Service (STS) and the BigQuery data-transfer service.

Upskilling. Invest in training on GCP and relevant GCP services. If internal expertise is limited, consider engaging Google Cloud Consulting or GCP partners for migration assistance.

See Also

"Best Practices of Migrating Hive ACID Tables to BigQuery" (*https://oreil.ly/_1k-G*) (Google Cloud blog post)

14.2 Data Migration Strategies

Problem

How can organizations effectively migrate large datasets from diverse sources like Amazon S3, Microsoft Azure, and HDFS to GCS? What factors should be considered when choosing the optimal migration strategy (batch vs. incremental), selecting the right tools, and ensuring a smooth transition while maintaining data integrity and security?

Solution

Effectively migrating data to GCS from various sources requires understanding your data's characteristics and key requirements. You'll need to choose between batch migration for large initial transfers or incremental migration for ongoing synchronization, using tools like gsutil for basic transfers, STS for managed migrations, or a Transfer Appliance for very large datasets. It's important to optimize network performance, ensure data integrity, and implement robust security measures throughout the process.

Discussion

For any migration job, it is critical to understand the data that you are going to migrate. Having a detailed view of the data's characteristics enables you to choose the most suitable transfer approach, optimize for performance, and minimize downtime or interruptions. Here's a breakdown of the key considerations:

Total data volume (GB/TB)
> Knowing the total size of the data to transfer is essential for selecting the right migration strategy and infrastructure. Large datasets may require network tuning, adjustments to maximize throughput, and a clear estimate of how long the transfer might take.

Number of objects
> The number of objects (files) affects the migration's complexity and the potential time required. Managing a large number of small files can introduce more overhead compared to fewer bigger files due to metadata handling and initiation of multiple connections.

Object size
> The size of objects (both average size and size range) affects transfer performance. Large files can maximize throughput, while small files can create more overhead due to multiple read/write operations, affecting speed and efficiency.

Read and write patterns (frequency and volume)

Understanding the access patterns of the data provides insight into how the migration may affect users or systems relying on the data. If the data is read or written frequently, transferring it all at once may cause disruptions.

Data format

Are there any specific data formats or compression techniques used in the source data?

Network connectivity

When planning your data migration to GCS, it's essential to assess your existing network connectivity to optimize transfer speed and security. Start by determining how your Google Cloud project connects to the source of your data. Are you using the public internet, or do you have a dedicated connection like Direct Peering, Cloud VPN, or Cloud Interconnect in place? These private connectivity options offer higher bandwidth, lower latency, and enhanced security compared to the public internet. Understanding your current setup will help you make informed decisions about potential network enhancements and ensure a smooth, efficient data transfer to GCS.

Additional questions you may need to answer include:

- Does the data contain sensitive information requiring encryption or specific security measures during and after the migration?
- Are there any regulatory or compliance requirements that need to be considered during the migration?

Once you have a handle on the key considerations we just covered, you'll need to select your migration strategy. There are a number of options:

Batch migration

This approach is ideal for datasets that are static or can be taken offline temporarily. It simplifies transfer by moving all the data at once, which is suitable for workloads with downtime tolerance. Example tools include Google Cloud STS, gsutil, and Storage Transfer API.

Incremental migration

This approach is best for frequently updated datasets that require minimal downtime. It allows data to be migrated in stages with continuous synchronization until the final cutover. Example tools include gsutil, rsync, and Transfer Appliance for initial bulk transfer, followed by Cloud STS for syncing updates.

Hybrid approach

Start with a bulk migration (batch) to move the bulk of the data, followed by incremental updates to capture recent changes. This approach balances speed with minimal downtime, especially for large, frequently accessed datasets. Example tools include STS configured with one-time refresh and incremental updates at regular intervals.

Table 14-1 provides further guidance to help you determine the best tools and solution for your needs.

Table 14-1. Recommended solution matrix based on source, destination, and size

Source	Destination	Data size	Recommended solution	Description
Another cloud storage provider	GCS	Any size	STS	For transfers from external providers (e.g., AWS, Azure) to GCS with automated scheduling and management
On premises (filesystem)	GCS	Less than 1 TB	gcloud storage	Suitable for smaller direct transfers from on premises to GCS
On premises (filesystem)	GCS	More than 1 TB	STS	Best for large on-premises transfers due to optimizations and error handling
GCS	GCS	Less than 1 TB	gcloud storage	Effective for small interregion transfers within GCS
GCS	GCS	More than 1 TB	STS	Ideal for large interregion transfers within GCS for improved reliability and control
GCS	On premises (filesystem)	Small dataset (less than 1 TB)	gcloud storage rsync	Straightforward and efficient for smaller transfers between GCS and filesystem storage
GCS	On premises (filesystem)	Large dataset (greater than 1 TB)	STS	Use with a locally installed agent in a Docker container to efficiently handle large transfers

Here's a summary of the different solution types:

gsutil
> A command-line tool for interacting with GCS, gsutil offers a wide range of commands for managing buckets and objects (uploading, downloading, deleting, moving, etc.). It provides options for scripting and automation, making it suitable for batch transfers and custom migration tasks. It can be used for both one-time transfers and ongoing synchronization with features like rsync.
>
> Here is a sample command for running gsutil:
>
> ```
> gsutil cp <source_path> <destination_path>
> ```

rsync
> Rsync is a command-line tool for efficient file synchronization. When used with gsutil (e.g., gsutil rsync), it enables incremental transfers by copying only new or changed files. This helps minimize data transfer time and costs by avoiding unnecessary copies. It is useful for keeping local directories and GCS buckets in sync or performing regular backups.

Here is a sample command for running gsutil with the rsync option:

```
gcloud storage rsync <source_path> <destination_path>
```

STS

A fully managed service for large-scale data transfers to GCS, STS supports various sources, including AWS S3, Azure, HTTP/HTTPS locations, and on-premises filesystems. It offers features like scheduling, error handling, and monitoring for reliable, automated transfers. It can perform both one-time batch migrations and incremental synchronizations. It provides a web UI and API for easy management and integration with other GCP services. More details about this solution are covered in Recipe 14.3.

DistCp (distributed copy)

This is a distributed copy tool for large inter- and intracluster copying within Hadoop environments (like HDFS). It uses MapReduce to distribute the copy process, providing parallelism and fault tolerance. It efficiently copies large datasets between HDFS clusters or within the same cluster. It can be used to move data from HDFS to a GCS connector that exposes GCS as an HDFS-compatible filesystem. It can be executed only from a machine that has Hadoop client libraries installed, typically a node within a Hadoop cluster.

Here is a sample command to copy data from HDFS to GCS using DistCp:

```
hadoop distcp hdfs://<hdfs_namenode>:<hdfs_port>/<source_path_in_hdfs> \
    gs://<your_gcs_bucket>/<destination_path_in_gcs>
```

14.3 Migrating Data with STS

Problem

The enterprise needs to migrate data from an AWS S3 bucket to a GCS bucket, ensuring seamless transfer with minimal disruption. The migration must include an initial full data transfer, followed by continuous synchronization of any new or modified files in the source S3 bucket during the migration process.

Solution

Google Cloud STS offers an efficient way to migrate data from various sources, including AWS S3, Azure, local filesystems, and HDFS, to a GCS bucket. It supports common migration patterns like one-time transfers and incremental synchronizations while also providing robust error handling.

The STS migration process involves these key steps:

1. Define the location of your data source and the target GCS bucket.
2. Ensure that STS can access the source data by providing the necessary authentication credentials.
3. Choose between a one-time transfer of all data or an incremental sync that only transfers new or updated files.
4. Customize the migration process with additional options such as:
   - Maintaining source bucket names
   - Changing file prefixes
   - Filtering source files for selective transfer

Discussion

Migrating large volumes of data can be a complex, time-consuming task, often requiring careful planning around file transfers, source and target connections, naming conventions, error handling, and retries. Google Cloud's STS streamlines this process by providing a fully managed solution for large-scale data migrations. It simplifies the transfer of data from a variety of sources, including AWS S3, Azure, on-premises filesystems, and GCS, to your target destination with minimal effort. With built-in automation, error handling, and the ability to scale according to your needs, STS lets you focus on your core applications while handling the intricacies of data movement efficiently and reliably.

STS currently supports a wide range of data sources for your migration needs, including:

- Migrating data between Cloud Storage buckets
- Transferring data from Amazon S3 buckets
- Moving data from Azure Blob Storage
- Migrating data from Azure Data Lake Storage Gen2
- Transferring data accessible via HTTP/HTTPS links
- Migrating data from on-premises or cloud-based filesystems
- Transferring data from HDFS

Currently, STS supports these destinations:

- Cloud Storage buckets
- On-premises or cloud-based filesystems

Google continuously updates STS with new features and supported sources and destinations. For the most up-to-date list, refer to the official documentation (*https://oreil.ly/V0K6Z*).

Let's consider a scenario where your source data is in an AWS S3 bucket and you are migrating it to a destination GCS bucket. Here are the steps to achieve this.

Preparing the source

Before you can set up the transfer, you need to gather the details of the source bucket in Amazon S3. These include:

S3 bucket name
  The name of the S3 bucket from which you'll transfer data.

AWS credentials
  To authenticate with AWS and access the S3 bucket, you need the access key and secret key for an AWS IAM user with appropriate permissions to read from the source S3 bucket.

Then, create an AWS credential file in the following format:

```
{
  "accessKeyId": "AWS_ACCESS_KEY_ID",
  "secretAccessKey": "AWS_SECRET_ACCESS_KEY"
}
```

Preparing the target

You also need a destination GCS bucket where the data will be copied. You should have the following details:

GCS bucket name
  The name of the target bucket in Google Cloud where data will be transferred.

Google Cloud credentials
  Ensure that you have a service account key or sufficient IAM permissions to write to the target GCS bucket.

Creating a storage transfer job

Use the following gcloud command to initiate a transfer from the S3 bucket to your GCS bucket:

```
gcloud storage transfer jobs create \
  --source-s3-bucket=<S3_BUCKET_NAME> \       ❶
  --source-aws-credentials-file=credentials.json \    ❷
  --destination-bucket=<GCS_BUCKET_NAME> \       ❸
  --do-not-run \    ❹
  --include-new-files=true \    ❺
  --description="S3ToGCS Transfer"    ❻
```

Let's walk through what each of these flags means:

❶ The name of the source Amazon S3 bucket

❷ The file containing your AWS access key and secret key

❸ The name of the target GCS bucket

❹ A --do-not-run flag to create a job without triggering or scheduling it

❺ Ensures that the transfer service will automatically detect and transfer new files that are added to the source S3 bucket after the initial transfer

❻ A description of the transfer job

In this example, the transfer job is configured to handle new files by setting --include-new-files=true. This means that after the initial bulk transfer of existing files, STS will automatically detect and transfer any new files added to the source S3 bucket according to the schedule you define (e.g., every 24 hours).

Triggering a transfer job manually

To manually trigger a transfer job, first obtain the job's ID by listing existing transfer jobs:

```
gcloud transfer jobs list
```

Here's a sample output of the list command:

```
dataproc-samples$ gcloud transfer jobs list
NAME                    LATEST_OPERATION_NAME
15503943997150141848    transferJobs-15503943997150141848-5925737718442033964
10202453196674762951    transferJobs-10202453196674762951-7879775533468794291
```

Once you have the ID (extracted from the Name column) of the STS job, use the following gcloud command to trigger it:

```
gcloud transfer jobs run 15503943997150141848
```

To monitor the progress of your transfer job, use the following command:

```
gcloud transfer operations list --job-id=15503943997150141848
```

Here's a sample output of the command for monitoring a transfer job:

```
dataproc-samples$ gcloud transfer operations monitor transferOperations/transfer
Jobs-15503943997150141848-66632930690318057
Operation name: transferJobs-15503943997150141848-66632930690318057
Parent job: 15503943997150141848
Start time: 2024-11-07T02:42:57.641574845Z
SUCCESS | 0% (0B of 0B) | Skipped: 1.7GiB | Errors: 0
End time: 2024-11-07T02:43:08.200843230Z
```

Scheduling a transfer job

Use gcloud command to run the transfer job every four hours:

```
gcloud storage transfer jobs create \
  --source-s3-bucket=<S3_BUCKET_NAME> \
  --source-aws-credentials-file=credentials.json \
  --destination-bucket=<GCS_BUCKET_NAME> \
  --schedule="every 4 hours" \ ❶
  --include-new-files=true \
  --description="S3ToGCS Transfer" \
  --schedule "start-time=YYYY-MM-DDThh:mm:ss+00:00,repeat-interval=4h"
```

❶ --schedule="every 4 hours" specifies how often the transfer should occur. You can modify the schedule depending on your needs (e.g., daily, hourly, or a custom schedule).

14.4 Accessing AWS S3 Data Using BigLake Tables

Problem

Your data is in AWS S3, and you want to query the data from GCP without moving the data.

Solution

You can use BigLake to achieve this by following these steps:

1. Create a dataset in one of the BigLake-supported cross-cloud regions (aws-us-east-1, aws-us-west-2, aws-ap-northeast-2, aws-ap-southeast-2, aws-eu-west-1, or aws-eu-central-1) where the S3 bucket is created, as shown in Figure 14-1.

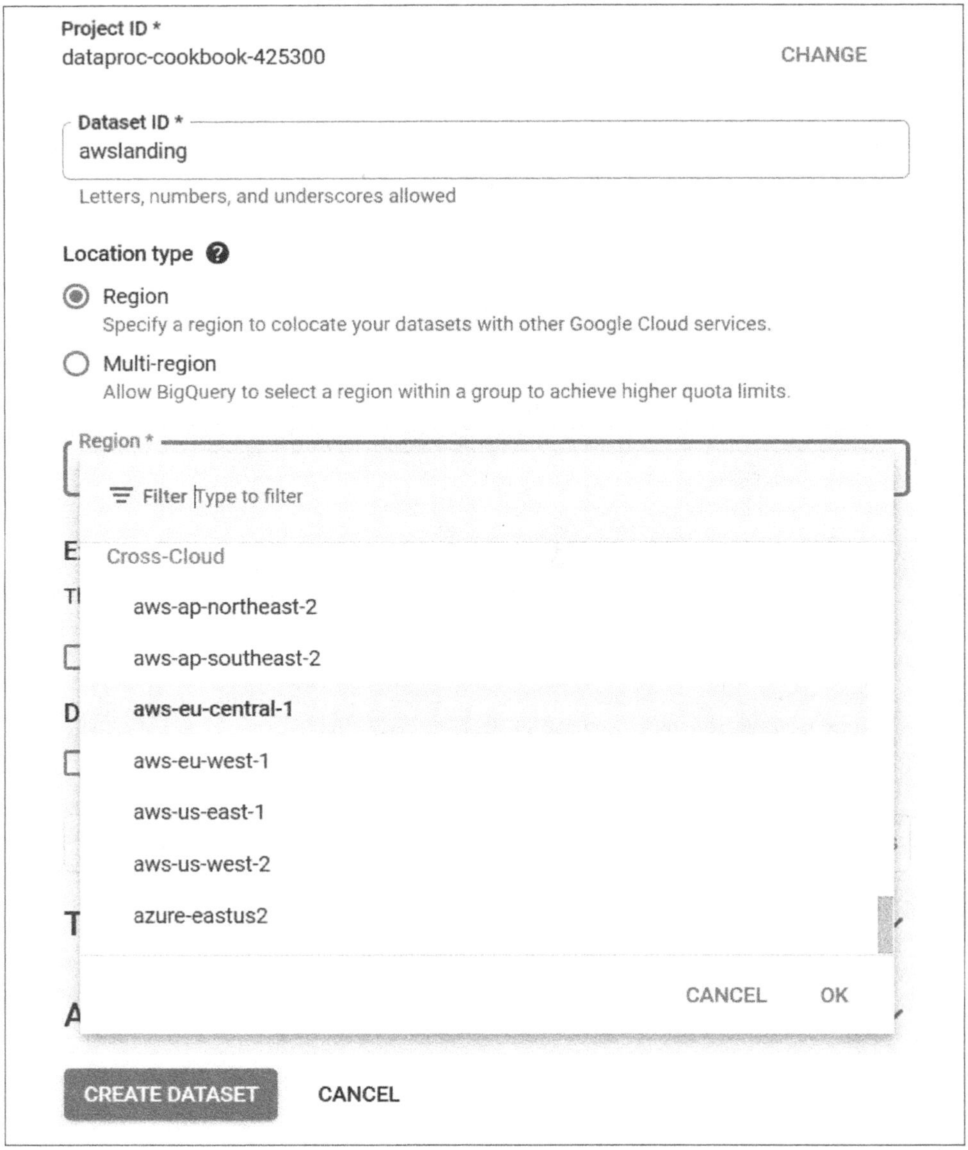

Figure 14-1. Creating a dataset in a BigLake-supported AWS region

2. Create a table in this dataset with the S3 path of the source data, as shown in Figure 14-2.

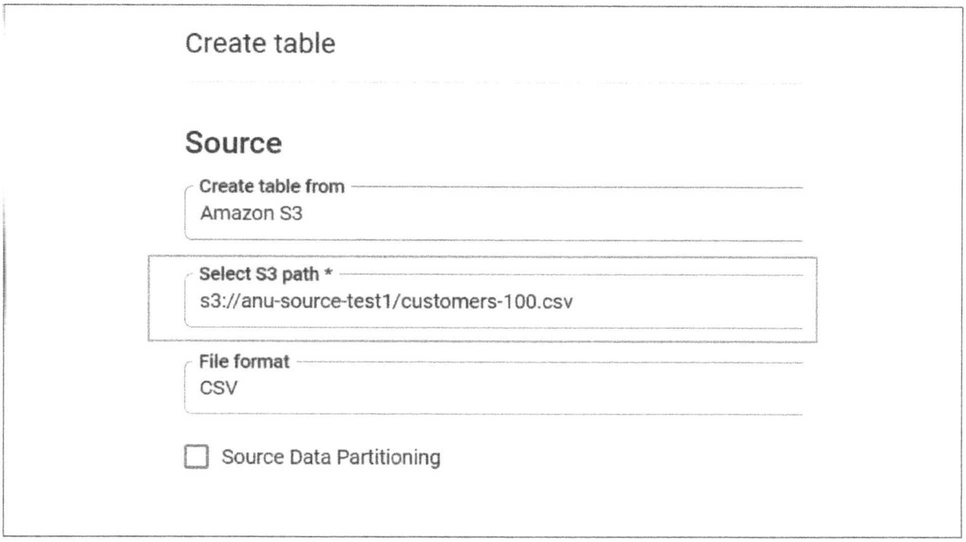

Create table

Source

Create table from
Amazon S3

Select S3 path *
s3://anu-source-test1/customers-100.csv

File format
CSV

☐ Source Data Partitioning

Figure 14-2. Creating a table referencing an S3 location

Discussion

Let's consider the AWS configuration first.

You must have an existing S3 bucket or create a new S3 bucket in one of the BigLake supported regions (us-east-1, us-west-2, ap-northeast-2, ap-southeast-2, eu-west-1, or eu-central-1) with the data to be queried from BigLake. Follow these steps:

1. Create a sample bucket:

```
aws s3api create-bucket --bucket anu-source-test1 --region us-east-1
```

2. Create an AWS IAM policy for read-only access to a specific S3 bucket and its contents. The following permission statements in JSON format allow the user to list the files in the specific bucket and download them, but they cannot upload, delete, or modify any files:

```json
{
    "Version": "2012-10-17",
    "Statement": [
        {
            "Effect": "Allow",
            "Action": [
                "s3:ListBucket"
            ],
            "Resource": [
                "arn:aws:s3:::anu-source-test1"
            ]
        },
        {
            "Effect": "Allow",
            "Action": [
                "s3:GetObject"
            ],
            "Resource": [
                "arn:aws:s3:::anu-source-test1",
                "arn:aws:s3:::anu-source-test1/*"
            ]
        }
    ]
}
```

3. Create an identity provider for *accounts.google.com*, as shown in Figure 14-3.

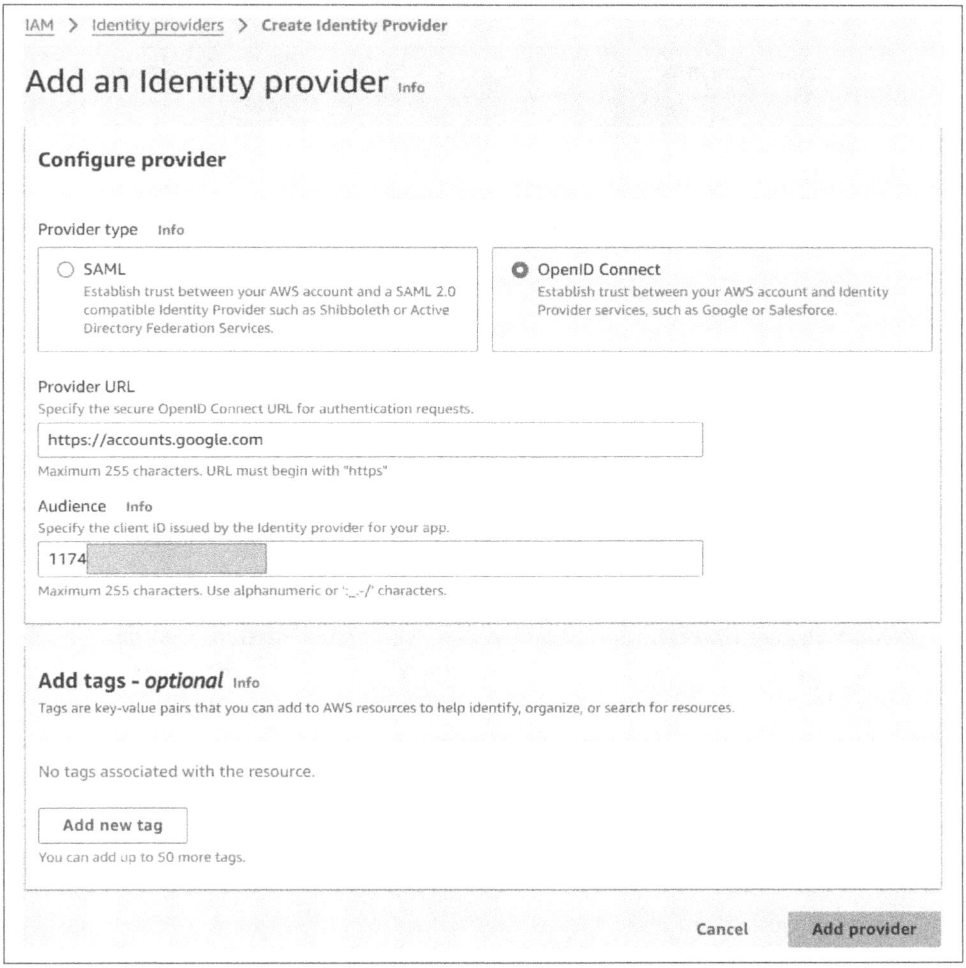

Figure 14-3. *Creating an identity provider for Google accounts*

4. Add the audience as a BigQuery Google identity value from the BigLake external connection, as shown in Figure 14-4.

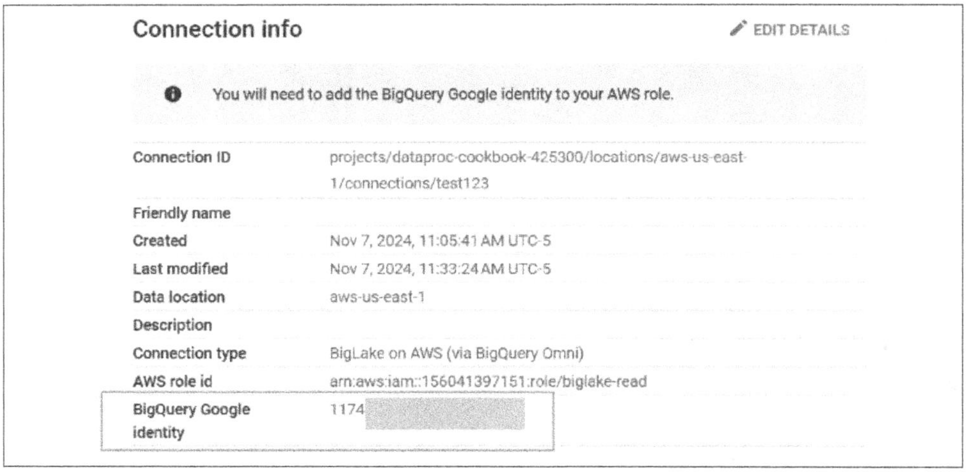

Connection info	✏ EDIT DETAILS
ℹ You will need to add the BigQuery Google identity to your AWS role.	
Connection ID	projects/dataproc-cookbook-425300/locations/aws-us-east-1/connections/test123
Friendly name	
Created	Nov 7, 2024, 11:05:41 AM UTC-5
Last modified	Nov 7, 2024, 11:33:24 AM UTC-5
Data location	aws-us-east-1
Description	
Connection type	BigLake on AWS (via BigQuery Omni)
AWS role id	arn:aws:iam::156041397151:role/biglake-read
BigQuery Google identity	1174

Figure 14-4. Retrieving the BigQuery Google identity value for the BigLake connection

5. Create a role with the trusted entity type "Web identity" and select the *accounts.google.com* identity provider that we created in step 3. When you hit Next, you will be asked to add the policy created in step 2, as shown in Figure 14-5.

Step 1
Select trusted entity

Select trusted entity Info

Step 2
Add permissions

Trusted entity type

Step 3
Name, review, and create

○ **AWS service** Allow AWS services like EC2, Lambda, or others to perform actions in this account.	○ **AWS account** Allow entities in other AWS accounts belonging to you or a 3rd party to perform actions in this account.	● **Web identity** Allows users federated by the specified external web identity provider to assume this role to perform actions in this account.
○ **SAML 2.0 federation** Allow users federated with SAML 2.0 from a corporate directory to perform actions in this account.	○ **Custom trust policy** Create a custom trust policy to enable others to perform actions in this account.	

Web identity

Allows users federated by the specified external web identity provider to assume this role to perform actions in this account.

Identity provider

accounts.google.com ▾   ↻   Create new ⤢

Audience

1174 ▾

Condition - *optional*

Add condition

Cancel   **Next**

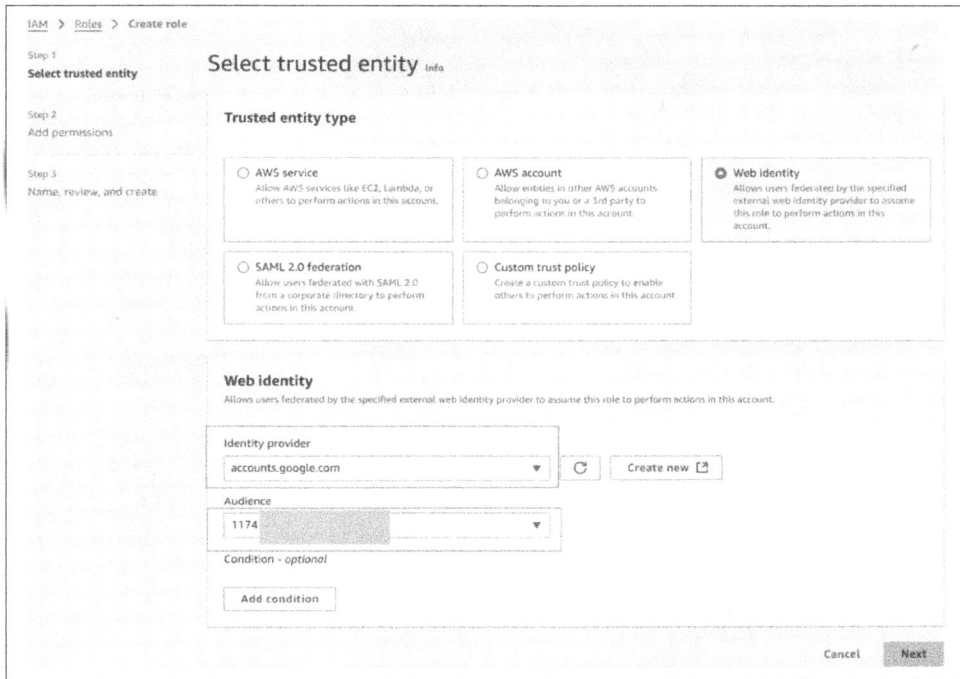

Figure 14-5. Creating a role and assigning the BigQuery Google identity as an audience

6. Once the role is created, copy and keep the AWS role ID Amazon Resource Name (ARN), as shown in Figure 14-6. This ARN has to be added when creating a BigLake connection.

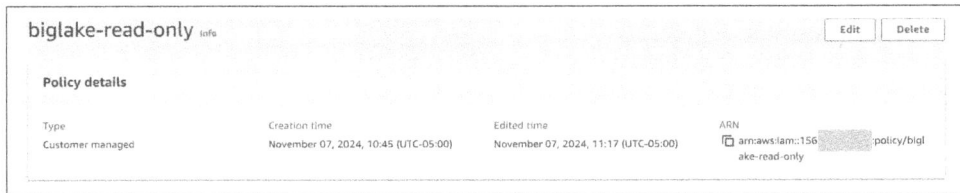

biglake-read-only Info

Edit   Delete

Policy details

Type	Creation time	Edited time	ARN
Customer managed	November 07, 2024, 10:45 (UTC-05:00)	November 07, 2024, 11:17 (UTC-05:00)	⧉ arn:aws:iam::156 policy/bigl ake-read-only

Figure 14-6. Retrieving the ARN of the created role

Now, let's go over the steps for creating the BigLake connection:

1. Navigate to BigQuery and click Add, as shown in Figure 14-7. Then, choose external connections, create a new connection in the region where the S3 bucket exists, and add the AWS role ID copied in step 6 previously.

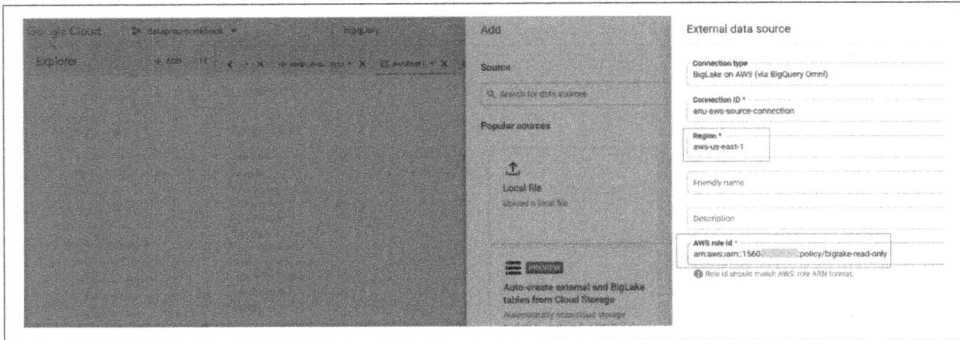

Figure 14-7. Creating a BigLake connection on AWS and adding the AWS role ID

2. Create the dataset and table, as shown in the "Solution" on page 401. While creating a table, you will see a drop-down menu under connection ID with the external configuration you configured. Select that, as shown in Figure 14-8. If you have configured everything correctly, the dataset and table will be successfully created.

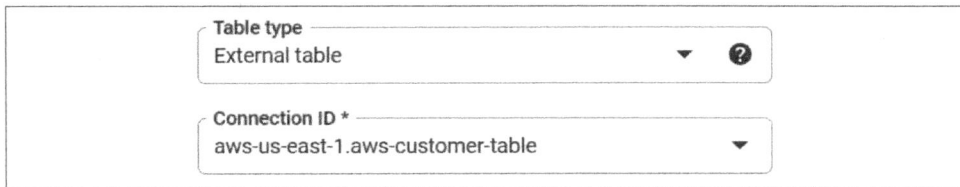

Figure 14-8. Selecting the connection details when creating the table

Then, you can query the table. It will query the S3 data and display the results in the console, as shown in Figure 14-9.

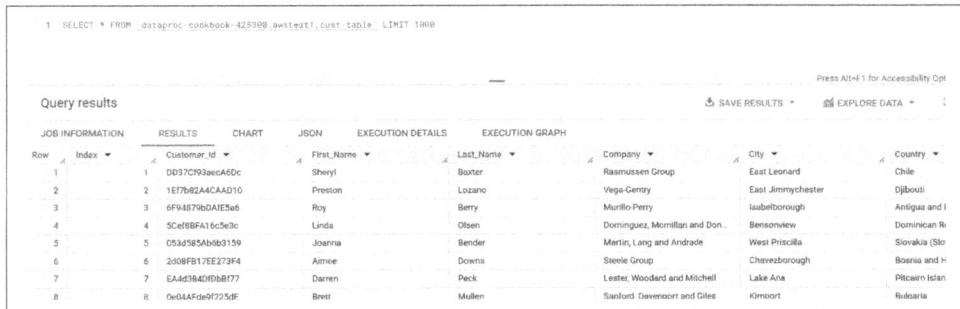

Figure 14-9. Querying the table after successfully creating it

14.5 Migrating Metadata

Problem

You want to migrate metadata from source systems (AWS Glue Data Catalog or Hive Metastore) to the Google Cloud environment.

Solution

Migrating metadata from the AWS Glue Data Catalog or Hive Metastore to Google Cloud can be achieved by exporting it to a generic format that the target system understands. Both source systems allow exporting metadata as SQL files, which can then be imported into Dataproc Metastore. For more complex scenarios, consider exporting to Avro format, which offers a schema-based approach for better compatibility and data validation. Alternatively, leverage tools like AWS Database Migration Service (DMS) or custom scripts to extract metadata and transform it into a format compatible with Google Cloud's Dataproc Metastore.

Discussion

To establish a centralized metadata repository during system migrations, it is essential to migrate metadata from databases like the Hive Metastore, which store details such as table names, columns, and data statistics.

Organizations use various solutions to store metadata, including generic options like Hive Metastore and specialized services like Glue Data Catalog in AWS or Unity Catalog in Databricks. These systems typically support exporting metadata into a generic, importable format. For example, you can export metadata from a MySQL-backed Hive Metastore into an SQL file.

Choosing a target service in GCP

When migrating, consider the available GCP options for your metastore. For Hadoop and Spark workloads, these options stand out:

DPMS
   A managed metastore service integrated with other GCP services

BigQuery
   A data warehousing service that can also function as a metastore

Standalone Hive Metastore server
   A self-managed Hive Metastore instance on a GCP Compute Engine

Table 14-2 compares the options for migrating metadata from the source to GCP.

Table 14-2. Key characteristics of metastore solutions on GCP

Feature	DPMS	BigQuery	Standalone Hive Metastore server
Type	Managed service	Data warehousing service with metastore capabilities	Self-managed
Integration with GCP services	Seamless integration with Dataproc	Native to GCP; integrates with other services	Requires manual integration
Scalability	Highly scalable; automatically managed by GCP	Highly scalable; automatically managed by GCP	Requires manual scaling and configuration
Maintenance	Choose the outage window, and GCP will handle upgrades and fixes	No maintenance required	Requires ongoing maintenance and updates
Cost	Pay-as-you-go pricing	Pay-as-you-go pricing	Costs associated with Compute Engine, database instances, and storage
Security	GCP IAM role-based security	GCP IAM role-based security	Requires manual security configuration but offers granular control and the ability to implement custom security measures
Suitability	Organizations seeking a managed, scalable, and integrated metastore solution	Organizations with data warehousing needs that want to leverage BigQuery's metastore capabilities	Organizations requiring full control and customization of their metastores

Choosing the migration strategy

Depending on the target system where you are migrating the metadata to in GCP, you can consider the following strategies:

Export to a generic format and import to the target system
> Export metadata from your source system (Hive Metastore, AWS Glue Data Catalog, etc.) into a portable format like Avro, JSON, or CSV. Then, use the target system's import functionality (e.g., Dataproc Metastore import) to load the data. If your source metastore uses a database (e.g., MySQL for Hive Metastore), use native database tools (like mysqldump) to create a file with all the metadata.

Use Spark Job
> Use Spark or an equivalent compute framework to connect to your source metastore database, extract schema information, and create corresponding tables and partitions in your target metastore (e.g., Dataproc Metastore) using its API.

Back up and import at database level
> This approach is similar to a database dump, but it is specifically suitable when both your source and your target metastores use the same or compatible database systems (e.g., both use MySQL). You can directly move the database backup with minimal changes.

Restore from data files of a type holding metadata

File formats like Parquet, Avro, and Optimized Row Columnar (ORC) contain schema information along with data. Tools like Dataplex and BigQuery can automatically infer schema and partitions when loading data in these formats. This can simplify metadata migration.

Restore the OS-level image

Create an image of the entire operating system containing your source metastore. Restore this image to a GCP VM. Then, migrate the metadata from this restored instance to your target metastore.

The best approach for migrating your metadata to GCP depends on various factors, including your source system, target metastore, data volume, and desired downtime.

> Metadata often includes the location of your data files. When migrating your metadata, make sure to update these locations to reflect the paths in your target environment. This ensures that your queries and applications can correctly access the underlying data after the migration.

14.6 Migrating Applications to Google Cloud

Problem

You have one hundred Spark applications on Spark version 3.5 that are connected to 30 external Hive tables running on a long-running Amazon EMR cluster that you want to migrate to Google Cloud with minimal code changes. How can you do this?

Solution

To migrate Spark 3.5 workloads and associated Hive tables with minimal code changes, use a lift-and-replatform strategy by moving workloads to the Dataproc 2.2 release version (*https://oreil.ly/CM1Fh*) on Google Cloud. Here's how:

1. Migrate data from AWS S3 to GCS (refer to Recipe 14.2 and Recipe 14.3).

2. Extract Hive DDLs from AWS EMR using *hive_ddl_extractor.py* from Google-CloudPlatform/dataproc-templates (*https://oreil.ly/Y8kw_*).

3. Migrate custom Spark dependency libraries from S3 to GCS (refer to Recipe 14.2 and Recipe 14.3).

4. Benchmark and select the appropriate Dataproc configurations (e.g., machine type, cores, memory, SSDs) to run your jobs.

5. Update the Spark code to use GCS paths instead of S3 paths, and run the Spark applications on Dataproc 2.2.

Discussion

Two primary migration strategies can guide the transition based on the organization's short-term and long-term goals:

Lift and replatform
> A quick migration approach that minimizes code changes and risks

Lift and modernize
> A modernization strategy that optimizes costs and fully leverages Google-native services

Assess the current environment (on premises or cloud) to determine the compute requirements and specific big data workloads (e.g., Spark, Hive, MapReduce) to be migrated. You'll want to gather details such as workload versions, application specifications, and resource-utilization metrics.

After selecting a migration strategy and discovering current workloads, identify the appropriate target services in Google Cloud for each workload, listed in Table 14-3. If you choose to lift and replatform Spark jobs, Hive, or MapReduce, Dataproc would be the target service in GCP. For notebooks, consider Dataproc with JupyterLab enabled or Dataproc Workbench. Similarly, if you are replatforming Flink or HBase, Dataproc remains the target. If the customer is open to modernization, then the target remains Dataproc for Spark, Flink, and MapReduce, with the option to upgrade MapReduce jobs to Spark for improved performance. For Hive, the optimized target would be BigQuery or BigLake, and for HBase, Bigtable is recommended. For each workload, identify the specific versions and replatform to an equivalent version in Spark. It's generally advisable to choose the latest version for added benefits and optimizations.

Table 14-3. Target services in Google Cloud for each workload and migration type

Migration type	Workload	Target service in GCP
Lift and replatform	Spark jobs	Dataproc
	Hive	Dataproc
	MapReduce	Dataproc
	Notebooks	Dataproc with JupyterLab, Vertex AI Workbench, and Google Colab
	Flink	Dataproc
	HBase	Dataproc
Lift and modernize	Spark batch	Dataproc, Dataflow, or BigQuery
	Flink	Dataproc
	MapReduce	Dataproc (consider upgrading to Spark)
	Hive	BigQuery or BigLake
	HBase	Bigtable
	Spark Streaming	Dataflow

You can also consider serverless options for cost savings. For Spark, Dataproc Serverless is available, and for Hive workloads, BigQuery and BigLake offer serverless functionality.

The migration steps are as follows:

1. Transfer the data from on premises or other clouds to GCS.

2. Migrate Hive DDLs to GCS. For modernization, consider using BigQuery's batch translation service to convert Hive DDLs to BigQuery DDLs.

3. Choose the compute configuration in GCP, then migrate and validate the applications.

> When migrating from other cloud providers or on-premises infrastructure to GCP, avoid selecting an exact compute equivalent. Instead, assess Dataproc and the specific throughputs of various compute and machine types to select the best fit. For example, if you currently use Amazon EMR with m8g.24xlarge instances (64 vCPUs, 256 GB), you don't necessarily need to choose an equivalent configuration like 96 vCPUs in GCP Dataproc. On Amazon EC2, the m8g.24xlarge instance (*https://oreil.ly/QlXNk*) offers a network bandwidth of 30 GBps. In GCP, an alternative might be the n2d-highmem-32 machine type (*https://oreil.ly/SlGuz*) (32 vCPUs, 256 GB), which has an egress bandwidth of 32 GBps. Consider throughput, discount options, consumption, and other features to make informed choices for GCP migration.

If your Spark notebooks are currently in Databricks, start by transferring the data from the source storage (Azure, AWS, or GCP) to GCS. Next, download any dependency libraries and upload them to GCS. Export the notebooks to *.ipynb* format and replace the dbutils commands with equivalent gsutil commands. The simplest way to run these notebooks in Google Cloud is by using Dataproc with JupyterLab enabled.

Index

machine types, 18
VPC (virtual private cloud)
    Compute Engines, 9
    creating, 9-11
    IP address range for Dataproc, finding, 162
    service controls, securing perimeter bound-
        aries, 241-247
    subnet ranges, 320

W

Web Interfaces, navigating to, 290
widgets
    chart, adding to dashboards, 217
Windows, installing gcloud CLI, 4
worker nodes, adding local SSDs, 73-74
worker role, granting to users, 30
workers
    autoscaling policy, configuration, 67-68
    environment size and, 339-340
workflow templates, orchestrating jobs,
    349-351
workflows, orchestration requirements, 320
workload identity, 124, 125
workloads
    autoscaling policy, 64-68
    cluster size considerations, 280
    Kubernetes clusters, machine types, 121
    performance, improving, 103-106
    scaling, Cloud Composer, 338-341

write_to_cloud_logging function, 194-196

X

XML data, converting to Parquet
    PySpark, 50-52
    Serverless Spark, 91-95
    Spark Scala, 46-49

Y

YAML configuration file, autoscaling policy, 64
YARN
    configuration, autoscaling policy, 66
    logs, 183
    memory chart, 207
    NodeManagers chart, 207
    pending memory chart, 207
    pending requests, creating alerts, 219-225
YARN ResourceManager (RM) (see RM (YARN
    ResourceManager))
YARN UI, 306-307
    opening, 38
    RM, 307-310

Z

Zeppelin, notebook development environment,
    354
zonal high availability, ensuring, 132-134
ZooKeeper, logs, 183

About the Authors

Narasimha Sadineni is a senior data engineer at Google, bringing over 15 years of experience in data architecture and analytics. Through his work on the professional services teams at Google and Cloudera, he has successfully guided more than 50 organizations in solving their most complex big data challenges.

A passionate advocate for knowledge sharing, Narasimha is known for mentoring data professionals to help them master difficult concepts. This book is an extension of that passion, written to share his expertise with the broader data community.

Anuyogam Venkataraman is a Senior Program Manager. She previously served as a Data Lake Engineer at Google, accumulating extensive experience in data technologies. Anu assists customers in migrating large-scale distributed systems to the cloud. She finds joy in speaking at universities and contributing technical blogs and videos to the Data community, aiming to expedite customers' journeys to the cloud. Anu played a key role as one of the leads for the Professional Services Tech Talk playlist on the Google Cloud Tech YouTube channel. She holds a master's degree in Electrical and Computer Engineering from Ryerson University, specializing in Medical Image Processing and Machine Learning.

Colophon

The animal on the cover of *Dataproc Cookbook* is the common degu (*Octodon degus*), a small rodent native to the semi-arid scrublands along the western slopes of the Andes Mountains in central Chile.

Degus are roughly the size of a large gerbil or small guinea pig. They have a round, compact body covered in soft brown fur with a lighter, yellowish underbelly and a long tail that ends in a tuft of fur. Their large eyes and ears help them remain alert to predators.

Degus prefer areas with loose, sandy soil for easier burrowing. They often live in colonies, constructing extensive tunnels and communal nests. Degus are highly social animals and exhibit cooperative behaviors, such as grooming and raising young. These rodents are herbivores with a diet consisting mainly of herbs, grasses, leaves, and bark, which they hoard in underground caches.

The common degu is currently listed as "Least Concern" by the International Union for Conservation of Nature (IUCN), though habitat destruction and hunting may impact some local populations. Many of the animals on O'Reilly covers are endangered; all of them are important to the world.

The color illustration is by Jose Marzan, based on an antique line engraving from *Cassell's Natural History*. The series design is by Edie Freedman, Ellie Volckhausen, and Karen Montgomery. The cover fonts are Gilroy Semibold and Guardian Sans. The text font is Adobe Minion Pro; the heading font is Adobe Myriad Condensed; and the code font is Dalton Maag's Ubuntu Mono.

O'REILLY®

Learn from experts.
Become one yourself.

60,000+ titles | Live events with experts | Role-based courses
Interactive learning | Certification preparation

**Try the O'Reilly learning platform
free for 10 days.**

www.ingramcontent.com/pod-product-compliance
Lightning Source LLC
Chambersburg PA
CBHW080137220326
41598CB00032B/5094